Advancing Body Fluid Diagnostics with Flow Cytometry

Advancing Body Fluid Diagnostics with Flow Cytometry

A Comprehensive Guide to Effusions and CSF

Edited by

Namrata P. Awasthi

Professor, Department of Pathology,
Dr. Ram Manohar Lohia Institute of Medical Sciences (Dr. RMLIMS),
Lucknow, Uttar Pradesh, India

ELSEVIER

ACADEMIC PRESS

An imprint of Elsevier

elsevier.com/books-and-journals

Academic Press is an imprint of Elsevier

125 London Wall, London EC2Y 5AS, United Kingdom
50 Hampshire Street, 5th Floor, Cambridge, MA 02139, United States

ISBN: 978-0-443-27426-8

For Information on all Academic Press publications visit our website at https://www.elsevier.com/books-and-journals

Publisher: Megan Ball
Acquisitions Editor: Patricia Osborn
Editorial Project Manager: Suman Deshwal
Production Project Manager: Rajeswari R
Cover Designer: Mark Rogers

Typeset by Aptara, New Delhi, India

Dedication

To the memory of my beloved mother, Mrs. Nirmala Dubey, whose blessings are eternal; to my dear father, Dr. Ram Shringar Dubey, for his lifelong guidance; my loving husband, Punit and precious daughters, Pranavi and Prakriti, for their unwavering love, patience, and support that made this work possible.

Namrata P. Awasthi

Contents

Contributors

Firoz Ahmad
Molecular Genomics, Unipath Speciality Laboratories Limited, Ahmedabad, Gujarat, India

Namrata P. Awasthi
Department of Pathology, Dr. Ram Manohar Lohia Institute of Medical Sciences (Dr. RMLIMS), Lucknow, Uttar Pradesh, India

Jasmita Dass
Department of Hematology, All India Institute of Medical Sciences, New Delhi, India

Venkatesh Dhanasekaran
Department of Pathology, All India Institute of Medical Sciences, Madurai, Tamil Nadu, India

Gauri Gaur
Department of Pathology, Dr. Ram Manohar Lohia Institute of Medical Sciences (Dr. RMLIMS), Lucknow, Uttar Pradesh, India

Nalini Gupta
Department of Cytology and Gynecologic Pathology, Postgraduate Institute of Medical Education and Research (PGIMER), Chandigarh, India

Parikshaa Gupta
Department of Cytology and Gynecologic Pathology, Postgraduate Institute of Medical Education and Research (PGIMER), Chandigarh, India

Ruchi Gupta
Department of Hematology, Sanjay Gandhi Post Graduate Institute of Medical Sciences, Lucknow, Uttar Pradesh, India

Aditya Jandial
Department of Clinical Hematology and Medical Oncology, Post Graduate Institute of Medical Education and Research (PGIMER), Chandigarh, India

Rashmi Kushwaha
Department of Pathology, King George's Medical University, Lucknow, Uttar Pradesh, India

Nabhajit Mallik
Department of Hematology, Postgraduate Institute of Medical Education and Research (PGIMER), Chandigarh, India

Sumaira Qayoom
Department of Pathology, King George's Medical University, Lucknow, Uttar Pradesh, India

Khaliqur Rahman
Department of Hematology, Sanjay Gandhi Post Graduate Institute of Medical Sciences, Lucknow, Uttar Pradesh, India

Thulasi Raman Ramalingam
Apollo Hospitals, Chennai, Tamil Nadu, India

Praveen Sharma
Department of Hematology, Postgraduate Institute of Medical Education and Research (PGIMER), Chandigarh, India

Sowmya Sharma
Clin Labs Bella Vista, Anatomic Pathology, Bellavista, NSW, Australia

Mona Singh
Department of Hematology, Sanjay Gandhi Post Graduate Institute of Medical Sciences, Lucknow, Uttar Pradesh, India

Saksham Singh
Department of Medical Oncology, Dr. Ram Manohar Lohia Institute of Medical Sciences (Dr. RMLIMS), Lucknow, Uttar Pradesh, India

About the Editor

Dr. Namrata Punit Awasthi, MD, FICH, MNAMS, is a Professor in the Department of Pathology at the Dr. Ram Manohar Lohia Institute of Medical Sciences (Dr. RMLIMS), Lucknow, India. With over 15 years of experience in academic pathology, she has established herself as a leading authority in diagnostic hematopathology and advanced flow cytometry.

While her foundational expertise is in hematology, Dr. Awasthi has developed a distinguished research niche in the nonhematological applications of flow cytometry. Her work in this area is expansive, ranging from the evaluation of radiation-induced DNA damage and DNA ploidy in cervical cancer to the use of immune biomarkers such as monocyte HLA-DR and neutrophil CD64 in sepsis and COVID-19. Her pioneering paper on circulating tumor cells in gallbladder carcinoma by flow cytometry was also recognized for its significant contribution to clinical science. Notably, her research on the diagnostic accuracy of flow cytometry in detecting malignant epithelial cells in serous effusions—published in the *Journal of the American Society of Cytopathology*—underlines her commitment to integrating this technology into routine cytomorphology.

A prolific researcher with publications in indexed national and international journals, Dr. Awasthi has received numerous accolades, including The Cytometry Society (TCS) Award, the Berend Houwen Award from the International Society of Laboratory Hematology (ISLH), and the Director's Merit Award (2025) from Dr. RMLIMS. She is an active member of several national and international professional bodies and a founding member of the Uttar Pradesh Hematology Group.

Dr. Awasthi is a Fellow of the Indian College of Hematology and a Member of the National Academy of Medical Sciences (India). Driven by a vision for diagnostic excellence, she is committed to empowering pathologists and cytomorphologists with the technical expertise needed to utilize flow cytometry as a standard diagnostic modality.

Preface

The diagnosis of malignant effusions and central nervous system involvement remains one of the most challenging frontiers in diagnostic cytology. For decades, the gold standard has been the morphological assessment of smears and cell blocks. While morphology remains the bedrock of our practice, it has some inherent limitations—particularly in cases where the morphology is not sufficient for a conclusive categorization into definite positive or negative diagnoses. This leaves the oncologist without an actionable certainty. Identification of low-burden lymphoid infiltrates in cerebrospinal fluid (CSF) is another challenging area.

This book was born out of a simple realization: flow cytometry, a technology traditionally housed within the specialized domain of hematopathology, is an underutilized powerhouse in the cytology laboratory. It is a robust, objective, and highly reproducible tool that can provide rapid diagnostic clarity where morphology alone may leave us in a "gray zone."

My goal is to demystify this technology for the practicing cytologist and general pathologist. I believe that the ability to adapt flow cytometric protocols for nonheme fluids—specifically effusions and cerebrospinal fluid (CSF)–should no longer be an alternative method, but a core competency in the modern diagnostic armamentarium.

Within these chapters, the text starts from the basics of effusion and CSF cytology and moves toward flow cytometry basics and its application on malignant effusions and CSF. Beyond theoretical discussions, the chapters provide practical, evidence-based guidance. Technical nuances of sample processing, the selection of antibody panels, and the interpretative pitfalls unique to fluid samples are all comprehensively addressed.

This text has been designed to serve as both a technical manual and a diagnostic guide, bridging the gap between the fluidics of the flow cytometer and the microscope slide.

As we move further into the era of precision medicine, the boundaries between laboratory subspecialties are blurring. I hope that this book empowers cytologists to integrate this technology seamlessly into their workflows to enhance diagnostic accuracy and, ultimately, improve patient outcomes.

Namrata P. Awasthi

Acknowledgments

The completion of this book was a journey that relied on the collective effort, patience, and expertise of many individuals.

First and foremost, a heartfelt thanks to **Dr. Parikshaa Gupta** for coming on board during a crucial phase of this journey. Her extraordinary commitment, from expanding our team of contributors to guiding the final manuscript revisions, has played an invaluable role in the successful completion of this book.

I wish to express my deepest gratitude to the contributing authors. Your specialized knowledge and commitment to excellence are the true foundation of this work. Thank you for your patience through the various stages of revision and for sharing your vision for the future of cytopathology.

I am sincerely grateful to Elsevier and the entire editorial and production team. In particular, I wish to express my deepest appreciation to Patricia Osborn, Senior Acquisitions Editor, and Suman Deshwal, Editorial Project Manager, whose professional guidance and belief in the importance of this subject matter have been instrumental in transforming a manuscript into a finished volume.

On a professional level, I would like to acknowledge the leadership at the Dr. Ram Manohar Lohia Institute of Medical Sciences, Lucknow, Uttar Pradesh, India, for providing an environment that encourages academic growth and clinical innovation. I am sincerely grateful to the **Director**, Prof. Chandra Mani Singh, the Dean, Prof. Pradyumn Singh, and the **Head of Department (Pathology)**, Prof. Nuzhat Husain, for their administrative support and constant encouragement.

I express my sincere appreciation to senior cytopathologist **Prof. Nalini Gupta** for her contribution and masterly guidance, which provided a strong foundational bridge between conventional cytology and advanced fluidics. I owe a special debt of gratitude to **Prof. Anurag Gupta**, who was my primary collaborator during the initial experiments. His early scientific inputs and cytology insights provided the vital foundation upon which this work was built. A special thanks to **Mr. Abhishek Bajpai**, for being the backbone of our flow cytometry experiments and ensuring the precision of every run.

Finally, and most importantly, I thank my family. To my parents, for their lifelong blessings and for the values they instilled in me. Special thanks to my father-in-law, Dr. Surendra Nath Awasthi, sisters-in-law Puja and Kasturee, and brother Prateek, whose belief in my potential and constant encouragement have been a vital source of strength throughout my career. To my spouse and children, thank you for your unwavering support and for the many hours of "family time" you sacrificed so that I could pursue this project. This book is as much a result of your patience as it is of my effort.

Namrata P. Awasthi

1

Cytology of effusion fluids: An introduction

Parikshaa Gupta and Nalini Gupta

DEPARTMENT OF CYTOLOGY AND GYNECOLOGIC PATHOLOGY, POSTGRADUATE INSTITUTE OF MEDICAL EDUCATION AND RESEARCH (PGIMER), CHANDIGARH, INDIA

1.1 Introduction

Body cavity effusions develop as a consequence of abnormal accumulations of fluid in the body cavities, such as the pleural, peritoneal, and pericardial spaces. Cytologic examination is a critical diagnostic tool for the assessment as well as management of body cavity effusions. Etiopathogenetically, these effusions can result from a wide range of causes, including infections, inflammatory diseases, and malignancies (Min, 2020). The nature of the effusion can often be determined based on its gross appearance, biochemical characteristics, and cytological findings (Shidham & Layfield, 2021).

Based on their location, body cavity effusions are classified as follows:

- *Pleural effusion*: Fluid accumulation in the pleural space surrounding the lungs.
- *Peritoneal effusion/ascites*: Fluid accumulation in the peritoneal cavity within the abdomen.
- *Pericardial effusion*: Fluid accumulation in the pericardial space around the heart.

Gross examination of effusion fluid samples plays a key role in the diagnosis by providing immediate insights into the nature and potential underlying causes of the effusion (Li et al., 2018). This initial gross examination helps prioritize the need for more detailed cytomorphologic analysis, thereby facilitating timely management decisions.

Depending upon their biochemical composition, body cavity effusions can be of two types, transudates and exudates. Transudates result from systemic fluid imbalances, while exudates arise from local pathological conditions (Malik & Datta, 2012). Understanding these distinctions helps guide appropriate management strategies, ensuring effective treatment of the underlying causes.

Cytologic examination plays a vital role in the diagnostic evaluation of body cavity effusions. The primary goal of cytologic examination is to provide a diagnosis by evaluating the cellular content of the effusion. By analyzing the cellular components of effusion fluids, cytology helps in determining the cause of the effusion, detecting malignancies, assessing inflammatory conditions, prognostication, and monitoring therapeutic responses (Michael, 2021).

Advancing Body Fluid Diagnostics with Flow Cytometry: A Comprehensive Guide to Effusions and CSF.
DOI: https://doi.org/10.1016/B978-0-443-27426-8.00004-1

1.1.1 Applications of cytologic examination in effusion cytology

The major applications of cytologic examination of body cavity effusions include:

- *Determining the cause of effusion*: Effusions can result from a variety of causes, including infections, trauma, or autoimmune diseases. Cytologic examination complemented by biochemical tests can often help differentiate between these causes (Chen et al., 2024). For example, the presence of lymphocytes might suggest a tuberculosis-related effusion, while eosinophils might indicate an allergic or parasitic cause.
- *Diagnosing infections*: High numbers of neutrophils in the effusion might suggest bacterial infection, while predominance of lymphocytes could point toward a chronic inflammatory process, such as in tuberculosis or systemic lupus erythematosus.
- *Detecting malignancy*: Cytology is particularly useful for detecting cancer cells in effusions. For instance, pleural effusions associated with lung cancer or breast cancer can demonstrate malignant cells, indicating metastasis. Similarly, peritoneal fluid analysis can help detect metastatic ovarian cancer or other abdominal malignancies (Kassirian et al., 2023).
- *Prognostication of the diseases*: Cytologic examination can also be employed to prognosticate various diseases by using ancillary tests, including immunocytochemistry (ICC) and various molecular diagnostic procedures (Michael, 2021).

1.1.2 Advantages of cytologic examination in effusion cytology

Cytologic examination offers several advantages in the evaluation of body cavity effusions (Michael & Davidson, 2016):

- *Minimally invasive*: It provides valuable diagnostic information with relatively simple procedures compared to more invasive biopsy techniques.
- *Quick and cost-effective*: It is generally quicker and less expensive than other diagnostic methods, such as imaging studies or tissue biopsies.
- *Prompt diagnosis*: It allows for rapid diagnosis and timely management decisions, which is crucial in acute situations.

1.1.3 Limitations of cytologic examination in effusion cytology

Despite its benefits, cytologic examination has some limitations (Shidham, 2021):

- *Sample quality*: The quality of the sample can significantly impact the accuracy of the results. For example, insufficient or contaminated samples often lead to inconclusive results.
- *False negatives*: Cytology may miss small clusters of malignant cells or low-grade malignancies, leading to false-negative results.
- *Limited specificity*: While cytology can suggest a diagnosis, in challenging cases, it may not provide a definitive etiology. In such cases, additional tests such as histopathological examination, molecular studies, or imaging may be needed for a complete diagnosis.

Despite its limitations, cytology remains an essential, cost-effective, and minimally invasive diagnostic tool that provides crucial insights into a wide range of underlying pathologies. For a complete diagnosis, cytology is often used in conjunction with other ancillary diagnostic methods, ensuring a comprehensive approach to patient management.

1.2 Collection of effusion samples

The entire process of collecting and transporting the effusion samples to the cytology laboratory must be performed under all aseptic conditions and with precision to ensure accurate diagnosis (Dey, 2023; Shidham, 2022).

- *Patient consent*: Before the procedure, explain the entire process to the patient to ensure their comfort and cooperation. Obtain informed consent, addressing any potential risks associated with the procedure.
- *Sterilization and equipment*: Ensure that all equipment, including syringes, needles, and collection containers, is sterile.
- *Patient preparation*: Prepare the patient's skin with an antiseptic solution to minimize the risk of infection.

1.2.1 Collection techniques

1.2.1.1 Thoracentesis (pleural effusion)

Positioning: The patient is made to sit upright, if possible, or in a semifowler's position. This allows for optimal fluid collection and minimizes the risk of complications.

Procedure: Locate the effusion site using clinical examination or imaging guidance. Insert a sterile wide-bore needle into the pleural cavity, typically at the posterior axillary line in the seventh–nineth intercostal space. Attach the needle to a syringe and aspirate the fluid. Collect an adequate volume for cytological examination, usually 30–50 mL.

1.2.1.2 Paracentesis (peritoneal effusion/ascitis)

Positioning: The patient may be positioned supine or in a semireclining position.

Procedure: The needle is inserted into the peritoneal cavity, usually in the left lower quadrant or a region away from major organs, and the fluid is aspirated. Around 50–100 mL of fluid should be collected for cytological analysis.

1.2.1.3 Pericardiocentesis (pericardial effusion)

Positioning: The patient is often placed in a semirecumbent position or sitting up.

Procedure: Using echocardiographic or fluoroscopic guidance, the optimal entry point in the anterior chest wall is identified, typically in the left xiphoid area or subxiphoid region. The needle is then inserted into the pericardial space, and the fluid is aspirated.

1.2.2 Collection tubes

For cytologic examination, the body cavity effusion fluid should preferably be collected in an anticoagulant-containing tube (Dey, 2023). The preferred anticoagulants include ethylene diamine tetraacetic acid (EDTA), acid citrate dextrose, 1% ammonium oxalate (ratio of 9 parts fluid:1 part anticoagulant), and heparin (3 units/mL). This helps in preventing coagulation, preserving cellular morphology, and limiting bacterial contamination.

The addition of a fixative is not recommended, as there is no significant alteration in cell morphology if the sample is processed within 4–12 h. The addition of formalin is not recommended as it prevents cells from adhering to the slide and interferes with Papanicolaou staining (Naylor, 2008; Shidham & Atkinson, 2007). Nonanticoagulated tubes can be used for sample collection for serological or microbiological analysis.

1.2.3 Fluid volume

Generally, around 20–30 mL of fluid is collected for cytologic examination. For cell block (CB) preparation, at least 30–50 mL of fluid is ideal. In general, fluid samples with low cellularity require a larger volume of fluid (Keebler et al., 2008).

1.3 Handling of effusion samples at the site of collection

- *Visual inspection*: The fluid should be immediately examined grossly after collection. The color, clarity, and any abnormal characteristics should be noted on the requisition/ case record form (e.g., hemorrhagic and purulent).
- *Labeling*: The sample container should be correctly labeled with the patient's identification details as well as the date, time of collection, and the site from which the sample was obtained.
- *Storage*: For cytological analysis, the fluid samples should be collected in sterile anticoagulated containers/ collection tubes. If the sample cannot be transported immediately to the cytopathology processing laboratory, it may be refrigerated at 4°C to slow the bacterial growth and preserve cell morphology. However, freezing of the sample should never be done as it can damage the cell membrane, cause cell rupture, and hence, hinder the cytomorphologic interpretation (Dey, 2023).

1.4 Transportation of the effusion samples to the cytology laboratory

The effusion fluid samples should be transported to the cytology laboratory as soon as possible. The timing of transport is crucial for preserving cell viability and ensuring accurate cytomorphological interpretation (Shidham, 2022).

- *Packaging in secure containers*: The sample containers should be placed in leak-proof biohazard bags. This prevents contamination and protects laboratory staff during transportation and handling.

- *Temperature control*: If immediate transport is not possible, use cool packs or ice packs to maintain the sample temperature at 4°C. Ensure that the sample remains at a stable temperature to prevent cellular degeneration and bacterial growth.
- *Requisition form*: A requisition form detailing the patient information, including clinical history, and the suspected diagnoses should be sent along with the appropriately labeled sample tube/container.
- *Long-distance transport*: If transport over long distances is required, a reliable courier service that can ensure optimal transport conditions for the fluid samples should be employed. A fixative can be added to the fluid sample to ensure cellular preservation and no bacterial contamination. An equal volume of 50% ethanol is commonly used as a fixative for such samples (Ali & Cibas, 2012).

1.5 Sample reception in the cytology laboratory

Upon arrival of the sample at the cytology laboratory, the following steps are essential (Dey, 2023):

- *Verification of the sample details*: Laboratory staff should verify and compare the sample details with the accompanying requisition form to ensure matching and accuracy.
- *Sample check*: The laboratory staff should check the sample for any signs of leakage or contamination.
- *Unique accession number*: A unique laboratory ID is assigned to each sample.
- Recording gross details of the sample.
- *Sample storage/processing*: The sample should be promptly processed in the laboratory. If a delay is anticipated, the same should be stored in the refrigerator at 4°C.

1.6 Gross examination of the effusion samples

Gross examination of effusion fluid samples plays a key role in the diagnosis by providing immediate insights into the nature and potential underlying causes of the effusion. During this preliminary evaluation, the medical laboratory professionals, or the cytopathologists themselves, assess and record the volume, color, viscosity, and clarity. These parameters can offer valuable diagnostic clues (Porcel & Light, 2006). For instance, a clear, pale yellow fluid suggests a transudative effusion, while a cloudy or turbid fluid generally indicates an exudative effusion. Bright red discoloration of the fluid samples often indicates blood in the effusion fluid (hemothorax, hemoperitoneum, or cardiac tamponade) and could suggest trauma, malignancy, or a bleeding disorder as the potential etiology.

Additionally, any unusual odors or suspensions observed in the fluid can further aid in diagnosis and can help in performing targeted diagnostic tests. However, sometimes the gross appearances can be deceptive; for instance, septic exudates may vary from being clear to opaque. Nonetheless, this initial gross examination helps prioritize the need for more detailed cytomorphological analysis to identify the root cause of the effusion and facilitates timely management decisions.

Table 1.1 Commonly encountered gross features of effusion fluid samples and their possible etiologies.

Gross characteristics	Potential etiologies
Thick yellowish-green, foul-smelling	Suppurative infections and tuberculosis
Orange to deep red	Trauma and malignancies
Black to chocolate brown	Melanoma
Milky	Lymphatic blockade and lymphoma
Thick mucoid	Mucinous neoplasms

Table 1.1 lists the gross features of fluid samples and the possible etiologies associated with them (Shidham & Atkinson, 2007).

1.7 Transudates and exudates

Effusions in body cavities are classified into two primary categories: transudates and exudates. Correctly distinguishing between these types is critical for diagnosing the underlying cause and determining the most effective treatment.

Transudates: These are fluids that accumulate in body cavities due to systemic factors affecting fluid dynamics, particularly changes in hydrostatic or oncotic pressures. They generally result from conditions that alter the equilibrium between fluid compartments rather than from localized pathological processes. Transudates are typically clear, pale yellow, and have low protein content.

Exudates: These fluids result from local pathological processes such as inflammation, infection, or malignancy. They are produced due to increased capillary permeability or lymphatic obstruction, leading to the leakage of high-protein fluid into body cavities. Exudates are often cloudy, turbid, or bloody and have a higher protein concentration.

1.7.1 Diagnostic characteristics

The differentiation between transudates and exudates is primarily based on specific physical and biochemical characteristics of the effusion fluid. Table 1.2 highlights the major differences between transudates and exudates (Mercer et al., 2019).

1.8 Cytologic processing techniques for effusion samples

Effusion fluid, collected from body cavities such as the pleural, peritoneal, or pericardial spaces, needs to be processed effectively to yield reliable diagnostic information. Various cytologic processing techniques, detailing their advantages and applications in diagnostic cytology, are as follows:

Table 1.2 Major differences between transudates and exudates.

Characteristics	Transudate	Exudate
Pathogenesis	Ultrafiltration of plasma; increased hydrostatic pressure with normal vascular permeability	Increased vascular permeability; inflammation induced
Gross appearance	Clear to straw-colored	Yellow to brown to deep red
Clot formation	Mostly not seen	Often noted due to high fibrinogen levels
Specific gravity	<1.018	>1.018
pH	>7.3	<7.3
Protein content	Low; <3 g/dL	High, >3 g/dL
Lactate dehydrogenase levels (LDH)	Low; <200 U/L	High; >200 U/l
Effusion LDH/Serum LDH	<0.6	>0.6
Glucose levels	Same as plasma	Low; <60 mg/dL
Light's criteria	Effusion protein/serum protein <0.5	Effusion protein/Serum protein >0.5
Cellularity	Low	High
Type of cells	Lymphocytes and mesothelial cells	Neutrophils, lymphocytes, and histiocytes, with or without malignant cells
Bacterial growth	Absent	May be noted
Common causes	Cirrhosis, nephrotic syndrome, congestive cardiac failure, constrictive pericarditis, pulmonary embolism, and protein-energy malnutrition	Bacterial infections, tuberculosis, connective tissue disorders, malignancies, and pancreatitis

1.8.1 Sediment smears

Sediment smear preparation is one of the most used techniques for the cytologic examination of effusion fluid samples. The process involves centrifugation (1500 rpm for 10 min) of the fluid sample to obtain a cell-rich sediment at the bottom of the centrifugation tube. The sediment is then spread onto the glass slides. After spreading, a minimum of one slide is air-dried, and the other is immediately fixed using alcohol based fixatives to preserve the cell morphology.

The air-dried smear can then be stained using May Grunwald Giemsa (MGG) or Leishman stain, and the wet-fixed smear can be stained using various stains, such as Hematoxylin & Eosin (H&E), Giemsa, or Papanicolaou stain, which highlight different cellular and background components. Additional smears can also be prepared from the sediment for performing special stains to aid in making a precise cytologic diagnosis. This technique, however, is less suitable for fluids with low-volume and low cellularity (Srinivasan et al., 2020).

1.8.2 Membrane filtration

Membrane filtration is an effective method for concentrating cells from fluid samples in cytologic analysis, especially when dealing with large volumes of fluid. This technique involves passing a

Table 1.3 Different types of filters commonly used in effusion fluid cytology.

Filter type	Pore size	Advantages
Nylon filters	0.45–5 μm	High strength and flexibility
Polycarbonate filters	0.2–5 μm	Highly durable
Cellulose acetate filters	0.45–8 μm	Rapid filtration and better cell retention
Polytetrafluoroethylene filters	0.45–5 μm	High resistance to chemicals and high temperatures

fluid sample through a membrane filter, which retains the cells while allowing the fluid to pass through.

1.8.2.1 Advantages of membrane filtration

- *Concentration of cells*: Membrane filtration is ideal for concentrating cells from large volumes of fluid, making it easier to detect and analyze rare or low-concentration cells. This is particularly useful for fluids like pleural, peritoneal, or cerebrospinal fluid.
- *Improved sample clarity*: By removing excess fluid, debris, and other unwanted materials, the membrane filter helps improve the clarity of the sample, helping in better cytomorphologic assessment.
- *Preservation of cellular integrity*: The filtration process minimizes potential cell damage compared to methods like centrifugation. Cells are retained in their natural state, which is crucial for accurate evaluation.

1.8.2.2 Limitations

- A significant limitation of membrane filtration is the risk of filter clogging, especially with viscous fluids or highly cellular samples. Selecting the appropriate filter pore size is essential to prevent this.

Different types of filters that can be used in effusion fluid processing are tabulated in Table 1.3 (Gill, 2012; Shidham, 2022).

1.8.3 Cytocentrifugation

Cytocentrifugation is a technique used to concentrate cells from low-volume and low cellularity effusion samples onto a slide. The fluid is placed in a special centrifuge chamber where it is spun at a constant speed. The centrifugal force causes cells to settle onto the surface of the glass slides, forming a monolayer (Gill, 2012; Stokes, 2004).

1.8.3.1 Procedure

- The labeled slides, filter card, sample chamber, and slide clip are placed into the cytospin head.

Table 1.4 Characteristic features of SurePath and ThinPrep liquid-based preparations.

Characteristics	SurePath preparation	ThinPrep preparation
Fixative	Ethanol-based	Methanol-based
Principle	Density gradient centrifugation	Membrane filtration
Cellularity and cell distribution	Cell concentration with uniform distribution, with a smear diameter of 13 mm	Cell concentration with uniform distribution, with a smear diameter of 20 mm
Monolayer	Scattered three-dimensional cell aggregates are noted	True monolayered preparations, with minimum to no cellular clumping
Cellular morphology	Better preservation of cellular morphology	Relatively more cellular shrinkage, flattening, and fragmentation of large cellular sheets and nuclear chromatin patterns noted in comparison to SurePath preparations
Preservation artifacts	Minimum	No to minimum
Automation	Compatible	Compatible
Immunocytochemistry	Can be easily performed	Can be easily performed
Molecular testing	Can be easily performed	Can be easily performed

- Around 6–10 drops of the fluid sample are put into the chamber.
- The lid is closed and the sample is centrifuged at 1000 rpm for 5 min.
- The clip assembly and slide are removed.
- The slides are then air-dried and fixed in 95% alcohol, stained accordingly, and coverslipped.

1.8.3.2 Advantages

1. *Cell concentration*: Cytocentrifugation effectively concentrates cells from dilute samples, which enhances the detection of abnormal cells.
2. *Improved cell preservation*: Cells are better preserved in a monolayer, allowing for clearer visualization under the microscope.

1.8.4 Liquid-based cytology

Liquid-based cytology involves processing effusion fluids through a filtration system or multiple centrifugations to obtain a uniform cell suspension. The cells are then deposited onto a slide in a thin, even cellular monolayer. The slide is then fixed and stained using the Papanicolaou stain, similar to the routine conventional cytologic smears.

Two commonly employed FDA-approved liquid-based cytology systems are available for cytologic sample processing. These include SurePath and ThinPrep. Both these systems have considerable advantages over conventional smears and have almost similar efficacy for diagnostic interpretation of effusion samples. The characteristic features of SurePath and ThinPrep liquid-based preparations (LBPs) are listed in Table 1.4 (Makde & Sathawane, 2022).

Table 1.5 Preferred cytopreparation method based on the physical properties of the samples.

Fluid sample properties	Sample processing technique
Low-volume and clear fluid	Cytocentrifugation and liquid-based preparation
Low-volume and mucoid fluid	Mucolytic fixative, sediment smears, and liquid-based preparation
High-volume and clear fluid	Sediment smears
High-volume and hemorrhagic fluid	Hemolyzing fixative, cytocentrifugation, and liquid-based preparation
High-volume and turbid fluid	Sediment smears

1.8.4.1 Advantages of liquid-based cytology techniques

1. *Cell concentration*: Helps in cellular concentration, especially useful for low cellularity samples.
2. *Uniform cell distribution*: Provides a uniform distribution of cells.
3. *Minimal obscuration*: Reduced likelihood of cell overlapping and clumping, thereby improving diagnostic accuracy.
4. *Minimal artifacts*: Minimizes artifacts caused by air-drying or uneven spreading.
5. Various ancillary tests can be performed, including molecular testing.

To summarize, the best method of cytopreparation may be selected based on the physical properties of the sample, as shown in Table 1.5 (Koss, 1992; Keebler et al., 2008).

1.9 Cytopreparation

An accurate cytodiagnosis relies on proper sample collection, thorough evaluation of the sample, correct fixation, preparation, and staining of the cells. The core principle in cytopreparation is to ensure the preservation of the cellular morphology, as close as possible to their natural form, for accurate interpretation and diagnosis.

1.9.1 Cytologic fixatives

The primary purpose of cytologic fixatives is to preserve the cytomorphologic characteristics and essential cytochemical elements of the cell as precisely as possible. An ideal cytologic fixative should rapidly penetrate the cells, minimize cellular shrinkage, and maintain their morphological integrity. It must also deactivate autolytic enzymes, replace cellular water, and facilitate the diffusion of dyes across cell membranes. Additionally, it should promote and not inhibit cell adhesion to the slide surface, and prevent microbial growth, ensuring reliable and accurate diagnostic outcomes (Dey, 2023; Gill, 2012; Keebler et al., 2008).

1.9.2 Fixation methods

- *Wet fixation*: Wet fixation involves immediately immersing the cytologic smear in a fixative solution, where it remains until it reaches the laboratory for accessioning and staining. Throughout this process, the cells are not exposed to air. Alcohols such as 95% ethanol, isopropanol, and methanol are all effective cytologic fixatives (Dey, 2023; Keebler et al., 2008).

- *Wet fixation with air-drying*: Wet fixation with subsequent air-drying involves immediately immersing the cell sample in a fixative solution for a specified duration, after which the sample is removed from the fixative, air-dried, and then placed in a container for transport to the laboratory. Upon receipt in the laboratory, the slide is again immersed in 95% ethanol or a similar cytologic fixative solution before staining (Dey, 2023; Gill, 2012; Keebler et al., 2008).
- *Spray fixation*: Spray fixation involves immediately applying a commercially available spray fixative to the wet cytologic smear, which is then air-dried and placed in a container for transport. This method is often used for fixing smears before long-distance transport. The spray fixatives typically consist of an alcohol base combined with a waxy substance, such as carbowax, which is sprayed onto the slide from a distance of 3–10 inches. This provides a thin protective coating over the cells. In the laboratory, the waxy coating of the spray fixative must be removed before staining by rinsing the slide 2–3 times in 95% ethanol solutions, for a minimum of 15–30 min. If the carbowax is not completely removed, cellular staining is altered, leading to foggy nuclei and pale cytoplasm (Keebler et al., 2008).
- *Air-drying*: Air-drying of the smears helps in better delineation of the cytoplasmic characteristics. The air-dried sample is stained with methods such as MGG, Wright-Giemsa, or Diff-Quik (Dey, 2023; Keebler et al., 2008).

1.9.3 Staining of the cytologic smears

The commonly used staining methods in the cytology laboratory include Papanicolaou staining, MGG, Leishman, Wright-Giemsa, Diff-Quick, and Hematoxylin and Eosin. A variety of special stains, including Ziehl-Neelson (ZN) stain, Periodic acid Schiff stain, Wade-file stain, Modified ZN stain, Oil Red O, and Perl's Prussian Blue, can also be performed on the cytologic smears.

1.10 Processing of hemorrhagic samples

Processing hemorrhagic cytologic fluid samples requires careful processing to ensure that diagnostic cells are preserved and readily identifiable with minimal obscuration. Specialized methods such as red blood cell (RBC) lysis, membrane filtration, and density gradient centrifugation are commonly used to concentrate cells, remove excess RBCs, and improve sample quality to ensure accurate diagnostic interpretation (Gill, 2012; Srinivasan et al., 2020).

Various methods that can be employed for processing hemorrhagic fluid samples include:

1.10.1 Methods to be used before sample processing

- *RBC lysis*: This method involves using a lysing solution to remove RBCs from the sample before further processing, to improve the clarity of the sample. Various hemolyzing solutions that can be used for this purpose are tabulated in Table 1.6.

Table 1.6 Various hemolyzing solutions that can be used for processing of hemorrhagic fluid samples.

Hemolyzing solution	Composition
Freshly prepared Carnoy's fixative	Absolute ethanol, chloroform, and glacial acetic acid (6:3:1)
Modified Carnoy's fixative	95% ethanol, chloroform, and glacial acetic acid (7:2.5:0.5)
Clarke's solution	Absolute ethanol and glacial acetic acid (3:1)
CytoLyt	Ethanol, isopropanol, acetic acid, buffered salts, and water
CytoRich red	Ethanol, acetic acid, buffered salts, water, and stabilizing agents

1.10.2 Methods to be used during sample processing

- *Membrane filtration and ThinPrep technique:* This method helps filter the RBCs and concentrate the other cells, which are then transferred to slides for staining.
- *Ficoll-hypaque-based density gradient centrifugation and SurePath technique*: This method separates the cells based on their density using a gradient medium. The sample is layered on a density gradient medium (Ficoll-Hypaque) and centrifuged, leading to the separation of cells into different layers. The desired layer can then be used for smear preparation.

1.10.3 Methods to be used after sample processing

- *Preparation of "thick and thin" and "fish-tail" smears*: The thicker areas of the smear ensure cell concentration, and the thinner areas can be used for better evaluation of cell morphology (Gill, 2012; Stokes, 2004; Bales & Durfee, 1992).
- *Use of saponin solution (0.1%–1%)*: Saponin is a natural surfactant derived from plants that can be used in cytologic fluid processing. It functions as a mild detergent to selectively disrupt the cell membrane of RBCs, causing them to lyse without affecting other types of cells in the sample. The smear may be dipped in saponin solution for a short time to ensure effective RBC lysis after slide preparation (Gill, 2012; Stokes, 2004; Bales & Durfee, 1992).
- *Use of urea solution (5–8 Molar)*: Urea solution works by selectively disrupting the osmotic balance of RBCs, leading to their lysis, while maintaining the integrity of other cells. A quick dip in a 5–8 M urea solution for a few seconds can help in RBC lysis after smear preparation (Gill, 2012; Stokes, 2004; Bales & Durfee, 1992).

1.11 Cell block preparation

While the routine preparation of CBs is not universally practiced in all cytopathology laboratories, it is strongly recommended. CBs often provide significant insights to pathologists, particularly those who are more familiar with histologic sections. The process of preparing a CB involves adding a coagulating/solidifying agent, such as thrombin, gelatin, or agar, to the effusion fluid sediment to ensure entrapment of the cell sediment. Once this is achieved, it is embedded in paraffin wax and processed as a histopathologic tissue block. Thin sections are obtained onto glass slides, and staining for subsequent microscopic examination (Torous et al., 2023).

Table 1.7 Various methods available for cell block preparation in effusion fluid cytology.

Cell block preparation method	Chief reagents	Principle
Plasma-thrombin method	Pooled plasma and thrombin	Involves the coagulation of plasma using thrombin to form a solid clot in which the concentrated cells from the fluid sample become entrapped. The cells are preserved in an intact form within the clot and are then processed into a paraffin-embedded block for further sectioning.
Sodium alginate method	Calcium chloride and sodium alginate	Calcium chloride solution is added to the mixture of sodium alginate and concentrated cells. The calcium ions cross-link the sodium alginate chains, resulting in a firm, three-dimensional gel structure (ionotropic gelation) that entraps the cells.
HistoGel	Commercially available HistoGel specimen processing gel is used	Involves using Histogel, a proprietary gel medium, to entrap and preserve cells concentrated from cytological samples.
Collodion bag method	Collodion: ether-based solvent	Celloidin is a viscous, gelatinous substance that helps preserve the cells in their native form and also provides a firm embedding medium that allows the cytologic specimen to be processed into a block for sectioning.
Agar embedding method	2% molten agarose	Involves embedding the concentrated cells in a gel-like matrix made from agar, a natural polysaccharide derived from seaweed, to preserve their morphology and facilitate sectioning.
Gelatin embedding method	2% molten gelatin	A warm gelatin solution is mixed with the cell pellet, solidifies on cooling leading to entrapment of the cells, and also provides a firm structure for further sectioning.
Automated cell block systems	Shandon cytoblock and cellient automated cell block systems	

A variety of methods are available for CB preparation and can be used by cytology laboratories depending on the available resources. The commonly used CB methods have been listed in Table 1.7 (Nambirajan & Jain, 2018; Shidham, 2019).

1.11.1 Advantages of cell blocks

1. *Cellular architecture*: CBs preserve the three-dimensional architecture of the tissue, providing a more comprehensive view of cellular and stromal relationships.
2. *Histological analysis*: Enables the use of histological stains and immunohistochemical techniques for detailed cellular analysis.

3. *Multiple sections*: Multiple sections can be obtained for performing additional microbiological or histochemical stains.
4. *Molecular testing*: CBs can also be used for nucleic acid-based molecular analyses to supplement or confirm the cytologic diagnoses.

1.12 Precautions to minimize sample cross-contamination during processing

- Laboratory technicians must carefully aspirate samples into pipette tips to avoid any sample entering the pipette barrel. If contamination occurs, the pipette should be removed, cleaned with alcohol and water, dried, and tested for contamination before reuse in processing (Bales & Durfee, 1992).
- Pipette tips must be replaced for each new sample.
- Nondisposable cytochambers and holders should be soaked for at least 30 min, washed, and dried before reuse.
- Staining solutions should be either filtered or replaced after each staining run with proper documentation.
- All water rinses should be discarded after use.
- Special staining should be carried out in separate Coplin jars, preferably with fresh reagents prepared daily.
- Cytoscreeners and cytopathologists are required to report any suspected contamination to the supervisor, who should document the incident and initiate corrective actions.
- If cross-contamination is suspected and no residual specimen sediment is available, the test should be canceled, the requesting clinician must be contacted, and if possible, a repeat sample may be requested.

1.13 Normal cellular constituents in effusion cytology

The exact cellular composition of effusion samples varies with the underlying disease etiology (Ali & Cibas, 2012; Shidham, 2021; Shidham & Atkinson, 2007). However, the most commonly encountered cellular constituents include:

1. *Mesothelial cells*: These cells line the body cavities and are usually the most abundant cells in a body cavity effusion sample. Their morphology can vary with the underlying cause of effusion; however, the typical cytomorphologic features include:
   a. Seen as singly scattered or flat sheets of cells or occasionally as papillary clusters
   b. Round to cuboidal in shape with characteristic "windows" in between the cells
   c. Round, central to eccentrically placed nuclei, fine chromatin, absent to small nucleoli
   d. Dense cytoplasm with peripheral skirting or fine vacuolations
   e. Reactive mesothelial cells may show nuclear enlargement, small nucleoli, and a single large vacuole imparting a "signet-ring"-like appearance (Figs. 1.1 and 1.2).

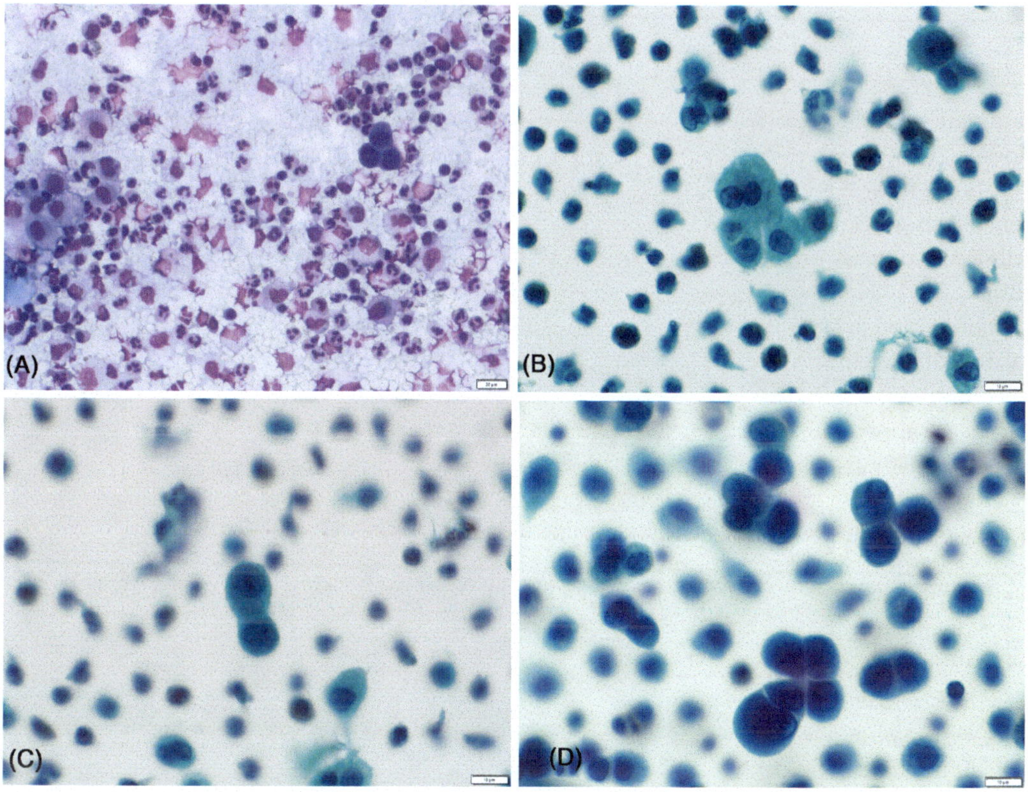

FIGURE 1.1 Benign mesothelial cells. (A) Conventional sediment smears showing groups of benign mesothelial cells with dense cytoplasm and intercellular windows, background shows neutrophils (MGG; 40×); (B–D) Liquid-based cytology preparation showing aggregates of benign mesothelial cells (Papanicolaou; B,C:40×, D:100×).

2. *Lymphocytes*: These are the second most common cell types seen in effusion fluids. Lymphocytes are usually small, round cells with a small nucleus, clumped chromatin, inconspicuous nucleoli, and scant cytoplasm. Effusions with predominant lymphocytes are generally reported as lymphocyte-rich effusions, and the common causes include tuberculosis, viral infections, chronic inflammatory diseases, autoimmune diseases, chylothorax, and lymphomas (Fig. 1.3).
3. *Macrophages*: These cells, which originate from monocytes, are generally present in small numbers in effusion fluids. They may exhibit variable cytomorphologic features and hence are often misinterpreted as malignant cells. The typical cytomorphologic features of macrophages include:
   a. Round to epithelioid cells, mostly scattered singly and rarely in small aggregates
   b. Fuzzy to ill-defined cell borders, round to oval, eccentrically placed, vesicular nuclei, and small nucleoli
   c. Abundant vacuolated cytoplasm with phagocytosed debris (Fig. 1.4)

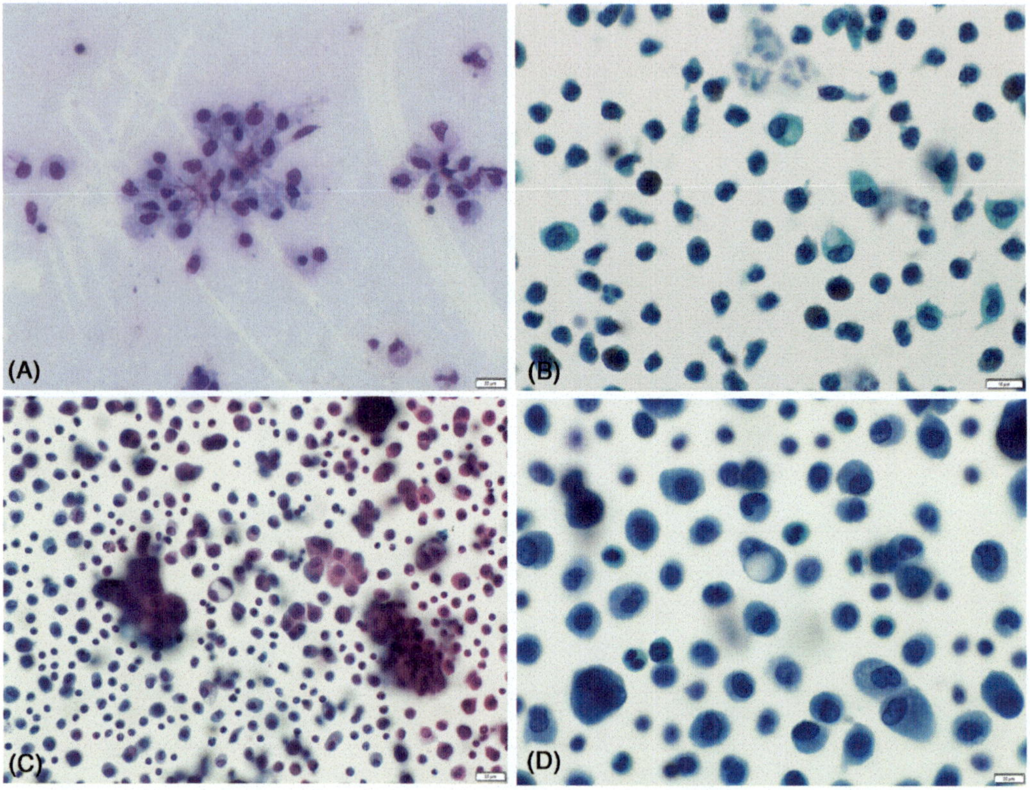

FIGURE 1.2 Reactive mesothelial cells. (A) Conventional sediment smear showing aggregates of reactive mesothelial cells with vacuolated cytoplasm (MGG; 40×); (B–D) Liquid-based cytology preparation showing singly scattered as well as clusters of reactive mesothelial cells, occasional binucleated cells also noted (Papanicolaou; B,C:40×, D:100×).

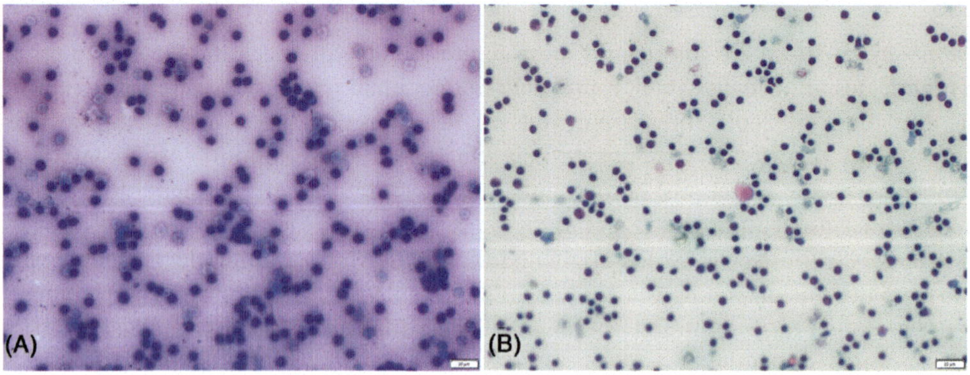

FIGURE 1.3 Lymphocytes in effusion fluid. (A) Conventional sediment smear showing dispersed population of small mature lymphocytes (MGG; 40×); (B) Liquid-based cytology preparation showing dispersed lymphocytes with occasional mesothelial cells (Papanicolaou; B:40×).

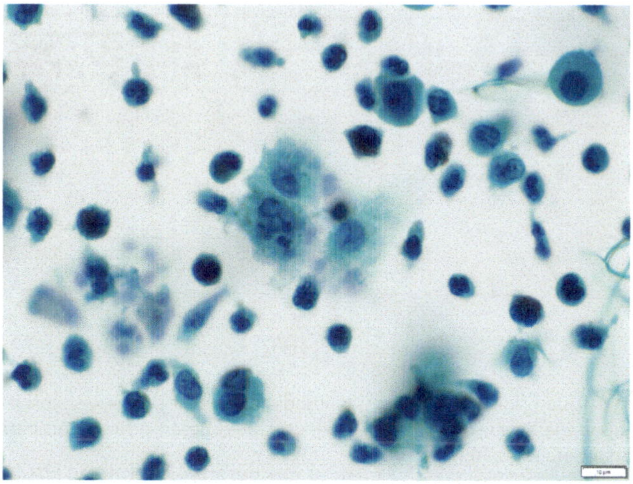

FIGURE 1.4 Macrophages in effusion fluid. Liquid-based cytology preparation showing aggregates of macrophages in the center with benign mesothelial cells, eosinophils, and neutrophils in the background (Papanicolaou; 100×).

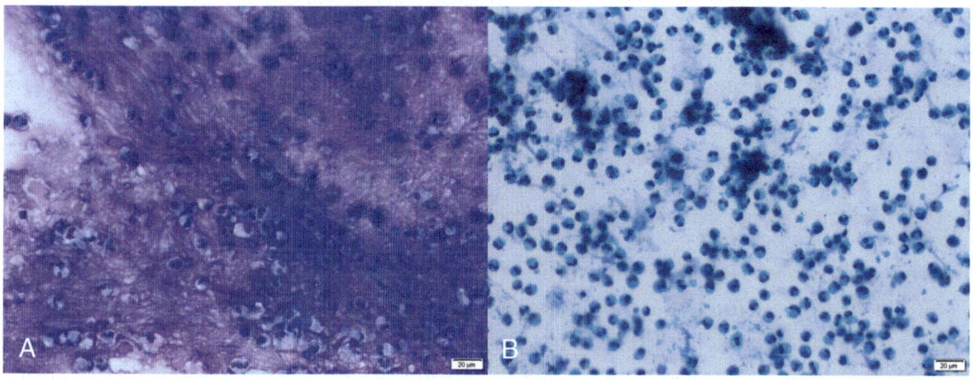

FIGURE 1.5 Neutrophils in effusion fluid. (A) Conventional sediment smear showing sheets of neutrophils with necrotic background (MGG; 40×); (B) Liquid-based cytology preparation showing dispersed neutrophils (Papanicolaou; B:40×).

4. *Neutrophils*: A small number of neutrophils, also known as polymorphonuclear cells, are commonly seen in effusion fluids. Higher numbers of neutrophils are often seen in conditions like tuberculosis, suppurative bacterial infections, acute inflammatory diseases, posttrauma, pancreatitis, and other diseases with superadded suppuration (Fig. 1.5).
5. *RBCs*: Generally, if there is no trauma or active bleeding, only a small number of RBCs are seen in effusion samples, or they may be completely absent. Significantly higher numbers of RBCs indicate hemorrhagic effusions, which may be associated with malignancies, trauma, bleeding disorders, vascular diseases, or infections.
6. *Eosinophils*: These cells are typically rare in effusion samples; however, their numbers are significantly higher in effusions associated with parasitic infestations, allergic reactions, or certain autoimmune diseases (Fig. 1.6).

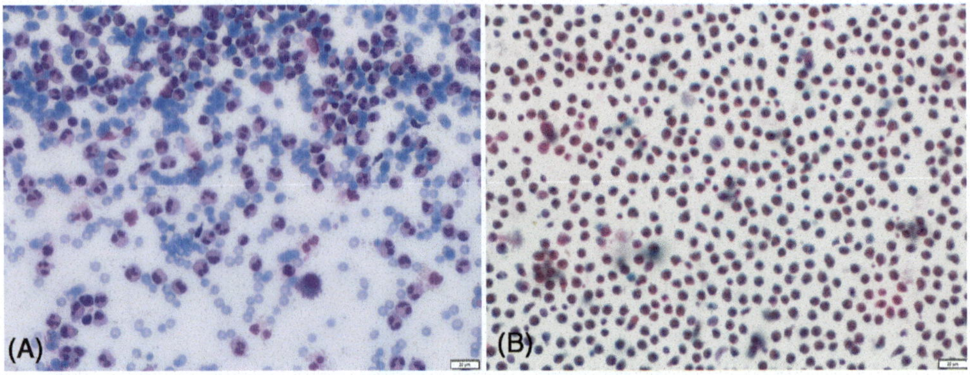

FIGURE 1.6 Eosinophils in effusion fluid. (A) Conventional sediment smear showing sheets of eosinophils with occasional mesothelial cells in the background (MGG; 40×); (B) Liquid-based cytology preparation showing dispersed eosinophils (Papanicolaou; B:40×).

1.14 Reporting systems in effusion cytology

Two standardized systems are available for reporting serous fluid cytopathology. Broadly, the two systems provide similar guidelines regarding the collection, processing, and reporting of the effusion fluid samples to ensure uniformity and reproducibility.

1.14.1 International system for reporting serous fluid cytopathology

The international system for reporting serous fluid cytopathology, first published in 2020, provides standardized guidelines that help cytopathologists interpret as well as communicate their findings to the treating clinicians in a clear and structured manner.

Various diagnostic categories and the associated risk of malignancy provided by the ISRSFC are presented in Table 1.8 (Chandra et al., 2020; Pinto et al., 2020) (Figs. 1.7 and 1.8).

1.14.2 Indian academy of cytologists guidelines for collection, preparation, interpretation, and reporting of serous effusion fluid samples

In December 2019, the Indian Academy of Cytologists published guidelines to ensure a standardized format for processing and reporting effusion cytopathology. These guidelines broadly cover all the aspects related to sample collection, preparation, and diagnostic categories of effusion specimens to ensure uniformity in cytologic reporting and better communication of the results to the clinicians.

The recommended diagnostic categories, associated risks of malignancy, and the recommendations provided by the IAC guidelines are presented in Table 1.9 (Kundu et al., 2021; Srinivasan et al., 2020).

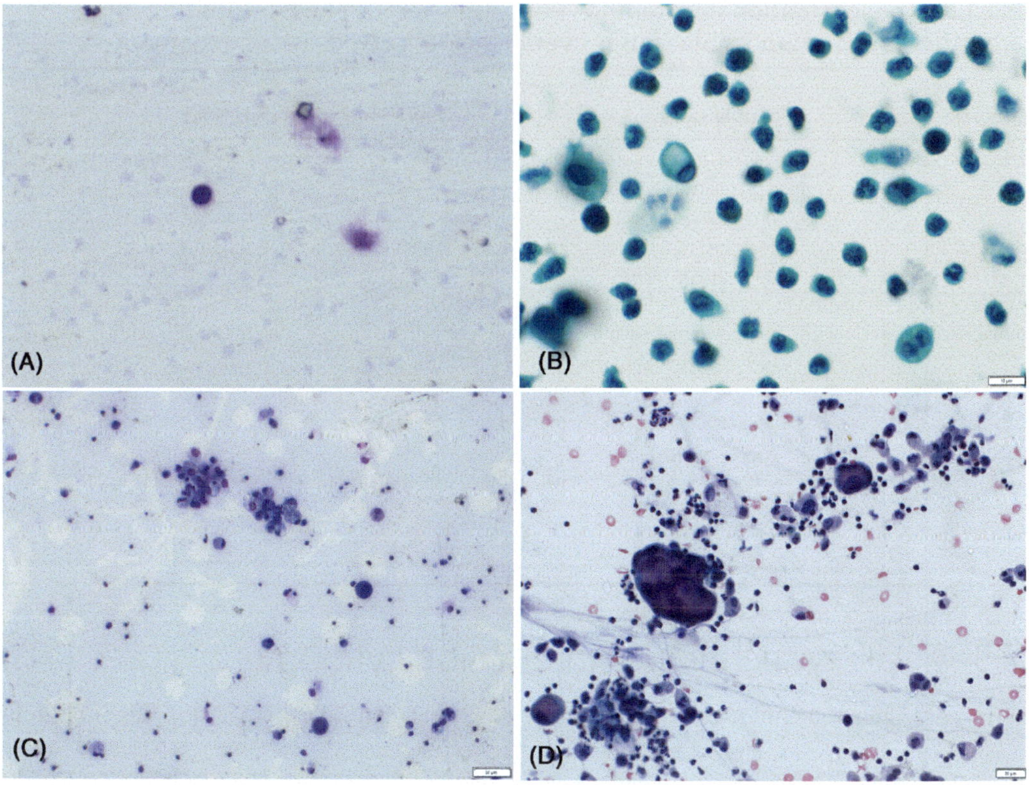

FIGURE 1.7 Cytological preparations showing various diagnostic categories. (A) Nondiagnostic sample: Conventional sediment smear showing an occasional lymphocyte and degenerated cells in a fluidy background (MGG; 40×); (B) Negative for malignancy: Liquid-based cytology preparation showing reactive mesothelial cells and few neutrophils (Papanicolaou; B:40×); (C) Atypical cells: Occasional scattered large atypical cells in a background showing reactive mesothelial cells (MGG; 20×); (D) Suspicious for malignancy: Conventional sediment smear showing an occasional cluster and singly scattered atypical cells, which are suspicious for malignancy (Papanicolaou; 40×).

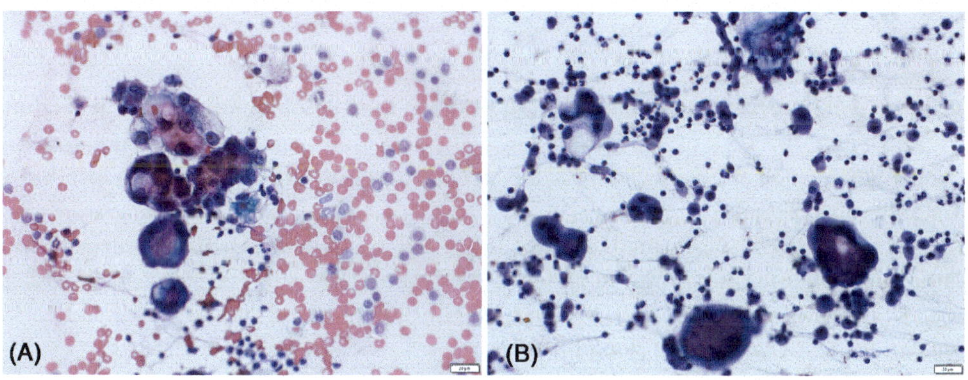

FIGURE 1.8 Malignant effusion. (A, B) Conventional sediment smears showing clusters of tumor cells with marked cellular pleomorphism, round to irregular nuclei, coarse chromatin, prominent nucleoli, and abundant vacuolated cytoplasm, indicating metastatic adenocarcinoma in the effusion sample (Papanicolaou; 40×).

Table 1.8 The International System for Reporting Serous Fluid Cytopathology: Diagnostic categories, possible causes overall incidence and risk of malignancy.

Diagnostic category	Possible causes	Overall incidence (%)	Risk of malignancy (%)
I. Nondiagnostic	Acellular/paucicellular sample; significant cellular obscuration; processing artifacts	0.9	17.4 ± 8.9
II. Negative for malignancy	Infections, renal failure, cirrhosis, and cardiac failure	70–80	21 ± 0.3
II. Atypia of undetermined significance	Reactive atypia; Benign tumors with cytologic atypia; well-differentiated malignant tumors	0.6–1.6	66 ± 10.6
IV. Suspicious for malignancy	Malignant tumors; rarely marked reactive atypia	3.6–6.3	82 ± 4.8
V. Malignant	Malignant tumors (epithelial, mesothelial, germ cell, mesenchymal, melanoma, and lymphoma)	30–50	99 ± 0.1

1.15 Ancillary techniques in effusion cytology

A variety of ancillary techniques can be employed on effusion fluid cytologic samples, including:

1.15.1 Special cytochemical stains

Special cytochemical stains are commonly employed in effusion cytology to help differentiate between various etiologies, such as infections, malignancies, or inflammatory diseases. Table 1.10 highlights various special stains that can be performed on the effusion fluid samples for reaching a definitive diagnosis (Fig. 1.9) (Dey, 2023; Shidham & Layfield, 2021).

1.15.2 Immunocytochemistry

ICC involves using specific antibodies to either effusion fluid cytology smears or CB sections to detect and characterize various cellular antigens. The antibodies are typically conjugated with a detectable marker, such as an enzyme or a dye, which, after binding to the antigen, produces an interpretable visual signal.

Samples: ICC can be performed reliably using CB sections, cytospin smears, ThinPrep, SurePath LBPs, membrane filtration smears, and sediment smears (Figs. 1.10 and 1.11) (Dey, 2023; Shidham & Layfield, 2021; Srinivasan et al., 2020).

Applications of ICC in effusion cytology:

- *Diagnosis of malignancy*: Especially useful in diagnosing malignant effusions in cases with morphological ambiguity.

Table 1.9 The Indian Academy of Cytologists guidelines for reporting serous fluid cytopathology: Diagnostic categories, overall incidence, and risk of malignancy.

Diagnostic category	Cytopathologic diagnoses and remarks	Overall incidence (%)	Risk of malignancy (%)
I. Unsatisfactory for evaluation/ nondiagnostic	Acellular sample; significant cellular obscuration; extensive cellular degeneration; processing artifacts	2.6	20
II A-No malignant cells detected II B-Benign changes seen	Reactive mesothelial cells, Inflammatory cells seen Lymphocyte-rich effusion Specific infections: Tuberculosis, microfilaria, fungal infection, hydatid cyst, and others	71.2	16.7
III. Atypical cells, not otherwise specified	Repeat cytology correlate clinically and with imaging studies ancillary technique (Optional)	1.3	50
IV. Atypical cells, suspicious for malignancy	Repeat cytology Ancillary techniques	4.4	94.4
V. Malignant cells seen	Mesothelial or nonmesothelial; subtype wherever possible	20.5	100

Table 1.10 Special stains commonly performed on effusion fluid samples.

Special stain	Purpose
Gram stain	Detection of gram-positive and negative organisms in infection-associated effusions
Ziehl-Neelson stain	Detection of acid-fast bacilli, including *Mycobacterium tuberculosis,* in suspected cases
Periodic acid schiff (PAS) stain	Detection of fungal organisms, glycogen, and neutral mucins
Alcian blue-periodic acid schiff (AB-PAS) stain	Detection of acidic mucins
Meyer's mucicarmine	Detection of acidic mucopolysaccharides, cryptococcus (highlights the capsule)
Gomori methenamine silver	Detection of fungal organisms and including Pneumocystis
Oil red O	Detection of lipid-containing cells
Perl's prussian blue	Detection of hemosiderin-containing cells
Hale's colloidal iron	Detection of acidic mucins and amoebiasis

- *Determining the primary source of metastasis and classifying malignant tumors*: Identifying the primary site of metastasis by delineating the cellular differentiation and differentiating between various types of malignancies.
- *Distinguishing between benign and malignant effusions*: In challenging cases, ICC can help in differentiating reactive mesothelial cells from malignant cells.

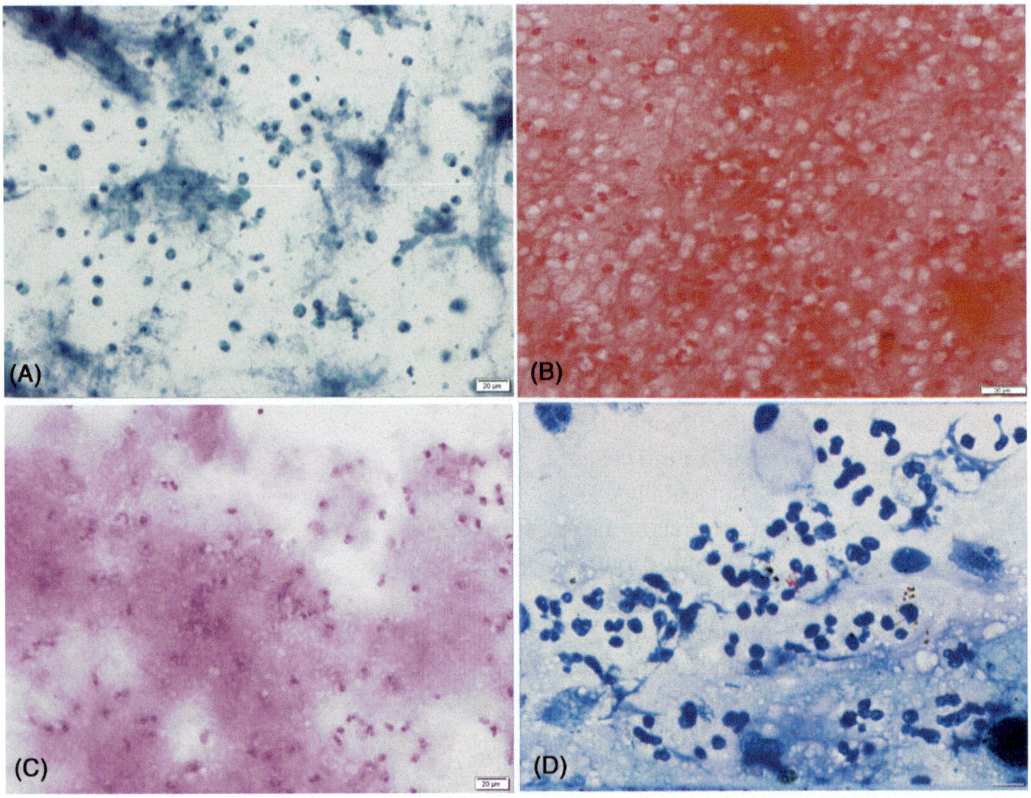

FIGURE 1.9 Special cytochemical stains. (A) Liquid-based cytology preparation showing scattered neutrophils in a necrotic background (Papanicolaou; 40×); (B) Gram-stained smear to look for the presence of gram-positive bacteria (Gram stain; 40×); (C) Periodic acid Schiff (PAS) stained smear to detect fungal profiles (PAS stain; 40×); (D) Ziehl-Neelson (ZN) stained smear highlighting the magenta-colored beaded acid-fast bacilli (ZN stain; 100×).

- *Identifying specific subtypes of malignancy*: ICC markers can help in identifying specific subtypes, like HER2, ER, and PR, in breast cancer.
- *Grading of neuroendocrine tumors*: Ki-67 ICC can be used in the grading of neuroendocrine tumors.
- *Prognostication of various malignancies*: PD-L1 ICC can help in the prognostication of various tumors, including lung cancer.

Table 1.11 lists various immunocytochemical markers that are commonly used in effusion cytology to characterize specific etiologies (Koss, 1992; Shidham & Atkinson, 2007; Shidham & Janikowski, 2022).

1.15.3 Flow cytometry

Flow cytometry in effusion cytology is a useful diagnostic tool that can help in the quantification as well as comprehensive phenotypic analysis of cells in effusion fluids. It is particularly useful

Table 1.11 Commonly used immunocytochemical markers in effusion fluid cytology to characterize specific etiologies.

Immunocytochemical Marker	Purpose
Pan-cytokeratin	Epithelial differentiation
Epithelial membrane antigen	Epithelial differentiation
Claudin-4	Epithelial differentiation
MOC 31	Epithelial differentiation
BerEP4	Epithelial differentiation
Keratin 7	Foregut epithelial tumors, lung adenocarcinoma, gastric cancer, fibrolamellar hepatocellular carcinoma, papillary renal cell carcinoma, and ovarian carcinoma
Keratin 20	Colorectal cancer, gastric cancer, and Merkel cell carcinoma (dot-like)
Keratin 5/6	Squamous cell carcinoma
Vimentin	Mesenchymal differentiation, sarcomas, endometrioid carcinoma (basolateral), and mesothelioma
CD 45/ leucocyte common antigen	Lymphocytic differentiation and lymphomas
CD 20	B- lymphocytic differentiation and B-cell lymphomas
PAX 5	B- lymphocytic differentiation and B-cell lymphomas, RS cells in classic type Hodgkin lymphoma
CD 3	T-lymphocytic differentiation and T-cell lymphomas,
CD 68 and CD 168	Histiocytic differentiation
CD 38 and CD 138	Plasma cell differentiation
ALK	Surrogate marker for *ALK* rearrangements, Anaplastic large cell lymphoma, lung adenocarcinoma
TTF1	Thyroid, thymic, and lung adenocarcinoma
ROS-1	Surrogate marker for *ROS-1* rearrangements in lung cancer
PD-L1	Prognostic marker in various tumors, including lung cancer
D2-40	Mesothelial and lymphatic differentiation
Calretinin	Mesothelial and sex cord differentiation
BAP1	Surrogate marker for *BAP1* deletion in mesothelioma
MTAP	Surrogate marker for *MTAP* deletion in mesothelioma
WT1	Mesothelial differentiation, serous ovarian tumors
PAX8	Thyroid, thymic, and Mullerian differentiation
p53	Surrogate marker for *TP53* mutation
p16	Surrogate marker for *CDKN2A* mutation
Chromogranin, synaptophysin, and CD56	Neuroendocrine differentiation
INSM1	Neuroendocrine differentiation
HER2	Prognostic marker in various cancers, including gastrointestinal, breast, and endometrial cancers
Estrogen receptor, progesterone receptor	Hormone receptors in various tumors, including breast cancer
Ki-67	Proliferation marker
HMB-45 and Melan A	Melanocytic differentiation
Desmin	Muscle differentiation
Smooth muscle antigen and h-caldesmon	Smooth muscle differentiation

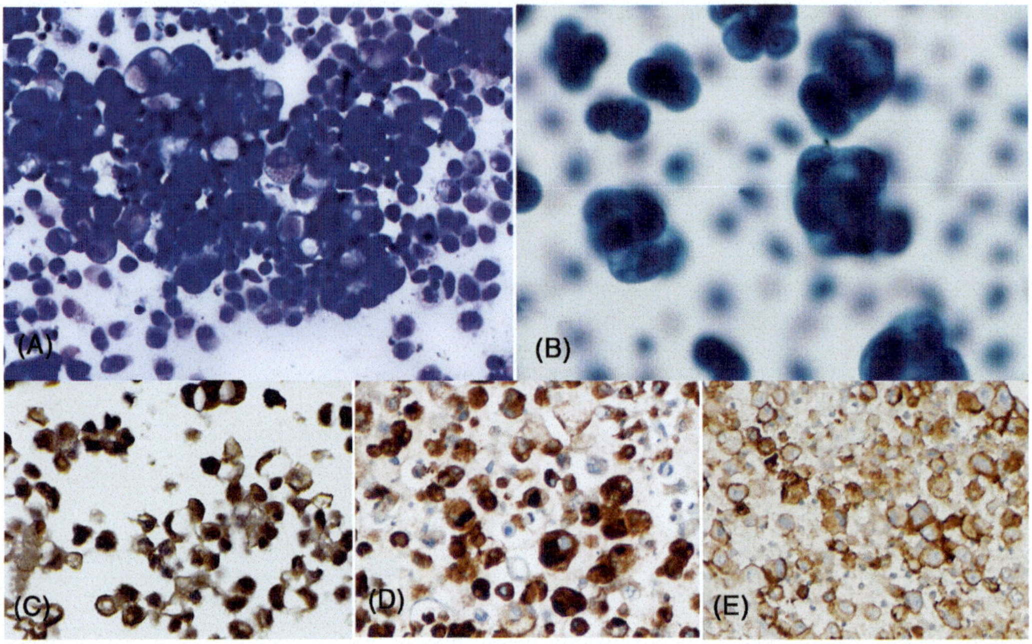

FIGURE 1.10 Immunocytochemistry. (A) Conventional sediment smears showing clusters of tumor cells indicating metastatic adenocarcinoma in the effusion sample (MGG; 40×); (B) Liquid-based cytologic preparation showing aggregates of tumor cells (Papanicolaou; 100×); (C–E) Immunocytochemistry showing tumor cells to be positive for TTF1 (C; 40×), ALK (D; 40×), and PD-L1 (E; 40×; Tumor proportion score >90%).

for characterizing cellular populations, diagnosing lymphoreticular as well as epithelial malignancies, and assessing the immunophenotype of suspicious or undifferentiated cells in various body cavity effusions. Flow cytometry provides much more detailed cellular information than cytomorphology alone, especially when combined with immunocytochemical and morphologic examination.

Flow cytometry involves passing specifically labeled cells in a single-cell suspension fluid stream through a laser beam. As cells pass through the laser, they scatter light and emit fluorescence, which is detected and analyzed by the flow cytometer (Wong-Arteta et al., 2020).

Table 1.12 lists various flow cytometric immunophenotypic markers that are commonly used in effusion cytology to characterize specific diseases (Gupta et al., 2021; Wong-Arteta et al., 2020).

Applications of flow cytometry in effusion cytology (Gupta et al., 2021; Wong-Arteta et al., 2020):

1. Detection and subtyping of hematologic malignancies
2. Cell subset identification and quantification (e.g., cancer stem cells)
3. Detection of minimal residual disease
4. DNA ploidy and cell cycle analysis
5. Detection of metastatic carcinomas

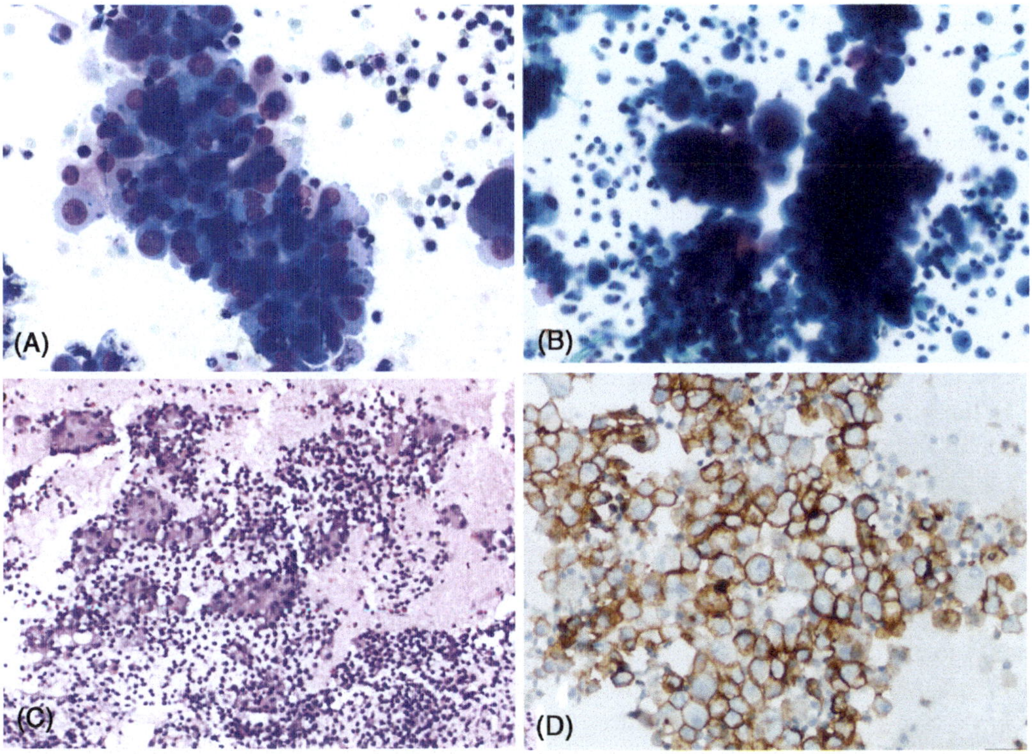

FIGURE 1.11 Immunocytochemistry on cell block. (A) Conventional sediment smears showing clusters of tumor cells indicating malignant mesothelioma in the effusion sample (MGG; 40×); (B) Liquid-based cytologic preparation showing similar three-dimensional balls of malignant mesothelial cells (Papanicolaou; 40×); (C) Cell block section showing similar tumor cells (Hematoxylin & Eosin; 20×); (D) Immunocytochemistry for D2-40 showing membranous positivity in the tumor cells (40×).

Table 1.12 List of commonly used flow cytometric immunophenotypic markers in effusion fluid cytology with their associated diseases.

Flow cytometric markers	Associated diseases
CD 45	Lymphomas
CD 19 and CD 20	B-cell lymphomas
CD 3	T-cell lymphomas
CD 5 and CD 23	Small cell lymphoma/chronic lymphocytic leukemia
CD 4, CD 7, and CD 8	T-cell lymphomas
CD 10	Precursor T-cell lymphoma and B-cell lymphoma
CD 13 and CD 33	Myeloid neoplasms
CD 117	Mast cell disorders and myeloid neoplasms
CD 23	Dendritic cell neoplasms
Epithelial membrane antigen	Anaplastic large cell lymphoma and epithelial malignancies
EPCAM and Cytokeratin	Epithelial malignancies
CD 38 and CD 138	Multiple myeloma
CD 30	Anaplastic large cell lymphoma, diffuse large B-cell lymphoma
CD 25, CD 103, CD 123, and CD 11c	Hairy cell leukemia

1.15.4 Electron microscopy

Electron microscopy (EM) is a highly specialized diagnostic tool in effusion cytology that can provide detailed, high-resolution images of cells at the ultrastructural level. Although EM is not routinely used in most clinical settings due to its technical complexity, time consumption, and high cost, it can be invaluable in certain specific situations where conventional cytology does not provide enough information (Shidham & Atkinson, 2007).

Applications of EM in effusion cytology(Shidham & Atkinson, 2007):

1. *Diagnosis of malignant effusions by identification of cellular differentiation of malignant cells*: EM can help identify specific ultrastructural features that differentiate malignant cells from benign cells, especially in challenging cases where conventional cytology is ambiguous. In mesotheliomas, EM helps in the identification of long microvilli, dense bodies, and tonofilaments within the cells, which are characteristic of mesothelial cells. In samples with metastatic neuroendocrine tumors, EM can help identify neurosecretory granules, which are diagnostic of neuroendocrine differentiation.
2. *Diagnosis of infectious organisms*: EM is particularly useful for detecting viruses, bacteria, and fungi that might be present in effusions. Viral particles can be directly visualized within the infected cells in effusions as intranuclear or intracytoplasmic inclusions. EM can reveal the ultrastructure of bacteria and fungi for precise diagnosis.
3. *Characterizing the individual tumor cells*: For poorly differentiated carcinomas or sarcomas, EM can reveal ultrastructural features that can help confirm a diagnosis of epithelial, mesenchymal, or neuroendocrine origin.

1.15.5 Molecular tests

In this era of personalized medicine, molecular testing in effusion cytology plays an indispensable role in diagnosing, characterizing, prognosticating, and monitoring diseases. Molecular tests can help in identifying the specific genetic profile of malignant cells, which helps in confirming the diagnosis, determining the primary tumor origin, and classifying the type of malignancy.

Samples: Cytoscrapes, supernatant fluid, LBC sediment, directly collected sample in RNAlater solution, and CBs.

Applications of molecular tests in effusion cytology:

1. *Diagnosis and characterization of malignant effusions*: Some cancers show characteristic mutations or gene amplifications that can be detected in effusion fluid using molecular techniques. For example, *EGFR* mutations in lung adenocarcinoma. *KRAS* mutations for identifying colon cancer. *HER2* amplification in breast cancer (Kundu *et al.*, 2021; Schmitt et al., 2021).
2. *Identification of the primary site of origin in metastatic effusions*: In malignant effusions with unknown primary malignancy, molecular tests, including next generation sequencing (NGS) and gene expression profiling, can often help identify the primary tumor origin by detecting its specific molecular genetic signatures (Awadallah et al., 2019).

Table 1.13 List of commonly used molecular tests in effusion fluid cytology.

Genetic abnormalities	Molecular tests	Associated conditions
EGFR mutations	Real-time polymerase chain reaction (PCR), next-generation sequencing (NGS)	Lung adenocarcinoma
ALK rearrangements	Fluorescence in situ hybridization	ALCL and lung adenocarcinoma
ROS 1 rearrangements	Fluorescence in situ hybridization (FISH)	Lung adenocarcinoma
BRAF mutations	Real-time PCR and NGS	Melanoma and papillary thyroid cancer
RAS mutations	Real-time PCR and NGS	Thyroid cancer
FOXO1 rearrangements	FISH and NGS	Alveolar rhabdomyosarcoma
SS18 rearrangements	FISH and NGS	Synovial sarcoma
NMYC amplification	FISH and NGS	Neuroblastoma
CMYC rearrangements	FISH and NGS	Burkitt lymphoma

3. *Detection of genetic mutations for targeted therapy*: Molecular testing can help identify genetic alterations that guide targeted therapy (Porcel, 2019). For example, the detection of *EGFR* mutations and anaplastic lymphoma kinase gene (*ALK*) rearrangements in patients with nonsmall cell lung cancer determines the use of specific tyrosine kinase inhibitors, which can show dramatic therapeutic responses in such patients.
4. *Detection of infectious diseases*: Molecular tests are also useful for diagnosing infectious effusions caused by various bacterial, viral, and fungal infections, or parasitic infestations. Molecular techniques like polymerase chain reaction (PCR) or NGS can help detect pathogen-specific DNA or RNA in the effusion fluids, thereby providing a rapid and highly sensitive way to diagnose infections that may not be detectable using conventional microbiological methods (Li et al., 2022). PCR-based assays can detect *M. tuberculosis* DNA in effusions caused by tuberculosis. PCR can also be used to detect viruses and fungi like cytomegalovirus, Epstein-Barr virus, human immunodeficiency virus, Aspergillus, Mucor, or Histoplasma in effusion fluids, thereby aiding in their rapid diagnosis.
5. *Molecular testing for prognostication and treatment monitoring*: Molecular tests like gene expression profiling and NGS can be used to identify specific molecular signatures that can help predict prognosis as well as in selecting appropriate therapy. Specific tests to detect various mutations that confer therapeutic resistance can also be performed during the disease (Wegmann et al., 2024).
6. *Detection of circulating tumor DNA*: Circulating tumor DNA (ctDNA) analysis can be performed on effusion fluid to identify specific genetic alterations present in tumor cell DNA circulating in the effusion fluid. ctDNA can be analyzed using NGS and can help not only in the early diagnosis of the disease but also play a role in the study of tumor evolution, disease monitoring, and prediction of relapse. This technique is highly sensitive and can be used to detect alterations that are not detectable by traditional molecular methods (Porcel et al., 2022).

Table 1.13 lists molecular tests that are commonly used in effusion fluid cytology (Oh et al., 2019; Sahu et al., 2021).

1.16 Conclusions

Effusion fluid cytology remains a cornerstone in the diagnostic evaluation of body cavity fluids, offering a rapid, minimally invasive, and cost-effective means of distinguishing benign from malignant diseases. The major strength of cytologic assessment lies in meticulous sample processing, optimal smear, and a structured interpretation. A standardized reporting approach, aligned with international guidelines, improves communication with clinicians and facilitates risk-based management. However, cytomorphology alone has inherent limitations, particularly in differentiating reactive mesothelial proliferations from malignancy and in accurately subtyping various malignancies. The incorporation of ancillary techniques has substantially enhanced diagnostic precision and clinical utility. Immunocytochemistry on cell blocks provides critical lineage-specific information, enabling reliable distinction between epithelial, mesothelial, and hematolymphoid malignancies. Flow cytometric immunophenotyping plays a pivotal role in identifying and characterizing hematolymphoid neoplasms, especially in morphologically ambiguous cases. Molecular assays further expand the diagnostic spectrum by enabling identification of actionable mutations, and prognostic stratification. Importantly, the interpretation of effusion cytology must always be contextual, integrating clinical history, imaging findings, and laboratory data, to avoid both overdiagnosis and false-negative reporting.

References

Ali, S. Z., & Cibas, E. S. (2012). Serous cavity fluid and cerebrospinal fluid cytopathology. *Essentials in Cytopathology, 11*. doi:10.1007/978-1-4614-1776-7.

Awadallah, S. F., Bowling, M. R., Sharma, N., & Mohan, A. (2019). Malignant pleural effusion and cancer of unknown primary site: A review of literature. *Annals of Translational Medicine, 7*(15), 353. doi:10.21037/atm.2019.06.33.

Bales, E.B., & Durfee, G.R. (1992). In Koss, L.G. (Ed.), Diagnostic cytology and its histopathologic bases: 2 (4th ed., pp. 1451-1531). Philadelphia: J.B. Lippincott.

Chandra, A., Crothers, B., Kurtycz, D., & Schmitt, F. (2020). *The international system for serous fluid cytopathology the international system for serous fluid cytopathology* (pp. 1–301). Springer. https://link.springer.com/book/10.1007/978-3-030-53908-5. doi:10.1007/978-3-030-53908-5.

Chen, H. H., Liu, X., & Wang, Q. (2024). Updates and challenges in serous fluid cytopathology. *Human Pathology Reports, 36*, 300738. doi:10.1016/j.hpr.2024.300738.

Dey, P. (2023). *Cytology sample procurement, fixation, and processing* (pp. 125–136). Springer Science and Business Media LLC. doi:10.1007/978-981-19-6616-3_13.

Gill, G. W. (2012). *Cytopreparation: Principles & practice. Essentials in Cytopathology*. Springer.

Gupta, P., Pandey, T., Gautam, U., Rajwanshi, A., Srinivasan, R., Gupta, N., Rohilla, M., Varma, N., & Dey, P. (2021). Lymphoreticular malignancies in serous effusions: Cytomorphologic, flow cytometric and immunocytochemical analysis. *Diagnostic Cytopathology, 49*(5), 647–656. doi:10.1002/dc.24729.

Kassirian, S., Hinton, S. N., Cuninghame, S., Chaudhary, R., Iansavitchene, A., Amjadi, K., Dhaliwal, I., Zeman-Pocrnich, C., & Mitchell, M. A. (2023). Diagnostic sensitivity of pleural fluid cytology in malignant pleural effusions: Systematic review and meta-analysis. *Thorax, 78*(1), 32–40. http://thorax.bmj.com/. doi:10.1136/thoraxjnl-2021-217959.

Keebler, oM., & Facik, M. (2008). Cytopreparatory technique, comprehensive cytopathology. *WB Saunders Co, 3*, 977–1012.

Koss, L. G. (1992). *Diagnostic cytology and its histopathologic bases* (4th ed.). Philadelphia: JB Lippincott. doi:10.1177/003591576205500348.

Kundu, R., Srinivasan, R., Dey, P., Gupta, N., Gupta, P., Rohilla, M., Gupta, S., Bal, A., & Rajwanshi, A. (2021). Application of Indian Academy of cytologists guidelines for reporting serous effusions: An institutional experience. *Journal of Cytology, 38*(1), 1–7. http://www.jcytol.org. doi:10.4103/JOC.JOC_224_20.

Li, Q. K. (2018). Serous effusion cytology. In X. Jing, M. Siddiqui, & Q. Li (Eds.), *Atlas of non-gynecologic cytology. Atlas of Anatomic Pathology* (1st ed.). Cham: Springer.

Li, Z., Sun, Q. i, Du, B., Jia, H., Dong, J., Lyu, L., Zhu, C., Xing, A., Yang, X., Wei, R., Chen, X., Zhang, Z., Pan, L., & Chen, X. (2022). Use of pleural fluid digital PCR analysis to improve the diagnosis of pleural tuberculosis. *Microbiology Spectrum, 10*(6). doi:10.1128/spectrum.01632-22.

Makde, M. M, & Sathawane, P. (2022). Liquid-based cytology: Technical aspects. *Cytojournal, 19*, 41. doi:10.25259/cmas_03_16_2021.

Malik, A., & Datta, V. (2012). Effusion cytology - A practical guide to cancer diagnosis. *Medical Journal Armed Forces India, 68*(4), 375. doi:10.1016/j.mjafi.2012.07.005.

Mercer, R. M., Corcoran, J. P., Porcel, J. M., Rahman, N. M., & Psallidas, I. (2019). Interpreting pleural fluid results. *Clinical Medicine, Journal of the Royal College of Physicians of London, 19*(3), 213–217. http://www.clinmed.rcpjournal.org/content/19/3/213.full.pdf+html. doi:10.7861/clinmedicine.19-3-213.

Michael, C. W. (2021). Serous fluid cytopathology: Past, present, and future. *Diagnostic Cytopathology, 49*(5), 577–581. http://onlinelibrary.wiley.com/journal/10.1002/(ISSN)1097-0339. doi:10.1002/dc.24663.

Michael, C. W., & Davidson, B. (2016). Pre-analytical issues in effusion cytology. *Pleura and Peritoneum, 1*(1), 45–56. http://www.degruyter.com/view/j/pp. doi:10.1515/pap-2016-0001.

Min, E. N (2020). *Approach to effusion cytology practical diagnostic approaches in non-gynaecologic cytology. 1.* Springer Singapore https://link.springer.com/book/10.1007/978-981-15-2961-0#bibliographic-information. doi:10.1007/978-981-15-2961-0.

Nambirajan, A., & Jain, D. (2018). Cell blocks in cytopathology: An update. *Cytopathology, 29*(6), 505–524. doi:10.1111/cyt.12627.

Naylor, B. (2008). Pleural, pericardial, and peritoneal fluids. In Bibbo, M. (Ed.), Comprehensive cytopathology (2nd ed., pp. 551–621). Philadelphia: W.B. Saunders.

Oh, S., Ronaghi, R., & Cai, G. (2019). Molecular testing on pleural fluid samples. *Seminars in Respiratory and Critical Care Medicine, 40*(03), 340–346. doi:10.1055/s-0039-1695702.

Pinto, D., Chandra, A., Crothers, B. A., Kurtycz, D. F. I., & Schmitt, F. (2020). The international system for reporting serous fluid cytopathology—diagnostic categories and clinical management. *Journal of the American Society of Cytopathology, 9*(6), 469–477. http://www.elsevier.com/journals/journal-of-the-american-society-of-cytopathology/2213-2945. doi:10.1016/j.jasc.2020.05.015.

Porcel, J. M. (2019). Diagnosis and characterization of malignant effusions through pleural fluid cytological examination. *Current Opinion in Pulmonary Medicine, 25*(4), 362–368. journals.lww.com/co-pulmonarymedicine/pages/default.aspx. doi:10.1097/MCP.0000000000000593.

Porcel, J. M., & Light, R. W. (2006). Diagnostic approach to pleural effusion in adults. *American Family Physician, 73*(7), 1211–1220. http://www.aafp.org/afp/20060401/1211.pdf.

Porcel, J. M., Sorolla, A., Parisi, E., Bielsa, S., Salud, A., & Sorolla, M. A. (2022). Cell-free DNA concentration and pattern fragmentation in pleural fluid and plasma to detect malignant effusions. *Annals of the American Thoracic Society, 19*(5), 854–856. https://www.atsjournals.org/doi/10.1513/AnnalsATS.202107-810RL. doi:10.1513/AnnalsATS.202107-810RL.

Sahu, S., Gupta, P., & Dey, P. (2021). Molecular testing on serous effusion: An update. *Cytojournal, 18*, 35. doi:10.25259/cytojournal_55_2020.

Schmitt, F., Di Lorito, A., & Vielh, P. (2021). Molecular testing on cytology for gene fusion detection. *Frontiers in Medicine, 8*. doi:10.3389/fmed.2021.643113.

Shidham, V. B. (2022). Collection and processing of effusion fluids for cytopathologic evaluation. *CytoJournal, 19*. https://cytojournal.com/collection-and-processing-of-effusion-fluids-for-cytopathologic-evaluation/. doi:10.25259/CMAS_02_14_2021.

Shidham, V. B., & Janikowski, B. (2022). Immunocytochemistry of effusions: Processing and commonly used immunomarkers. *CytoJournal, 19*. https://cytojournal.com/immunocytochemistry-of-effusions-processing-and-commonly-used-immunomarkers/. doi:10.25259/CMAS_02_15_2021.

Shidham, V. B., & Layfield, L. J. (2021). Approach to diagnostic cytopathology of serous effusions. *CytoJournal, 18*(1). https://cytojournal.com/approach-to-diagnostic-cytopathology-of-serous-effusions/. doi:10.25259/CMAS_02_03_2021.

Shidham, V. B., & Atkinson, B. F. (2007). *Cytopathologic diagnosis of serous fluids E-Book*. Elsevier Health Sciences.

Shidham, V. B. (2019). CellBlockistry: Chemistry and art of cell-block making – A detailed review of various historical options with recent advances. *CytoJournal, 16*(1), 12. doi:10.4103/cytojournal.cytojournal_20_19.

Shidham, V. B. (2021). Diagnostic pitfalls in effusion fluid cytology. *Cytojournal, 18*(1), 33. doi:10.25259/cmas_02_04_2021.

Srinivasan, R., Rekhi, B., Rajwanshi, A., Pathuthara, S., Mathur, S., Jain, D., Gupta, N., Gautam, U., Rai, N., Nijhawan, V. S., Iyer, V., Dey, P., Deb, P., & Prasoon, D. (2020). Indian academy of cytologists guidelines for collection, preparation, interpretation, and reporting of serous effusion fluid samples. *Journal of Cytology, 37*(1), 1–11. http://www.jcytol.org. doi:10.4103/JOC.JOC_157_19.

Stokes, B. O. (2004). Principles of cytocentrifugation. *Laboratory Medicine, 35*(7), 434–437. doi:10.1309/FTT5-9GWK-DWH6-9FB0.

Torous, V. F., Cuda, J. M., Manucha, V., Randolph, M. L., Shi, Q., & VandenBussche, C. J. (2023). Cell blocks in cytology: Review of preparation methods, advantages, and limitations. *Journal of the American Society of Cytopathology, 12*(2), 77–88. http://www.elsevier.com/journals/journal-of-the-american-society-of-cytopathology/2213-2945. doi:10.1016/j.jasc.2022.11.003.

Wegmann, R., Bankel, L., Festl, Y., Lau, K., Lee, S., Arnold, F., Cappelletti, V., Fehr, A., Picotti, P., Dedes, K. J., Franzen, D., Lenggenhager, D., Bode, P. K., Zoche, M., Moch, H., Britschgi, C., & Snijder, B. (2024). Molecular and functional landscape of malignant serous effusions for precision oncology. *Nature Communications, 15*(1). https://www.nature.com/ncomms/. doi:10.1038/s41467-024-52694-8.

Wong-Arteta, J., Rey, M., Aragón, L., Gil-Rodríguez, E., & Bujanda, L. (2020). The utility of flow cytometry in the diagnostic work up of malignant effusions due to nonhematopoietic neoplasms. *Cytometry Part B: Clinical Cytometry, 98*(6), 504–515. doi:10.1002/cyto.b.21886.

2

Cerebrospinal fluid—Physiology and cytological evaluation

Rashmi Kushwaha

DEPARTMENT OF PATHOLOGY, KING GEORGE'S MEDICAL UNIVERSITY, LUCKNOW, UTTAR PRADESH, INDIA

Cerebrospinal fluid (CSF) is a clear and colorless fluid created in the brain's ventricles primarily by the choroid plexus, a network of microscopic blood capillaries in the third and fourth ventricles. CSF is present in the cerebral ventricles, spinal canal, and subarachnoid space. It is mostly a plasma ultrafiltrate consisting of water, low concentrations of proteins, ions, and glucose, which is renewed several times a day (Kawthalkar, 2018; Wichmann et al., 2022).

CSF is produced from the choroid plexus and the ependymal cells in the cerebral ventricles. From the lateral ventricles, CSF flows to the third ventricle through the foramen of Monro. From here, it passes through the cerebral aqueduct of Sylvius into the fourth ventricle. CSF then enters the subarachnoid space via the median aperture of Magendie and the two lateral apertures of Luschka. From the Foramen Magnum, CSF enters the spinal subarachnoid space and the central spinal canal. Reabsorption of CSF occurs primarily through arachnoid granulations in arachnoid mater and through dural lymphatics (Wichmann et al., 2022; Yamada & Mase, 2023).

2.1 Normal composition of cerebrospinal fluid

- *Total volume*: 100–150 mL (10–60 mL in the newborn) (Kawthalkar, 2018; McPherson and Pincus, 2021; Wichmann et al., 2022; Yamada & Mase, 2023).
- *Opening pressure:* 90–180 mm of water (10–100 mm of water in infants and young children).
- *Density*: 1.003 to 1.008 g/cm^3.
- *Appearance*: Clear and colorless, no clots (viscosity similar to water).
- *White blood cell (WBC) count*:
- Adults: 0–5 cells/μL
- Infants: 0–30 cells/μL
- 1–4 years: 0–20 cells/μL

Advancing Body Fluid Diagnostics with Flow Cytometry: A Comprehensive Guide to Effusions and CSF.
DOI: https://doi.org/10.1016/B978-0-443-27426-8.00014-4

- 5–18 years: 0–10 cells/µL
- *WBC types*: Predominantly lymphocytes
- *Red blood cells (RBC) count:* Nil
- *Glucose*: 50–80 mg/dL (Normally, CSF glucose is two-thirds of the corresponding blood glucose level)
- *Proteins:* 15–45 mg/dL (Normally, CSF proteins are 1% of plasma proteins)
- *Oligoclonal bands*: Negative
- *Chloride*: 120–130 mEq/L (20 mEq/L more than serum level)
- *Bilirubin*: Absent
- *CSF lactate*: 1–3 mmol/L

2.2 Functions of cerebrospinal fluid

- *Acts as a shock absorber*: CSF provides the buoyant support to the brain and the spinal cord. Since CSF envelops the brain, it helps absorb and redistribute downward pressure, minimizing mechanical strain and protecting the brain's structural integrity. When present in sufficient volume, this buoyant support also reduces traction on the nerve roots as they exit the central nervous system (CNS). It provides a cushion to the brain against rapid acceleration and deceleration.
- CSF functions as an intermediary between the bloodstream and the brain, facilitating the delivery of nutrients and the removal of metabolic waste. It contains glucose, proteins, lipids, and electrolytes that help meet the metabolic needs of the CNS. The solute composition of CSF supports and supplements vascular nutrient transport, particularly in areas where blood-brain barrier (BBB) permeability is limited. Metabolic waste products diffuse into the CSF and are subsequently cleared through absorption at the arachnoid granulations into the venous circulation, with an additional portion drained via lymphatic pathways.
- CSF plays an important role in CNS immune defense. It contains immunoglobulins (Ig) and mononuclear cells that help monitor and respond to potential pathogens. By draining into the cervical lymph nodes, CSF supports immunosurveillance, allowing antigens from the CNS to be sampled and enabling immune cell trafficking between the CNS and peripheral lymphatic system.
- CSF supports homeostasis within the CNS by helping regulate temperature and maintain osmotic equilibrium. Its electrolyte composition controls osmotic pressure, thereby sustaining normal CSF pressure and promoting adequate cerebral perfusion (Kawthalkar, 2018; Wichmann, Damkier, & Pedersen, 2022; Yamada & Mase, 2023).

2.3 Collection of cerebrospinal fluid

CSF collection is done by lumbar puncture (LP). The procedure entails advancing a long needle (spinal or LP needle) into the lumbar subarachnoid space, usually between the third and fourth lumbar vertebrae, or between the fourth and fifth (L3-L4 or L4-L5), followed by extraction of CSF from the subarachnoid space.

Determining the opening CSF pressure is a critical step in the procedure. An increased opening pressure can suggest conditions such as hydrocephalus, an intracranial mass, or idiopathic intracranial hypertension (IIH). In patients with a space-occupying intracranial lesion, removal of CSF may trigger brain herniation. Such conditions must be ruled out by computed tomography (CT) of the head or magnetic resonance imaging (MRI) of the brain before performing the LP (Chowdhury & Sharma, 2018; Costerus et al., 2018; Kim, 2022; Yamada & Mase, 2023).

The patient lies on their side (lateral recumbent) with knees curled toward the chest and chin tucked down, or sits bending forward, to open the spaces between the vertebrae. The procedure is performed under strict aseptic precautions to avoid procedure-related infections. The area surrounding the intended needle entry site must be carefully disinfected. A concentric circular technique using an antiseptic solution such as povidone-iodine or chlorhexidine-containing solution is commonly recommended. Sterile drapes must be placed. Administration of local anesthesia is optional. A sterile LP needle, ideally 22 gauge, is advanced gradually. It passes sequentially through several anatomical layers: The skin, subcutaneous tissue, supraspinous ligament, interspinous ligament, ligamentum flavum, epidural space (including the internal vertebral venous plexus), dura mater, arachnoid, and finally the subarachnoid space. As it traverses the spinal ligaments and dura mater, increased resistance is encountered. Typically, the spinal needle is inserted about two-thirds of its length (approximately 4–5 cm) to reach the subarachnoid space. Entry into the subarachnoid space is typically signaled by a distinct loss of resistance. The stylet is then carefully withdrawn. Once CSF begins to flow, a preassembled manometer is connected to the needle to obtain the opening pressure measurement. Accurate assessment of opening pressure requires the patient to be positioned in the lateral recumbent position (Chowdhury & Sharma, 2018; Kamimura et al., 2022).

Once CSF starts to flow, it is safest to allow it to drain passively rather than aspirating, which can pose risks. The rate of CSF collection is influenced by the needle gauge, though gentle abdominal pressure on the patient can help speed up the process. Before withdrawing the needle, always reinsert the stylet. After removal, cover the puncture site with sterile gauze and apply light pressure to secure a dressing. If the initial opening pressure was elevated, a closing pressure measurement should be documented after fluid collection. Before collecting the sample, the clinician should determine the amount needed for the requested tests to ensure an adequate volume is obtained. Up to 20 mL of sample can typically be safely withdrawn. After positioning the three-way stopcock correctly, CSF is distributed into sterile, plain collection tubes in a specific sequence (Table 2.1). Usually, the first tube is for biochemical analysis. However, if the first tube has a hemorrhagic sample due to traumatic puncture, then avoid using it for protein analysis; instead, use tube 3's sample for the same (McPherson & Pincus, 2021).

The first tube should never be used for microbiological testing due to the chances of contamination with skin bacteria. CSF is typically collected in sterile, plain (additive-free) tubes, usually clear plastic polypropylene tubes. Plastic tubes are preferred over glass, especially for

Table 2.1 Sequence of collection of CSF.

Sequence of collection	–
Tube 1	**Biochemical analysis (glucose and protein) and serologic testing**
Tube 2	**Microbiological studies, including Gram stain, culture, and antibiotic sensitivity**
Tube 3	**Hematologic evaluation, including total and differential cell counts**
Tube 4	**Cytologic examination and any additional specialized investigations**

CSF, Cerebrospinal fluid.

Table 2.2 Indications of lumbar puncture.

Category	Indications/details
Diagnostic evaluation	Examination of CSF in suspected conditions: • CNS infections (bacterial, tuberculosis, viral, and fungal meningitis); viral encephalitis • Subarachnoid hemorrhage • Inflammatory and demyelinating disorders (multiple sclerosis and Guillain-barré syndrome) • CNS syphilis, infectious polyneuritis, paraspinal abscess, CNS neoplasms; Meningeal involvement from leukemia/malignancy
Measurement of CSF pressure	Diagnosing idiopathic intracranial hypertension Evaluating normal pressure hydrocephalus
Therapeutic indications	Intrathecal administration of: • Anesthetic agents • Antibiotics (e.g., amphotericin B for fungal meningitis) • Chemotherapy agents (e.g., methotrexate for acute lymphoblastic leukemia) removal of CSF to reduce pressure in **Idiopathic intracranial hypertension**
Radiologic studies	Introduction of contrast media for myelography

cell counts, as they reduce cell adherence. A venous blood sample should be drawn either just before or immediately after the LP to allow comparison of serum and CSF glucose levels.

2.3.1 Indications of lumbar puncture

Indications of LP can be broadly classified into four categories, i.e., for diagnostic evaluation, for therapeutic CSF pressure reduction, for drug administration, and for radiological studies (McPherson & Pincus, 2021) (Table 2.2).

2.3.2 Contraindications of lumbar puncture

LP is generally a safe procedure; however, it should be avoided or postponed in certain situations due to the risk of serious complications. In the presence of signs suggestive of increased intracranial pressure, such as altered consciousness, focal neurological deficits, recent-onset seizures, or papilledema, it is advisable to perform brain imaging with CT or MRI before proceeding with the LP (Kim, 2022). The common contraindications of LP are enumerated in Table 2.3. In patients with severe sepsis where an LP is essential to confirm or rule out the diagnosis (for example, suspected meningococcal sepsis), the procedure may be carried out regardless of the platelet count. However, if the platelet count is below 10,000/μL, a platelet transfusion should be given.

For patients receiving dual antiplatelet therapy (clopidogrel plus aspirin), prophylactic platelet transfusion is advised. In contrast, if the patient is taking only low-dose aspirin (100 mg), an LP can generally be performed without platelet transfusion, as the risk of bleeding remains low (Tumani et al., 2020).

2.3.3 Complications of lumbar puncture

A clear understanding of the possible complications and their pathophysiological basis is vital for enhancing the safety, procedural quality, and postprocedure care of LP. Table 2.4 enumerates various complications associated with LP (Kim, 2022).

2.4 Laboratory examination of cerebrospinal fluid

CSF analysis is a fundamental method used in the diagnosis of neurological disorders. After collection, the sample must be transported to the laboratory by hand without delay, as cellular elements rapidly deteriorate. CSF samples must not be transported through the pneumatic system. CSF must be examined within 1 h of collection, and cell count must be done within 30–60 min. During prolonged standing, cells disintegrate, glucose levels decline, and lactate concentrations rise (Aggarwal et al., 2021; Hrishi & Sethuraman, 2019).

If immediate analysis is not feasible, samples may be stored at 4–8°C for short periods or at −20°C for longer durations. However, only certain components, such as proteins and properly preserved RNA, remain suitable for analysis after storage (Deisenhammer et al., 2006).

For microbiological examination, the sample must be transported to the microbiology laboratory within 1 h for culture and analysis. Specimens intended for culture must be protected from refrigeration, extreme temperatures, and direct sunlight. They should be maintained at 20–35°C during transport, and CSF samples should be inoculated onto culture media within 1 h to ensure optimal results. Refrigeration should be avoided for culture specimens, as fastidious organisms (such as *Haemophilus influenzae* and *Neisseria meningitidis*) may not survive under these conditions. If that is not possible, then inoculate the CSF sample into trans-isolate medium (T-I medium). T-I serves both as a growth medium and as a medium for specimen storage and transport. It is a biphasic medium used for the primary culture of meningococci and other causative organisms of bacterial meningitis, such as *S. pneumoniae* and *H. influenzae*, from CSF.

Table 2.3 Contraindications of lumbar puncture.

Category	Contraindications/considerations	Notes/rationale
Structural/anatomical	• Arnold-Chiari malformation • Vertebral abnormalities • Developmental defects (e.g., myelomeningocele)	Structural anomalies can increase procedural risk or prevent safe needle insertion
Raised intracranial pressure	• Symptoms/signs suggesting elevated ICP: altered consciousness, focal neurologic deficits, new-onset seizures, immunocompromised state, malignancy	A brain CT or MRI is recommended to rule out mass effect before lumbar puncture
Local infection	• Soft tissue infection at or near the puncture site	Risk of introducing infection into the CNS and unreliable CSF results
Coagulation/hematologic	• Platelet count < 10,000/ μL is an absolute contraindication and < 50,000/μL is a relative contraindication (Tumani et al., 2020) • INR > 1.8	Low platelets or coagulopathy increase the risk of bleeding; ensure values are within a safe range before LP
Medications	• Anticoagulants • Antiplatelet agents	Considered relative contraindications; data are limited, but caution is advised
Clinical practice note	Clinicians must exercise due diligence in evaluating risks before a lumbar puncture	Proper neuroimaging, lab assessment, and patient evaluation minimize complications

Table 2.4 Complications of lumbar puncture.

Category	Complication	Notes / mechanism
Neurological/CNS	**Postlumbar puncture headache**	Most common; caused by CSF leakage from the puncture site, often positional, worse when upright
–	**Back pain/local discomfort**	Pain at the puncture site or along the nerve root trajectory; usually transient
–	**Nerve injury/radicular pain**	Rare; occurs if the needle contacts the nerve root
–	**Brain herniation**	Life-threatening; can occur if LP is performed in the presence of elevated intracranial pressure due to a mass lesion
Infectious	**Meningitis/CNS infection**	Rare; risk increases with poor aseptic technique or puncture through infected tissue
Bleeding/hematologic	**Epidural or spinal hematoma**	More likely in patients with coagulopathy or on anticoagulants; it can cause neurological deficits if compressive
CSF abnormalities	**Bloody or traumatic tap**	Needle may puncture vessels, causing blood contamination and complicating the interpretation of CSF results
Other/systemic	**Vasovagal reaction**	Fainting, hypotension, or bradycardia during procedure; usually self-limited
–	**Rare complications**	Cerebral venous thrombosis, subdural hematoma, or infection spreading via CSF (extremely uncommon)

If T-I is not available, incubate CSF at 35–37°C with ~5% CO_2. CSF is centrifuged at 1500–3000g for 15 min. Use the sediment for Gram stain, other stains (e.g., India ink), and inoculating culture media. The supernatant can be used for biochemistry tests (protein/glucose) or antigen testing (Ikken et al., 2021).

2.4.1 Basic cerebrospinal fluid examination includes

- CSF opening pressure
- CSF appearance
- Cell count
- Quantitative estimation of total protein
- Quantitative estimation of glucose
- Quantitative estimation of lactate
- Qualitative and quantitative cytological examination
- Spectrophotometry of CSF

2.4.1.1 Cerebrospinal fluid opening pressure

CSF opening pressure refers to the initial pressure of the CSF measured at the beginning of an LP procedure. It reflects the pressure within the subarachnoid space. Measurement of CSF opening pressure is commonly performed during LP using a manometer.

CSF opening pressure is measured when the patient is in the lateral recumbent (lying on the side) position with legs extended and neck in neutral position to avoid artificial increases in pressure. After inserting the spinal needle into the subarachnoid space, a sterile manometer is attached to the needle to measure the pressure of the CSF column. The level at which the CSF stabilizes in the manometer represents the opening pressure.

The normal CSF opening pressure in adults is generally 90–180 mm H_2O. In children, the normal range may be slightly lower, 10 to 100 mm H_2O, and reaches adult level by 6–8 years of age. Opening pressure of more than 250 mm H_2O is diagnostic of intracranial hypertension, and in such a situation, care must be taken not to remove more than 2.0 mL CSF during LP (McPherson & Pincus, 2021).

Clinical significance:

Increased CSF opening pressure

- Bacterial or fungal meningitis
- Intracranial tumors
- Subarachnoid hemorrhage
- Hydrocephalus
- Cerebral edema
- Benign intracranial hypertension (Pseudotumor cerebri)
- Thrombosis of venous sinuses
- Superior vena cava syndrome
- Hypo-osmolality

Decreased CSF opening pressure

- CSF leakage after trauma or surgery
- Dehydration
- Spinal block or obstruction
- After repeated LPs

2.4.1.2 Cerebrospinal fluid appearance

Under physiological conditions, CSF is clear, colorless, and water-like in consistency. This transparent appearance reflects its low protein content and minimal number of cells. Under normal physiological conditions, CSF is clear and colorless. The presence of any discoloration is considered pathological. Abnormal coloration most commonly results from hemoglobin, methemoglobin, or bilirubin.

a. *Variations in color*: Color changes usually indicate underlying pathology
    - Traumatic LP
    - Intracranial hemorrhage

    Traumatic tap can occur in 20% of patients undergoing LP. Blood may enter the sample artificially due to injury of local blood vessels during LP (a traumatic tap). In such cases, the CSF may appear pink to red, a finding described as an erythrochromic (sanguinolent) appearance. When CSF is collected sequentially into three separate tubes, a traumatic tap is suggested if the discoloration becomes progressively lighter from the first tube to the last. In such cases, spinning the sample in a centrifuge typically clears the fluid, as the RBCs settle at the bottom. By contrast, in intracranial hemorrhage, the discoloration remains consistent across all tubes collected and does not disappear after centrifugation. However, if the bleeding has occurred very recently, the supernatant may still appear clear despite the presence of blood. Moreover, hemosiderin-laden macrophages are observed in cases of intracranial hemorrhage (Aggarwal et al., 2021) [10].

b. *CSF supernatant color and associated conditions* (Aggarwal et al., 2021; Hrishi & Sethuraman, 2019) [9,10]:

- *Yellow (xanthochromia)*: A yellow colored CSF or xanthochromia is due to an older bleed and due to the presence of bilirubin originating from catabolism of hemoglobin. Xanthochromia is seen in:
- Subarachnoid hemorrhage (after several hours)
- Severe hyperbilirubinemia
- Markedly elevated protein levels (more than 150mg/dL)
- Traumatic taps (RBC >100,000 /μL)
- Complete spinal block, polyneuritis, and meningitis
- Due to oxyhemoglobin produced due to artifactual lysis of RBC caused by detergent contamination of the needle or collection tube
- Carotenoids in people with dietary hypercarotenemia (hypervitaminosis A)
- Rifampin therapy
- *Orange*:
- Blood breakdown product
- High carotenoids
- *Ochre yellow to brown*: The presence of methemoglobin in CSF causes a discoloration ranging from ochre yellow to brown
- *Greenish tint*:
- May appear in purulent infections (e.g., bacterial meningitis)
- Hyperbilirubinemia
- *Brown*: Meningeal Melanomatosis
- *Pink-red color*: Due to the presence of blood (RBC count > 6000/ μL). Usually seen in traumatic taps, subarachnoid hemorrhage, intracerebral hemorrhage, and cerebral infarcts

c. *Turbidity (clarity changes)*: Normal CSF is crystal clear. Cloudiness or turbidity suggests increased cellular or protein content:

- *Slightly cloudy*→Mild pleocytosis (*increased WBCs*)
- *Markedly turbid/purulent*→bacterial meningitis
- *Opalescent or viscous*→Very high protein concentration

Typically, visible turbidity occurs when the leukocyte count exceeds approximately 200–400 cells/μL. The presence of erythrocytes (> 400 cells/ μL) can also make the CSF turbid. The presence of microorganisms (such as bacteria, fungi, or amoebas), radiographic contrast media, aspirated epidural fat, or elevated protein levels above 150 mg/dL (1.5 g/L) can all contribute to varying degrees of cloudiness.

d. *Clot formation*: CSF normally does not clot. Clotting may occur when:
- Protein concentration is markedly elevated (>150mg/dL)
- Traumatic tap
- Froin syndrome (complete spinal block)
- There is significant fibrinogen leakage (e.g., inflammatory processes)
- A delicate fibrin web (fine cobweb-like clot) may form in tuberculous meningitis after standing
- Purulent meningitis

Clots may interfere with an accurate cell count by entrapping the cells in a fibrin meshwork (McPherson & Pincus, 2021).

2.4.1.3 Cytological examination of cerebrospinal fluid

Examination of CSF cellularity is a key component in the evaluation of infections, inflammatory disorders, hemorrhage, and malignancy of the CNS. Cytological examination should be carried out within 2 h of LP, and ideally within 30 min, as both red and *WBCs* begin to undergo lysis soon after collection (Deisenhammer et al., 2006).

CSF cell count (Ahmed et al., 1996; Fishman, 1992; Karchar & McPherson, 2022; Ropper et al., 2019):

Examination of CSF cellularity is a key component in the evaluation of infections, inflammatory disorders, hemorrhage, and malignancy of the CNS. Pleocytosis is defined as an increase in cell count in CSF. WBC count up to 200/cmm and red cell count up to 400/cmm are associated with a clear appearance of CSF; hence, it is important to do microscopic examination of all CSF samples (Kawthalkar, 2018)

a. *Normal CSF cell count* (Ahmed et al., 1996; Fishman, 1992; Karchar & McPherson, 2022):
Under physiological conditions:
- *Total leukocyte count*: Adults: 0–5 cells/μL
- Mostly lymphocytes and monocytes (ratio of 70% and 30% respectively)
- *Neutrophils*: Normally absent
- *RBCs*: Absent
- *Eosinophils/plasma cells*: Absent

In neonates, slightly higher leukocyte counts may be observed (up to 30 cells/μL), and few neutrophils may be present.

b. *Method of cell counting*:
- CSF cell count should be performed immediately after collection because leukocytes rapidly deteriorate. In addition, cells may adhere to the walls of glass tubes, leading to falsely low counts.
- The sample collected in the third tube is preferred for analysis to minimize the effect of possible contamination from a traumatic tap.

- CSF is generally examined without dilution. Dilution is required only when the fluid appears turbid and is suspected to contain a markedly elevated leukocyte count.
- Counting is usually performed using a manual hemocytometer method. Counting can also be performed on automated cell counters (where validated).
- Differential count is done on a centrifuged and stained smear (e.g., Wright or May-Grünwald-Giemsa stain).

Procedure of the manual hemocytometer method (Kawthalkar, 2018):

The CSF specimen must be thoroughly mixed before analysis. Clear samples are examined without dilution. If the fluid appears cloudy or turbid, a 1:20 dilution is prepared by mixing 0.05 mL of CSF with 0.95 mL of Turk's solution (prepared from 4 mL glacial acetic acid, ten drops of methylene blue, and distilled water up to 200 mL).

1. Clean the modified Neubauer chamber and coverslip.
2. Place the coverslip properly over the counting platform.
3. Using a pipette, fill the chamber by touching the tip to the edge of the coverslip.
4. Allow cells to settle for 2–3 min.
5. Avoid overfilling or air bubbles.
6. For the WBC's count, the corner four large squares
7. Follow the counting rule (include cells touching top and left borders; exclude bottom and right).
8. Count both sides of the chamber and take the average.

Calculation:

For WBC count: If Undiluted CSF is Used: Cells per μL=Cells counted×2.5

If Diluted CSF is used: Cells per μL=Cells counted×2.5×Dilution factor

For CSF cell counting, a Fuchs-Rosenthal chamber is preferred because its depth is twice that of the improved Neubauer chamber, allowing more accurate counting in fluids with low cellularity. Cells are counted in five large squares (the four corner squares and the central square).

When undiluted CSF is used, the total number of cells counted in the five large squares corresponds directly to the number of cells per cubic millimeter (cmm) of CSF. If the sample has been diluted, the counted value must be multiplied by the dilution factor (e.g., ×20).

a. *Cell counts in hemorrhagic CSF:*

The presence of blood in CSF, whether from a traumatic LP or a subarachnoid hemorrhage, can falsely elevate the leukocyte count. As a general estimate, one WBC is added for every 1000 RBC introduced into the sample. This correction can be applied when the patient's peripheral blood counts are within normal limits. However, if the patient has marked anemia or leukocytosis, the CSF leukocyte count should be adjusted using a correction based on the individual's peripheral blood cell counts (Kawthalkar, 2018; McPherson & Pincus, 2021).

$$\begin{aligned} &\textit{Corrected WBC count in CSF} = \textit{WBC count in CSF (cells/cmm)} \\ &\quad -(\textit{WBC count in blood} \times \textit{Red cell count in CSF} \ / \ \textit{Red cell count in blood}) \end{aligned}$$

b. *Automated cell counting in cerebrospinal fluid:*

This has increasingly supplemented and partially replaced the manual count using a hemocytometer. This shift is driven by the need for improved precision, faster turnaround time, and reduced observer variability, particularly in settings with high sample throughput (Barnes et al., 2004; Deisenhammer et al., 2006; Zimmermann et al., 2011). Modern automated analyzers that are flow cytometry-based systems use principles like fluorescence, impedance, and light scatter to identify and quantify cells. These systems can rapidly provide total WBC counts, RBC counts, and, in some cases, perform preliminary differentiation of leukocytes. Compared with manual counting, automated methods offer better reproducibility and are less affected by interobserver differences (Wick, Gross, Tumani, Wildemann, & Stangel, 2021; de Jonge et al., 2010). However, there are certain limitations, especially at very low cell counts, which are common in normal CSF. Many automated analyzers have reduced sensitivity below 5–10 cells/μL, which can lead to underestimation or imprecision in clinically critical ranges. In addition, abnormal and atypical cells (e.g., malignant cells, blasts, or microorganisms) may not be reliably identified by automated systems. For this reason, microscopic review remains essential in cases with abnormal flags, suspected infection, or malignancy.

Preanalytical factors such as sample handling, time to analysis, and cell lysis also affect automated counts. Prompt processing is crucial, as leukocytes and erythrocytes rapidly degrade in CSF. Some systems incorporate specialized body fluid modes designed to improve accuracy in low cellularity fluids, including CSF. Overall, automated CSF cell counting is a valuable screening and quantitative tool, but it should be complemented by manual microscopy when clinically indicated (Aune et al., 2014; Delanghe, 2014; Roccaforte et al., 2018).

c. *Causes of increased cell count in CSF*:

- Meningitis (bacterial/viral/fungal/ Tubercular)
- Parasitic infection
- Noninfectious inflammatory disorders (multiple sclerosis, autoimmune encephalitis, neurosarcoidosis, systemic lupus erythematosus (SLE) with CNS involvement)
- Subarachnoid or intracerebral hemorrhage
- Traumatic LP
- Miscellaneous (drug-induced meningitis, ventriculoperitoneal shunt reaction, and demyelinating disorders)

d. *Differential leukocyte count*

This is done to check the percentage of different kinds of WBC's present. For this centrifuge, the specimen is centrifuged at high speed (3000 g for 10 min), and then a smear is made from the sediment. However, if the cell count is high, then a smear can be made without centrifugation. The smear is stained with a Romanowasky stain and examined under the microscope.

However, this traditional method can have certain disadvantages if the cell count is low or if cells have undergone degeneration and fragmentation. In such situations, Cytospin preparation using a cytocentrifuge is recommended to prevent cellular damage and morphological distortion. Cytological distortion can further be minimized by adding two drops of 22% bovine serum albumin to the sample (Kawthalkar, 2018).

This technique employs high speed centrifugation to concentrate cells onto a slide in a uniform monolayer, enhancing cell recovery and maintaining cellular morphology more effectively. The cytocentrifuge method is rapid and requires minimal training. It has advantages such as only a small sample volume is required, improvement in diagnostic sensitivity, especially for malignancy. It is the recommended method for differential cell counts in all body fluids (Karchar & McPherson, 2022; Khan et al., 2025; Marcos, Santos, Marrinhas, Correia-Gomes, & Caniatti, 2016).

Procedure

1. A small volume of well-mixed CSF (usually 0.1–0.5 mL) is placed into a cytocentrifuge chamber fitted with a filter card and slide.
2. The sample is centrifuged at low speed (commonly 500–1000 rpm for 5–10 min).
3. Cells are sedimented onto the slide while excess fluid is absorbed by the filter paper.
4. The slide is then air-dried or fixed immediately.
5. Staining is performed using Romanowasky stains such as May-Grünwald-Giemsa, Wright stain, or Papanicolaou stain.

In healthy adults, the CSF differential typically consists of approximately 70% lymphocytes and 30% monocytes. In young children, the proportion of monocytes can be higher, sometimes reaching up to 70%.

Besides mature blood cells, CSF may contain immature hematopoietic cells, tissue-derived cells such as ependymal cells or pia-arachnoid mesothelial cells, and malignant cells. Increased CSF Polymorphonuclear cells (neutrophils) occur in many conditions, as mentioned in Table 2.5. A PMN count of more than 2000 cells/ μL is highly suggestive of bacterial meningitis. More than 60 % PMN's are seen in early viral meningitis. However, it changes to lymphocytic pleocytosis within 2–3 days. Neutrophilic meningitis persisting for more than 1 week may be noninfectious in origin or caused by less common pathogens, including *Nocardia, Actinomyces, Aspergillus,* and *zygomycetes* (Spanos et al., 1989).

Increased CSF lymphocytosis is seen in conditions mentioned in Table 2.5. Cases of early acute bacterial meningitis may be associated with increased lymphocytes in CSF when the CSF leukocyte count is less than 1000 cells/ μL. Table 2.5 shows the predominant cell types associated with various infections (Rifai et al., 2018; Ropper et al., 2019).

a. *CSF in malignant diseases:*

Malignancies most commonly showing CSF involvement include leukemia, lymphoma, breast carcinoma, lung carcinoma, melanoma, medulloblastoma, and other primary brain tumors. Leptomeningeal metastasis or carcinomatous meningitis is malignant involvement of the CSF that usually occurs when cancer spreads to the leptomeninges. This happens when tumor cells disseminate through the bloodstream or directly invade the meninges and subsequently circulate within the CSF pathways (Bönig et al., 2019; Torres & Naous, 2026).

Morphological examination of CSF for malignant cells has variable sensitivity, with the highest (70%) for leukemic patients, moderate (20%–60%) for metastatic adenocarcinoma, and the lowest (30%) for primary CNS malignancies. Sensitivity may be increased by performing

Table 2.5 CSF cell pattern interpretation.

Predominant cell type	Common cause	Associated findings
Polymorphonuclear cells (PMNs)	Acute bacterial meningitis, early viral meningitis, brain abscess, subdural empyema, amoebic encephalomyelitis, AIDS- related CMV radiculopathy, subarachnoid hemorrhage, repeated lumbar punctures, Introduction of anticancer drugs or contrast media in the subarachnoid space, Meningeal metastasis	Usually high protein, low glucose in bacterial cases
Lymphocytes	Viral meningitis, tuberculous meningitis, fungal meningitis, *Leptospiral* meningitis, parasitic infestations (cysticercosis, trichinosis, and toxoplasmosis), aseptic meningitis due to septic focus adjacent to meninges, multiple sclerosis, Subacute sclerosing panencephalitis, drug abuse encephalopathy, Guillain-Barré syndrome, acute disseminated encephalomyelitis , polyneuritis, and CNS periarteritis	Glucose is usually normal in viral; low in tubercular/fungal
Monocytes/macrophages	Chronic infections (tuberculosis and fungal), neurosyphilis, and the recovery phase of meningitis	Often, a chronic inflammatory process
Eosinophils	Parasitic infections (e.g., cysticercosis), fungal infections, drug reaction, reaction to foreign material (e.g., shunts)	Eosinophilic meningitis (Defined as more than 10 eosinophils per mm^3 or a total CSF cell count made up of more than 10% eosinophils)[17]
Increased TLC with mixed cells Neutrophils + Lymphocytes	Partially treated bacterial meningitis, tubercular meningitis	Transitional phase pattern
Mild increase in TLC lymphocytes	Multiple sclerosis and autoimmune disorders	Oligoclonal bands may be present
Plasma cells	Acute viral infections, Guillain-Barré syndrome, multiple sclerosis, parasitic CNS infestations, sarcoidosis, subacute sclerosing panencephalitis, syphilitic meningoencephalitis, tuberculous meningitis	–

CSF, Cerebrospinal fluid.

serial punctures in cases with strong suspicion or using liquid-based thin-layer methods for the detection of neoplastic cells (Marton & Gean, 1986).

Cytological changes in CSF in malignant diseases

In normal conditions, CSF contains very few cells, usually fewer than 5 cells/μL, mainly lymphocytes and monocytes. In malignant conditions, several cytological changes can be observed:

1. *Increased cellularity*: Malignant cells may appear singly or in three-dimensional clusters.
2. *Abnormal morphology*: Malignant cells have pleomorphic appearance, hyperchromatic nuclei, irregular chromatin distribution, prominent nucleoli, and increased nuclear-cytoplasmic ratio.
3. *Background changes*: The smear may show necrotic debris, proteinaceous material, or inflammatory cells, reflecting tumor invasion and tissue damage.
4. *Biochemical changes CSF in malignant conditions*: Associated biochemical changes are increased protein levels due to disruption of the BBB, reduced glucose levels due to increased metabolic activity of tumor cells, occasionally xanthochromia or blood-stained CSF due to tumor-related hemorrhage (Liu et al., 2009).
5. Increased opening pressure during LP (Liu et al., 2009).

Types of metastatic cells seen in CSF:

1. *Metastatic adenocarcinoma cells*: Adenocarcinoma is the most common metastatic tumor found in CSF. It commonly originates from the breast, lung, or gastrointestinal tract. These cells appear in clusters or gland-like arrangements with large pleomorphic nuclei, prominent nucleoli, and abundant cytoplasm. Sometimes mucin-containing vacuoles may be present (Ruotolo et al., 2023).
2. *Small cell carcinoma cells*: These cells usually originate from small cell carcinoma of the lung. They appear as small, round cells with scant cytoplasm and nuclear molding. They may resemble lymphocytes, but usually occur in clusters.
3. *Melanoma cells*: Metastatic melanoma can spread to the meninges and appear in CSF. These cells are typically large, pleomorphic, and may contain brown or black melanin pigment within the cytoplasm.
4. In pediatric patients, tumors such as medulloblastoma may shed tumor cells into the CSF. These cells are usually small, round cells with hyperchromatic nuclei and scant cytoplasm.
5. In leukemia, blast cells may infiltrate the CSF. These cells usually have lymphoid blast morphology in cases of acute lymphoblastic leukemia and myeloid blast morphology in cases of acute myeloid leukemia. Their presence is important for diagnosing CNS involvement in leukemia. Moreover, flowcytometric immunophenotyping will further confirm the immunophenotypic character of these cells. CSF analysis is particularly important in acute lymphoblastic leukemia (ALL) to assess CNS involvement. A WBC count exceeding 5/μL along with the presence of lymphoblasts is indicative of CNS infiltration (Ruotolo et al., 2023).
6. *Lymphoma*: Lymphoid cells may appear in CSF in cases of CNS lymphoma or systemic nonHodgkin lymphoma, like lymphoblastic lymphoma, diffuse large B-cell lymphoma,

Table 2.6 Causes of increased CSF protein.

Condition	Protein level	Mechanism
Acute bacterial meningitis	Markedly increased (100–500 mg/dL)	BBB damage + inflammation
Viral meningitis	Mild–moderate increase	Inflammation
Tuberculous / Fungal meningitis	Moderate–marked increase	Chronic Inflammation
Guillain-barré syndrome	Increased (often very high) with normal cell count	Albuminocytologic dissociation
Brain tumor, abscess, herniated disk, and loculated CSF effusion	Increased	Obstruction of CSF flow
Subarachnoid hemorrhage	Increased	Blood breakdown products and increased blood-CSF permeability
Multiple sclerosis, neurosyphilis, and subacute sclerosing panencephalitis	Mild Increase (oligoclonal band)	Increased intrathecal IgG synthesis
Collagen vascular diseases (e.g., lupus, periarteritis), chronic inflammatory demyelinating polyradiculopathy	Increased	Increased blood-CSF permeability

CSF, Cerebrospinal fluid.

Burkitt's lymphoma. In inflammatory conditions, the predominant cell type is T cells. However, most lymphomas, especially those occurring in immunocompromised patients, are of the B-cell type (DeAngelis et al., 1999; Bierman & Giglio, 2005).

Flow cytometric immunophenotyping for malignant cells, Polymerase chain reaction, and DNA sequence analysis are some of the high-end techniques to diagnose leukemia/lymphoma cells (Bromberg et al., 2007; Gupta et al., 2024; Rimelen et al., 2019). Presence of malignant cells in CSF produces characteristic cytological, biochemical, and physical changes, and careful microscopic examination, as well as immunophenotyping, can reveal various types of metastatic tumor cells. Therefore, CSF cytology remains a critical component in the diagnosis and management of malignancies affecting the CNS.

2.4.1.4 Quantitative estimation of cerebrospinal fluid total protein

CSF protein is much lower than plasma protein because of the BBB. Normal value in adults is 15–45 mg/dL and is slightly higher in newborns (up to ~150 mg/dL). Elevated CSF protein is seen in infections, intracranial hemorrhages, multiple sclerosis, Guillain-Barré syndrome, malignancies, some endocrine abnormalities, certain medication use, and a variety of inflammatory conditions. Table 2.6 shows various causes of increased CSF proteins.

Traumatic LP can falsely elevate the CSF protein content because of the presence of RBC's. However, this can be corrected by subtracting 1 mg/dL of protein for every 1000 RBC's per cu.mm. However, this correction should be done if the same tube is used for CSF cell count and protein estimation (Seehusen et al., 2003).

a. *Types of proteins in CSF*:
   1. Albumin (major fraction): Derived from plasma; increased when the BBB is damaged.
   2. *Globulins*: Especially IgG; increased in inflammatory and demyelinating diseases
   3. *Specific proteins*: Myelin basic protein (*MBP*) (demyelination); oligoclonal bands (seen in multiple sclerosis)

   Protein electrophoresis of concentrated normal CSF shows two key differences from serum: A prominent transthyretin (prealbumin) band and the presence of two transferrin bands. The relatively high level of transthyretin is due to its production by both the liver and the choroid plexus. The second transferrin band, known as β2-transferrin, migrates more slowly than the serum form because of neuraminidase-mediated removal of sialic acid residues in the brain (Delpech, 1985).

b. *Causes of decreased CSF proteins* (Greenberg, 2020; Seehusen et al., 2003):
   - Rare
   - CSF leak
   - Repeated LP
   - Acute water intoxication
   - Untreated hyperthyroidism
   - IIH (minority of patients)

   CSF proteins do not fall in hypoproteinemia

c. *Methods to measure CSF proteins*: The methods to measure CSF proteins are broadly classified into qualitative and quantitative methods.
   - Qualitative tests available are heat and acetic acid tests, and sulphosalicylic acid (SSA) test
   - Quantitative tests available are the turbidimetric method using trichloroacetic acid for precipitation of proteins. It is based on the principle that trichloroacetic acid, when added to the CSF, causes precipitation of proteins that make the CSF turbid. This turbidity is then measured with the turbidity of a known standard using a photoelectric calorimeter. Other methods available for quantitative protein estimation are the biuret method, Nephelometry, and protein electrophoresis
   - The permeability of the BBB can be assessed by measuring the CSF to serum albumin index

$$CSF/Serum\ albumin\ index = CSF\ albumin\ (mg/dL)/\ Serum\ albumin(g/dL)$$

An intact BBB has index values less than 9. Whereas values between 9 and 14 suggest mild impairment, and values between 14 and 30 suggest moderate impairment. Severe impairment has values above 30. The albumin index is invalid in cases with traumatic tap. The index is slightly elevated in infants due to an immature BBB (Delpech & Lichtblau, 1985).

a. Intrathecal immunoglobulin (Greenberg, 2020; Seehusen et al., 2003). The detection of intrathecal oligoclonal Ig in CSF is diagnostically important, particularly in patients with multiple sclerosis. Methods of detection are the isoelectric focusing (IEF) technique. Different oligoclonal band patterns may suggest various CNS disorders, including multiple sclerosis, SLE, Guillain-barré syndrome (GBS), multiple myeloma, and other neurological diseases.

Increased CSF to serum IgG Index suggests intrathecal IgG synthesis or increased plasma IgG crossover from breakdown of the BBB

$$CSF/serum\ IgG\ index:\ \ CSF\ IgG\ (mg/dL)/Serum\ IgG(\ /dL)$$

Normal range is 3.0 to 8.7

The following formula can be used to do correction for immunoglobulin derived from plasma crossover by dividing the CSF/ serum IgG index by the CSF/albumin index to yield the CSF IgG index

$$\text{CSF IgG index} = \frac{\text{CSF IgG (mg/dL) / Serum IgG (g/dL)}}{\text{CSF albumin (mg/dL) / Serum albumin (g/dL)}}$$

The reference range for the IgG index varies, with a normal upper range of 0.8. CSF IgG is normally 3% to 5% of total CSF protein. In multiple sclerosis and other inflammatory neurological diseases, the CSF IgG increases significantly.

a. CSF protein electrophoresis (Greenberg, 2020; Seehusen et al., 2003)

   CSF protein electrophoresis is used for the identification of oligoclonal bands and to determine whether fluid submitted for examination is CSF. CSF protein electrophoresis typically separates proteins into several fractions:

   1. Prealbumin (transthyretin)–A prominent band in CSF that originates mainly from the choroid plexus and is often more pronounced in CSF than in serum.
   2. Albumin–The major protein fraction in CSF, derived primarily from plasma through the BBB.
   3. Alpha-1 globulins–Present in small quantities and include proteins such as α1-antitrypsin.
   4. Alpha-2 globulins–Include haptoglobin and ceruloplasmin.
   5. Beta globulins–Include transferrin and complement proteins. CSF-specific β2-transferrin is particularly important in diagnosing CSF leakage.
   6. Gamma globulins–Mainly Ig (especially IgG). Increased gamma fraction may indicate intrathecal immunoglobulin production.

b. Clinical applications (Ahmed et al., 2022a; Cabrera, 2022; Lunding et al., 2000; Siddiqui et al., 2002).

   1. *MS*: Detection of oligoclonal IgG bands in CSF is a hallmark of MS. Oligoclonal bands indicate intrathecal synthesis of Ig.
   2. *CNS infections*: Increased gamma globulins may occur in chronic infections such as neurosyphilis or tuberculous meningitis.
   3. *BBB dysfunction*: Elevated albumin in CSF relative to serum suggests increased permeability of the BBB.
   4. *Neoplastic conditions*: Certain CNS tumors or metastatic diseases may alter the protein pattern.
   5. *CSF Leakage diagnosis*: Detection of β2-transferrin confirms the presence of CSF in body fluids such as nasal or ear discharge.

c. *Other proteins found in CSF*:
   1. *MBP*: It is a component of the nerve sheath and is released during demyelination. MBP in CSF acts as a surrogate marker to check acute exacerbations of MS. Increased values are also seen in Guillain- Barré syndrome, lupus erythematosus (LE), subacute sclerosing panencephalitis, and various brain tumors, following CNS irradiation and chemotherapy. It can act as a prognostic marker in head injury cases (Agrawal et al., 2004; Tumani et al., 2008).
   2. *α2 macroglobulin*: It does not cross the BBB because of its large size. Increased values seen in polyneuropathies, subdural hemorrhage, and bacterial meningitis (Reiber & Peter, 2001)
   3. *β2 microglobulin*: values above 1.8 mg/L are associated with CNS leukemia/lymphoma. Viral infections, inflammations, and various malignancies can also show increased CSF β2 Microglobulin (Fan et al., 2025; Reiber, 2001)
   4. *C-reactive protein*: normal CSF CRP levels almost rule out bacterial meningitis. An increase in CSF CRP is more common in Gram-negative bacterial meningitis when compared with Gram-positive bacterial meningitis. Correlation with CSF CRP levels is more specific than blood CRP levels (Heckenberg et al., 2014)
   5. *Fibronectin*: Elevated CSF levels of fibronectin are found in ALL, Burkitt lymphoma, metastatic solid tumors, astrocytoma, and bacterial meningitis. Decreased levels are found in viral meningitis and AIDS dementia complex (Delpech et al., 1985)
   6. *β- Amyloid protein 42 and Tau (τ) protein*: Decreased CSF levels of β- Amyloid protein 42 and increased levels of Tau (τ) Protein increase the accuracy of diagnosis of Alzheimer's disease (AD) (Blennow et al., 2010; Olsson et al., 2016)
   7. *Protein 14-3-3*: A group of regulatory proteins involved in cell signaling. In CSF, protein 14-3-3 is used as a marker of rapid neuronal damage. It is commonly elevated in conditions like Creutzfeldt–Jakob disease (CJD), but also in other acute brain injuries (Sanchez-Juan et al., 2006).

2.4.1.5 Quantitative estimation of glucose in cerebrospinal fluid

A definite normal range cannot be assigned to CSF glucose. Generally, in a normal adult, CSF glucose is approximately two thirds of the blood glucose level (45–80 mg/dL) measured within the previous 2–4 h (Greenberg, 2020; Seehusen et al., 2003). However, this proportion decreases as serum glucose levels rise. CSF glucose estimation must be done immediately after the sample collection, as the values fall due to glycolysis if the sample is kept stored. The glucose oxidase peroxidase method is the most preferred technique for glucose estimation. Decreased CSF glucose is seen when there is increased glucose consumption by pathogenic organisms or when glucose transport across the BBB is impaired. Conditions associated with reduced CSF glucose:

- Meningitis(bacterial, tubercular, fungal, and parasitic)
- Noninfectious causes like subarachnoid hemorrhage, meningeal carcinomatosis, leukemic infiltration, neurosarcoidosis, and chemical meningitis
- Severe hypoglycemia

CSF glucose is normal in viral meningitis

2.4.1.6 Lactate in cerebrospinal fluid

Lactate is produced during anaerobic glycolysis, a process in which glucose is broken down without sufficient oxygen to produce energy. Healthy brain tissue produces a small amount of lactate, which reflects the metabolic activity of the brain. Lactate does not cross the BBB, and unlike glucose, its concentration in CSF is independent of its level in plasma. Generally, the concentrations of lactate and glucose in CSF change in opposite directions. In the brain and nervous system, lactate levels increase when there is reduced oxygen supply, increased metabolic activity, or infection. Therefore, the measurement of lactate in CSF can reflect metabolic changes occurring in the brain (Baheerathan et al., 2020; Curti et al., 2020; Stephani et al., 2021; Vélez-Paez et al., 2022; Wang et al., 2023). Lactate in CSF is estimated by an enzymatic spectrophotometric method based on the Warburg optical test.

Normal levels: The normal concentration of lactate in CSF is 1.1–2.4 mmol/L (about 10–22 mg/dL)

Clinical significance: (Baheerathan et al., 2020; Curti et al., 2020; Stephani et al., 2021; Vélez-Paez et al., 2022; Wang et al., 2023).

1. *Bacterial meningitis*: CSF lactate levels are significantly increased in bacterial meningitis, usually above 3.5 mmol/L. This occurs due to increased anaerobic metabolism and bacterial activity.
2. *Viral meningitis*: CSF lactate levels generally remain normal or only slightly increased. This helps in differentiating viral meningitis from bacterial meningitis.
3. *Cerebral hypoxia*: When the brain does not receive enough oxygen, such as in stroke or severe respiratory disorders, lactate production increases, leading to elevated CSF lactate levels.
4. *Mitochondrial and metabolic disorders*: Some metabolic diseases that affect energy production in cells may also cause increased CSF lactate.
5. *Brain injury*: Trauma or bleeding in the brain can increase lactate levels due to tissue damage and altered metabolism.

2.4.1.7 Chloride in cerebrospinal fluid

Chloride is a major inorganic ion present in CSF (Dong et al., 2022; Fang et al., 2023). It plays an important role in maintaining the osmotic pressure, electrolyte balance, and acid–base equilibrium of the CNS. Its concentration in CSF is generally higher because of selective transport mechanisms across the BBB and the choroid plexus. The normal concentration of chloride in CSF ranges from 118 to 132 mmol/L.

Decreased CSF Chloride is seen in:

- Tubercular meningitis
- Fungal meningitis

Increased CSF chloride levels (less common but may occur in):

- Dehydration
- Renal disorders
- Metabolic acidosis

2.4.1.8 Enzymes in cerebrospinal fluid

Enzyme evaluation in CSF is a useful adjunct in diagnosing neurological diseases, particularly infections, malignancies, and tissue damage. Although not always definitive on their own, CSF enzyme levels provide supportive biochemical evidence.

- *Adenosine deaminase (ADA)*: Elevated in tuberculous meningitis (Gupta et al., 2010)
- Creatine *kinase (CK)*: Elevated CSF CK activity has been observed in a variety of CNS disorders, including hydrocephalus, cerebral infarction, primary brain tumors, and subarachnoid hemorrhage. CSF CK-BB activity below 5 U/L (the upper limit of normal) suggests minimal neurological damage. Levels between 5 and 20 U/L are associated with mild to moderate CNS injury, while values ranging from 21 to 50 U/L are often associated with a high risk of death. Levels exceeding 50 U/L are almost invariably linked to fatal outcomes (Johnsson et al., 1995; Davison et al., 2011)
- *Lactate dehydrogenase (LDH):* Increased in bacterial meningitis (especially neutrophilic response), CNS leukemia, lymphoma, metastatic carcinoma, and subarachnoid hemorrhage (Huy et al., 2010)
- *Lysozyme*: Although useful, CSF lysozyme is not disease-specific. Elevated levels may overlap in bacterial infections, tuberculosis, and some inflammatory disorders (Mishra et al., 2003)

2.4.1.9 Tumor markers in cerebrospinal fluid

1. *Carcinoembryonic antigen*: Elevated in metastatic carcinomas (e.g., lung, breast, and gastrointestinal tumors). Helps distinguish metastatic tumors from primary CNS tumors (Huang et al., 2025).
2. *Alpha-fetoprotein (AFP)*: A glycoprotein produced by the yolk sac component of the germ cell tumor. CSF AFP levels are used for diagnosis and monitoring of the response to therapy in patients with intracranial germ cell tumors (Nørgaard-Pedersen et al., 1978).
3. *Beta-human chorionic gonadotropin (β-hCG)*: Produced by choriocarcinoma and malignant germ cell tumors with a trophoblastic component. Used for diagnosis and monitoring of the response to therapy in patients with intracranial germ cell tumors.
4. *Placental alkaline phosphatase (PLAP)*: PLAP in CSF is useful in determining whether the pineal gland tumor is pinealoma or germ cell tumor.
5. *Neuron-specific enolase (NSE)*: elevated in neuroblastoma, metastatic small cell lung carcinoma (Isgro et al., 2015).

2.4.2 Microbiological examination of cerebrospinal fluid

Microbiological examination of CSF is mainly done to diagnose infections of the CNS, such as bacterial meningitis, Viral Meningitis, tuberculous meningitis, and cryptococcal meningitis

(Brouwer et al., 2010; Tunkel et al., 2004). The sample is usually collected by LP under sterile conditions.

2.4.2.1 Bacterial meningitis

The most common organisms for bacterial meningitis are:

In neonates (0–28 days):

- Group B *streptococcus* (most common)
- *Escherichia coli* (especially K1 strain)
- *Listeria monocytogenes*

Infants (1–3 months):

- *Streptococcus pneumoniae*
- *N. meningitidis*
- *H. influenzae*

Children & young adults:

- *N. meningitidis* (most common in this group)
- *S. pneumoniae*

Adults:

- *S pneumoniae* (most common overall)
- *N. meningitidis*

Elderly (>50 years):

- *S. pneumoniae*
- *L. monocytogenes*
- Gram-negative bacilli (e.g., *E. coli*) (Smets et al., 2020; Tunkel et al., 2004; Tunkel et al., 2004; van de Beek et al., 2006).

Direct microscopic examination: This is the initial step in detecting microorganisms.

a. *Gram stain*: Gram stain is done when the sample is purulent and has mostly neutrophils. Detects Gram-positive and Gram-negative organisms in CSF. However, one must note that Gram smear is positive in only 80% of cases of bacterial meningitis. Hence, a negative Gram stain does not rule out bacterial infection (Brouwer et al., 2010; Tunkel et al., 2004). Fig. 2.1 shows Gram-positive cocci arranged in pairs and short chains, suggestive of streptococci.

Table 2.7 shows various bacteria that can cause meningitis and can be detected by Gram stain.

2.4.2.2 Tuberculous meningitis

Abnormal CSF findings, particularly increased protein levels along with lymphocytic predominance, are characteristic and hallmark features of tuberculous meningitis. These changes reflect

Table 2.7 Bacteria commonly causing meningitis detectable by Gram stain.

Bacteria	Gram reaction	Morphology	Presentation and association
Streptococcus pneumoniae	Gram-positive	lancet-shaped diplococci	The most common cause of community-acquired meningitis in adults
Neisseria meningitidis (also called meningococcus)	Gram-negative	Kidney-shaped diplococci	Common in children and adolescents; may occur in outbreaks
Haemophilus influenzae	Gram-negative	Small pleomorphic coccobacilli	life-threatening meningitis and sepsis, particularly in children under five, especially in unvaccinated children
Listeria monocytogenes	Gram-positive	Rod-shaped bacterium	The bacterium poses a particular risk to pregnant women, newborns, the elderly, and immunocompromised individuals
Escherichia coli	Gram-negative	Rod-shaped bacilli	Common cause of neonatal meningitis
Streptococcus agalactiae also known as Group B Streptococcus (GBS)	Gram-positive	Cocci in a chain	The major cause in newborns

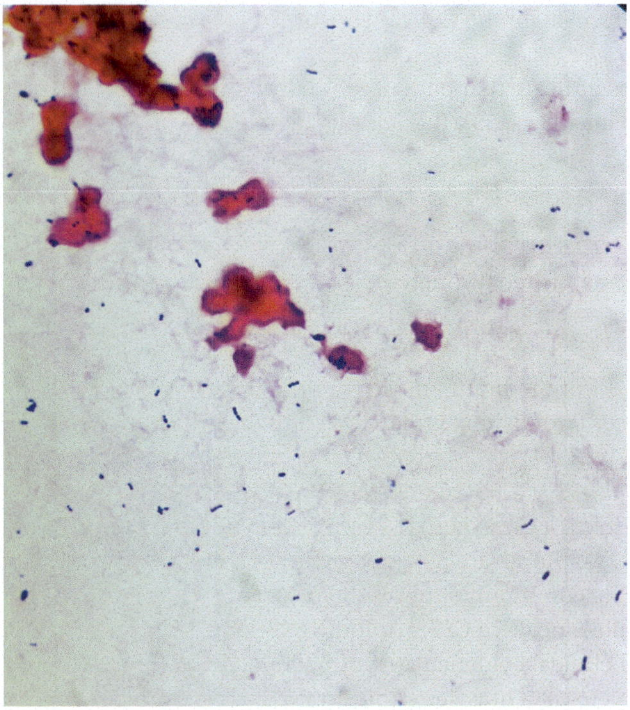

FIGURE 2.1 CSF Gram stain. CSF Gram stain showing numerous pus cells (neutrophils) with Gram-positive cocci arranged in pairs and short chains, suggestive of streptococci. *CSF*, Cerebrospinal fluid. *Source: Image courtesy Dr Prashant Gupta, King George's Medical University, Lucknow, India.*

the chronic inflammatory response typically seen in this condition. Elevated CSF ADA is supportive of TBM and helps differentiate TBM from viral meningitis.

The Ziehl-Neelsen (ZN) Stain is used to detect acid-fast bacilli. However, ZN staining is not highly sensitive for detecting *Mycobacterium tuberculosis* in CSF. In approximately 70% of cases of tuberculous meningitis, the Acid-Fast Bacilli (AFB) smear may be negative. Fluorescent auramine staining is more sensitive and therefore improves the detection of *M. tuberculosis* compared with the ZN method (Thwaites et al., 2004)[27] (Fig. 2.2).

Lowenstein-Jensen (LJ) culture medium can be used, but growth is slow, often requiring several weeks for growth, although it is highly specific. PCR-based nucleic acid amplification targeting specific DNA sequences has shown great promise for the rapid and accurate detection of *M. tuberculosis*, particularly in the diagnosis of tuberculous meningitis (Bahr & Boulware, 2012; Miftode et al., 2015; Rock et al., 2008).

2.4.2.3 Fungal meningitis

In fungal meningitis, CSF shows increased opening pressure, increased cell count predominantly lymphocytes, elevated protein, and low glucose. *Cryptococcus neoformans* is the most frequently

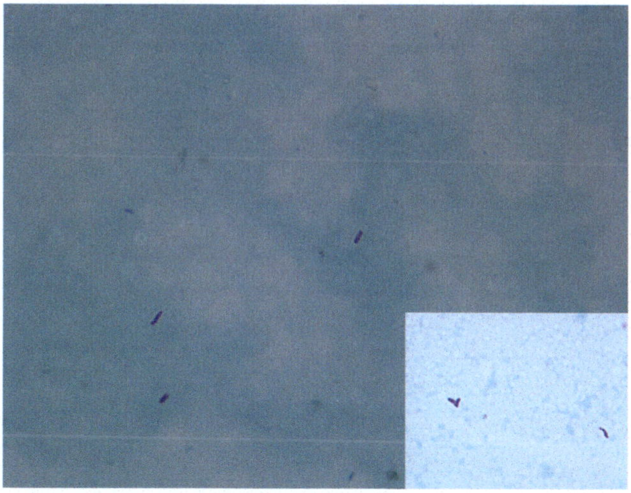

FIGURE 2.2 Ziehl-Neelsen stain revealing acid-fast bacilli in a case of Tuberculous meningitis. Bacilli are slender, beaded, and slightly curved or sometimes in an L-shaped arrangement (Inset). *Source: Image courtesy Dr Jyotsna Agarwal, Dr Anupam Das, Dr RMLIMS, Lucknow, India.*

isolated fungus that can be demonstrated by Indian Ink preparation (Perfect et al., 2010). It demonstrates the capsule of yeast cells (Fig. 2.3).

Although it has a sensitivity of only 25%, that may be increased by doing multiple LPs. It has a sensitivity of greater than 90% in untreated HIV patients. Fungal culture yields very slow growth. However, in the case of dimorphic or filamentous fungi, culture using a large quantity of CSF 15–20 mL) can yield growth. Detection of cryptococcal antigen (CrAg) using latex agglutination has a higher sensitivity of 60%–90%. PCR-based tests can be used to detect fungal DNA. Detection of CrAg in serum and CSF demonstrates very high diagnostic accuracy in adults with HIV who present with symptoms suggestive of cryptococcal meningitis, outperforming conventional methods such as fungal culture and India ink microscopy (Perfect et al., 2010; Temfack et al., 2021).

2.4.2.4 Spirochetal meningitis

A form of meningitis caused by spirochete bacteria, most commonly including *Treponema pallidum, Borrelia burgdorferi*, and *Leptospira* species. It is typically characterized by a subacute or chronic course and distinct CSF findings. CSF findings include lymphocytic pleocytosis, elevated protein, normal to slightly decreased glucose, and elevated Ig. Positive CSF VDRL and elevated protein are seen in neurosyphilis. Diagnosis of Lyme disease depends on serological methods, such as the ELISA/Western blot technique (Bharti et al., 2003; Djukic et al., 2012).

2.4.2.5 Primary amoebic meningoencephalitis

It is a rare, rapidly fatal brain infection caused by the free-living amoeba *Naegleria fowleri, Acanthamoeba species, and Balamuthia mandrillaris.* CSF shows neutrophilic pleocytosis, increased

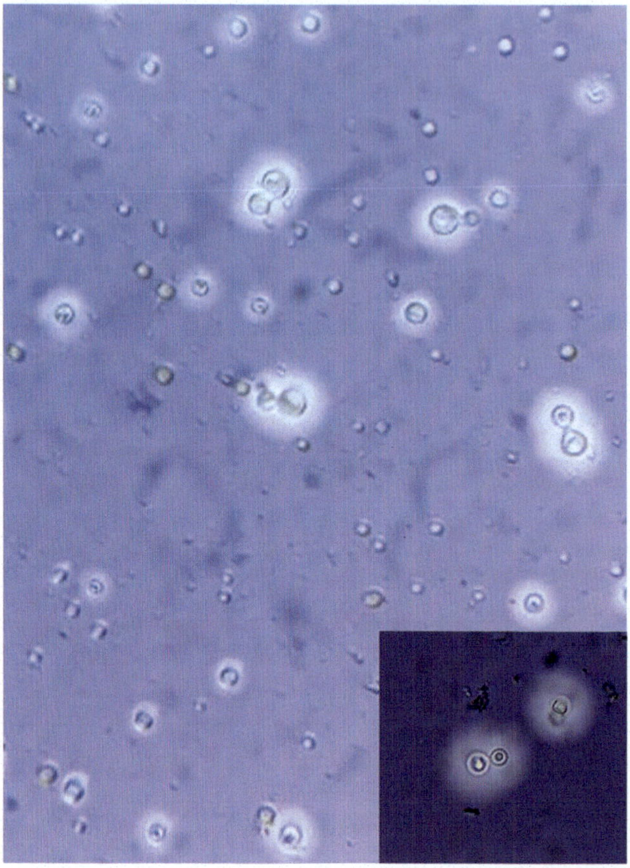

FIGURE 2.3 *Cryptococcus neoformans* in India ink preparation. Spherical budding yeast cells surrounded by a large, clear polysaccharide capsule against a dark background (Inset). *Source: Image courtesy Dr Prashant Gupta, King George's Medical University. Inset: Image provided by Dr Jyotsna Agarwal and Dr Anupam Das, Dr RMLIMS, Lucknow, India.*

proteins, and reduced glucose. Wet mount on light or phase contrast microscopy shows motile trophozoites. Gram stain is usually negative. Acridine orange stains the amoeba with a brick red color and leukocytes with a bright green color. PCR-based techniques are both sensitive and specific for the diagnosis (McPherson & Pincus, 2021).

2.4.3 Other diagnostic tests

2.4.3.1 Cerebrospinal fluid culture

CSF culture is done when bacteria are seen on Gram stain, or CSF has increased leukocytes and proteins. It is considered to be the gold standard for the diagnosis of bacterial meningitis. Culture helps in identifying the exact organism and performing antibiotic sensitivity testing (Brouwer et al., 2010; Gray & Fedorko, 1992; Tunkel et al., 2004). Common media used include:

Table 2.8 Summary of CSF parameters in various infections.

Parameter	Bacterial meningitis	Viral meningitis	Tuberculous meningitis	Fungal meningitis
Opening pressure	Markedly increased	Normal or mildly increased	Increased	Increased
Appearance	Turbid or purulent	Clear	Slightly cloudy or opalescent	Clear to slightly cloudy
Cell count	1000–5000 cells/μL	50–1000 cells/μL	50–500 cells/μL	20–500 cells/μL
Predominant cells	Neutrophils	Lymphocytes	Lymphocytes	Lymphocytes
Protein	Markedly increased (100–500 mg/dL)	Mildly increased (50–100 mg/dL)	Increased (100–500 mg/dL)	Increased (50–200 mg/dL)
Glucose	Decreased (<40 mg/dL)	Normal	Decreased	Decreased or low-normal
CSF/Serum glucose ratio	<0.4	Normal (>0.6)	<0.5	<0.5
Chloride	Normal or slightly decreased	Normal	Decreased	Normal or decreased
Gram stain/microscopy	Often positive for bacteria	Negative	Acid-fast bacilli may or may not be seen	Fungal elements may be seen
Culture	Positive for bacteria	Usually negative	Mycobacterium tuberculosis may grow	Fungal culture may be positive
Special tests	Latex agglutination, PCR	PCR for viruses	AFB stain, GeneXpert, PCR	India ink, cryptococcal antigen

CSF, Cerebrospinal fluid.

- Blood agar
- Chocolate agar
- MacConkey agar
- Sabouraud dextrose agar (for fungi)

2.4.3.2 Antigen detection tests

(Brouwer et al., 2010; Liesman et al., 2018; Tunkel et al., 2004) used when organisms are few or the culture is negative. Methods include:

- Latex agglutination test
- ELISA

These tests detect antigens of organisms like:

- *N. meningitidis*
- *Streptococcus pneumoniae*
- *H. influenzae*

2.4.3.3 Polymerase chain reaction

Detects microbial DNA/RNA. Very useful for diagnosing viral meningitis and tuberculous meningitis. PCR is a rapid, sensitive, and specific method that greatly improves the detection of pathogens in CSF, especially when traditional microscopy or culture methods are negative. PCR can detect pathogens even if they are dead, which is helpful when the patient has already started antibiotic therapy. Some pathogens are slow-growing or difficult to culture, such as *M. tuberculosis* or certain viruses; PCR can detect them easily(Ahmed et al., 2022b; de Almeida et al., 2019; Kitagawa et al., 2023; Welinder-Olsson et al., 2007).

Table 2.8 summarizes the changes in various parameters of CSF in different kinds of infections.

2.5 Conclusion

CSF represents a highly specialized extracellular milieu of the CNS, characterized by tightly regulated production, circulation, and reabsorption mechanisms that reflect the integrity of the blood-brain and blood-CSF barriers. Its physicochemical composition, low cellularity, and dynamic turnover are critical for maintaining neuronal homeostasis, facilitating metabolic exchange, and providing mechanical as well as immunological protection to the CNS.

A detailed understanding of CSF physiology underpins its diagnostic utility. Even subtle deviations in CSF parameters—such as alterations in opening pressure, cellular profile, protein concentration, glucose ratio, and metabolic markers—serve as sensitive indicators of pathological processes affecting the CNS. LP, when performed with appropriate clinical judgment and technical precision, remains the gold standard for CSF acquisition, enabling direct assessment of intracranial pathophysiology.

Cytological evaluation constitutes a cornerstone of CSF analysis, offering valuable insights into the nature of neurological disorders. Distinct cellular patterns, including neutrophilic, lymphocytic, eosinophilic, or malignant cell predominance, provide critical diagnostic clues for differentiating infectious, inflammatory, hemorrhagic, and neoplastic conditions. When integrated with biochemical and microbiological findings, CSF cytology significantly enhances diagnostic accuracy and disease stratification.

Biochemical analysis further elucidates underlying pathophysiological mechanisms, with parameters such as protein elevation indicating BBB dysfunction or intrathecal synthesis, decreased glucose reflecting increased metabolic consumption or impaired transport, and elevated lactate serving as a marker of anaerobic metabolism and tissue hypoxia. Advanced investigations, including immunoglobulin profiling, electrophoresis, tumor markers, and molecular diagnostics (e.g., PCR), have expanded the scope of CSF evaluation, enabling early and precise detection of CNS infections, demyelinating disorders, and malignancies.

Despite technological advancements, preanalytical variables—particularly prompt sample processing and appropriate handling—remain critical determinants of result reliability. Therefore, optimal interpretation of CSF findings necessitates a multidisciplinary approach that integrates clinical context with laboratory data.

In conclusion, CSF analysis is an indispensable, high-yield diagnostic modality in neurological practice. Its comprehensive evaluation provides profound insights into CNS pathophysiology, facilitating early diagnosis, prognostication, and targeted therapeutic intervention across a broad spectrum of neurological diseases.

References

Aggarwal, J., Singh, S., & Rawat, A. (2021). Cerebrospinal fluid (CSF) analysis and role in clinical neurology. *International Journal of Clinical Biochemistry and Research, 8*(2), 73–78. doi:10.18231/j.ijcbr.2021.016.

Agrawal, A., Timothy, J., Pandit, L., & Manju, M. (2004). Myelin basic protein in cerebrospinal fluid as a marker of demyelination. *Neurol India, 52*(2), 182–185.

Ahmed, A., Hickey, S. M., Ehrett, S., Trujillo, M., brito, F., Goto, C., Olsen, K., Krisher, K., & Mccracken, G. H. (1996). Cerebrospinal fluid values in the term neonate. *The Pediatric Infectious Disease Journal, 15*(4), 298–303. doi:10.1097/00006454-199604000-00004.

Ahmed, M. A., Askar, G. A., Farghaly, H. S., Ahmed, A. O., Kamal, D. T., Ahmed, S. S., & Mohamad, I. L. (2022a). Accuracy of cerebrospinal fluid C- reactive protein and multiplex polymerase chain reaction and serum procalcitonin in diagnosis of bacterial and viral meningitis in children. *Acta Neurologica Taiwanica, 31*(2), 61–71. http://www.ant-tnsjournal.com/Mag_Files/31-2/N20223116189.pdf.

Ahmed, M. A, Al Ghamdi, S., & Al-Habib, A. (2022b). Accuracy of cerebrospinal fluid multiplex PCR in diagnosis of bacterial and viral meningitis in children. *Acta Neurologica Taiwanica, 31*(1), 1–8.

Aune, M. W, Becker, J. L., & Brugnara, C. (2014). Automated cerebrospinal fluid cell counts using the Sysmex XN-series analyzers. *American Journal of Clinical Pathology, 142*(1), 81–88.

Baheerathan, A., Pitceathly, R. DS., Curtis, C., Davies, N. WS., & lactate, C. S. F. (2020). *Practical Neurology, 20*(4), 320–323. doi:10.1136/practneurol-2019-002191.

Bahr, N. C., & Boulware, D. R. (2012). Methods of diagnosing tuberculous meningitis. *Clinical Infectious Diseases, 55*(6), 749–755.

Barnes, P. W., Eby, C. S., & Shimer, G. (2004). An evaluation of the utility of performing body fluid counts on the coulter LH 750. *Laboratory Hematology, 10*(3), 127–131. doi:10.1532/LH96.04023.

Bharti, A. R., Nally, J. E., Ricaldi, J. N., Matthias, M. A., Diaz, M. M., Lovett, M. A., Levett, P. N., Gilman, R. H., Willig, M. R., Gotuzzo, E., & Vinetz, J. M. (2003). Leptospirosis: A zoonotic disease of global importance. *Lancet Infectious Diseases, 3*(12), 757–771. http://www.journals.elsevier.com/the-lancet-infectious-diseases. doi:10.1016/S1473-3099(03)00830-2.

Bierman, P., & Giglio, P. (2005). Diagnosis and treatment of central nervous system involvement in non-Hodgkin's lymphoma. *Hematology/oncology clinics of North. America, 19*(4), 597–v. https://doi.org/10.1016/j.hoc.2005.05.003.

Blennow, K., Hampel, H., Weiner, M., & Zetterberg, H. (2010). Cerebrospinal fluid and plasma biomarkers in Alzheimer disease. *Nature Reviews Neurology, 6*(3), 131–144. doi:10.1038/nrneurol.2010.4.

Bönig, L., Möhn, N., Ahlbrecht, J., Wurster, U., Raab, P., Puppe, W., Sühs, K. W., Stangel, M., Skripuletz, T., & Schwenkenbecher, P. (2019). Leptomeningeal metastasis: The role of cerebrospinal fluid diagnostics. *Frontiers in Neurology, 10*. http://www.frontiersin.org/Neurology. doi:10.3389/fneur.2019.00839.

Bromberg, J., Breems, DA., Kraan, J., Bikker, G., van der Holt, R., Sillevis Smitt, P., … Gratama, J. W. (2007). CSF flow cytometry greatly improves diagnostic accuracy in CNS hematologic malignancies. *Neurology, 68*(20), 1674–1679. https://doi.org/10.1212/01.wnl.0000261909.28915.83.

Brouwer, M. C., Tunkel, A. R., & Van De Beek, D. (2010). Epidemiology, diagnosis, and antimicrobial treatment of acute bacterial meningitis. *Clinical Microbiology Reviews, 23*(3), 467–492. http://cmr.asm.org/cgi/reprint/23/3/467. doi:10.1128/CMR.00070-09.

Cabrera, C. M. (2022). Oligoclonal bands: An immunological and clinical approach. *Advances in Clinical Chemistry, 109*, 129–163. http://www.elsevier.com/wps/find/bookdescription.cws_home/705213/description#description. doi:10.1016/bs.acc.2022.03.004.

Chowdhury, A., & Sharma, H. (2018). A review of 571 radiographs on Tuffier's inter-cristal line and its application in lumbar spinal surgery. *Journal of Spine Surgery, 4*(2), 383–387. doi:10.21037/jss.2018.06.10.

Costerus, J. M., Brouwer, M. C., & van de Beek, D. (2018). Technological advances and changing indications for lumbar puncture in neurological disorders. *The Lancet Neurology, 17*(3), 268–278. http://www.journals.elsevier.com/the-lancet-neurology/. doi:10.1016/S1474-4422(18)30033-4.

Curti, J. M., Queiroz, G. R., Pereira, P. F. V., Anjos, M. C., Flaiban, K. K. M. C., & Lisbôa, J. A. N. (2020). L-lactate in cerebrospinal fluid can be used as a biomarker of encephalitis in cattle. *Canadian Journal of Veterinary Research, 84*(2), 146–152. https://www.ingentaconnect.com/contentone/cvma/cjvr/2020/00000084/00000002/art00010.

Davison, D. L., Terek, M., & Chawla, L. S. (2011). Neurobiomarkers in acute brain injury. *Critical Care Clinics, 27*(2), 241–255.

de Jonge, R., Brouwer, R., de Graaf, Luitwieler, R. L., Fleming, C., … Lindemans, J. (2010). Evaluation of the new body fluid mode on the Sysmex XE-5000 for counting leukocytes and erythrocytes in cerebrospinal fluid and other body fluids. *Clinical Chemistry and Laboratory Medicine, 48*(5), 665–675. https://doi.org/10.1515/CCLM.2010.108.

de Almeida, S. M., Dalla Costa, L. M., Siebra, C., Arend, L. N. V. S., & Nogueira, K. d. S. (2019). Validation of multiplex PCR for the diagnosis of acute bacterial meningitis in culture negative cerebrospinal fluid. *Arquivos de Neuro-Psiquiatria, 77*(4), 224–231. http://www.scielo.br/pdf/anp/v77n4/1678-4227-anp-77-04-0224.pdf. doi:10.1590/0004-282x20190028.

DeAngelis, L. M. (1999). Primary central nervous system lymphoma: A clinical challenge. *Neurology, 52*(8), 1579–1586.

Deisenhammer, F., Bartos, A., Egg, R., Gilhus, N. E., Giovannoni, G., Rauer, S., & Sellebjerg, F. (2006). Guidelines on routine cerebrospinal fluid analysis. Report from an EFNS task force. *European Journal of Neurology, 13*(9), 913–922. doi:10.1111/j.1468-1331.2006.01493.x.

Delanghe, J. R., & Speeckaert, M. M. (2014). Preanalytical requirements of urinalysis and body fluid analysis (including cerebrospinal fluid). *Biochemia Medica (Zagreb), 24*(1), 89–104.

Delpech, B., et al. (1985). Fibronectin in cerebrospinal fluid in neurological and infectious diseases. *Annales de Biologie Clinique (Paris), 43*(3), 159–162.

Delpech, B., & Lichtblau, E. (1985). Serum–CSF albumin ratio as a measure of blood–brain barrier function. *Annales de Biologie Clinique, 43*(3), 159–162.

Djukic, M., Schmidt-Samoa, C., Lange, P., Spreer, A., Neubieser, K., Eiffert, H., Nau, R., & Schmidt, H. (2012). Cerebrospinal fluid findings in adults with acute Lyme neuroborreliosis. *Journal of Neurology, 259*(4), 630–636. doi:10.1007/s00415-011-6221-8.

Dong, Q., Huang, Z., Yu, P., Song, E., & Chen, Z. (2022). Feng Qin, Ventricular and lumbar cerebrospinal fluid analysis in 77 HIV-negative patients with Cryptococcal meningitis who received a ventriculoperitoneal shunt. *Scientific Reports, 12*(1). doi:10.1038/s41598-022-25742-w.

Fang, X., Lu, Y., Fu, Y., Liu, Z., Kermode, A. G., Qiu, W., Ling, L., & Liu, C. (2023). Cerebrospinal fluid chloride is associated with disease activity of relapsing–remitting multiple sclerosis: A retrospective cohort study. *Brain Sciences, 13*(6). http://www.mdpi.com/journal/brainsci. doi:10.3390/brainsci13060924.

Fan, Y., Huang, Y., Zheng, W., Su, Z., Hu, J., & Yuan, S. (2025). Cerebrospinal fluid β2-microglobulin as a diagnostic biomarker in central nervous system lymphoma: A single-center retrospective analysis. *Annals of Hematology, 104*(11), 5711–5718. https://doi.org/10.1007/s00277-025-06719-x.

Fishman, R. A. (1992). *Cerebrospinal fluid in diseases of the nervous system*. Saunders.

Gray, L. D., & Fedorko, D. P. (1992). Laboratory diagnosis of bacterial meningitis. *Clinical Microbiology Reviews, 5*(2), 130–145. https://journals.asm.org/journal/cmr.

Greenberg, M. S. (2020). Greenberg's handbook of neurosurgery. *Section on diagnostic lumbar puncture and CSF analysis* (9th ed.). Thieme;.

Gupta, B. K., Bharat, A., Debapriya, B., & Baruah, H. (2010). Adenosine deaminase levels in cerebrospinal fluid of tuberculous meningitis patients. *Journal of Clinical Medicine Research, 2*(5), 220–224.

Gupta, M., Bradley, J. D., Massaad, E., Burns, E. J., Georgantas, N. Z., Maron, G. E., Batten, J. M., Gallagher, A., Thierauf, J., Nayyar, N., Gordon, A., Jones, S. A. S., Pisapia, M., Sun, Y., Jones, P. S., Barker, F. G., Curry, W. T., Gupta, R., Romero, J. M., … Shankar, G. M. (2024). Rapid tumor DNA analysis of cerebrospinal fluid accelerates treatment of central nervous system lymphoma. *Blood, 144*(10), 1093–1100. doi:10.1182/blood.2024023832.

Heckenberg, S. G., Brouwer, M. C., & van de Beek, D. (2014). Bacterial meningitis: Current epidemiology and the role of CRP and other biomarkers. *Clinical Microbiology and Infection, 20*(5), 1–9.

Hrishi, A. P., & Sethuraman, M. (2019). Cerebrospinal fluid (CSF) analysis and interpretation in neurocritical care for acute neurological conditions. *Indian Journal of Critical Care Medicine, 23*, S115. https://www.ijccm.org/lov/IJCCM. doi:10.5005/jp-journals-10071-23187.

Huang, K., He, H., Jin, W., Liu, X., Weng, Y., Yan, D., … Chen, Q. (2025). Tumor Marker Test in Cerebrospinal Fluid for Leptomeningeal Metastasis Diagnosis and Response Assessment in Non-Small-Cell Lung Cancer. *European journal of neurology, 32*(6), e70245. https://doi.org/10.1111/ene.70245.

Huy, N. T, et al. (2010). Cerebrospinal fluid lactate concentration to distinguish bacterial from viral meningitis: A meta-analysis. *Crit Care, 14*(6), R240.

Ikken, Y., Charof, R., Elouennass, M., & Sekhsokh, Y. (2021). The novel biphasic medium for transport, culture and conservation at an ambient temperature of Neisseria meningitidis, Streptococcus pneumoniae and Haemophilus influenzae. *World Journal of Microbiology and Biotechnology, 37*(11). doi:10.1007/s11274-021-03149-5.

Isgro, M. A, et al. (2015). Neuron-specific enolase as a biomarker in neurological disorders. *Journal of the Neurological Sciences, 353*(1–2), 1–8.

Johnsson, P., et al. (1995). Increased cerebrospinal fluid creatine kinase BB activity after brain injury: Relationship to severity and outcome. *Acta Neurologica Scandinavica, 91*(3), 229–234.

Kamimura, Y., Nagai, A., Nakanishi, T., Tsuji, T., Tanaka, M., & Sobue, K. (2022). Intercristal line determined by palpation in parturients in the sitting and the lateral decubitus positions: A prospective observational study. *Journal of Anesthesia, 36*(2), 270–275. doi:10.1007/s00540-022-03045-8.

Karchar, D. S., & McPherson, R. A. (2022). Henry's clinical diagnosis and management by laboratory methods. In St. Louis (Ed.). *Elsevier; 2017. Chapter 29: Cerebrospinal, synovial, serous body fluids, and alternative specimens for testing*: 24 (23rd ed.). Elsevier.

Kawthalkar, SM. (2018). Examination of cerebrospinal fluid. In SM Kawthalkar (Ed.), *Essentials of Clinical Pathology* (pp. 93–103). New Delhi: Jaypee Brothers Medical Publishers.

Khan, M. S., Rajput, J. S., Bhardwaj, P., Dhawan, S., & Shailza (2025). Comparative evaluation of cytospin and centrifuge techniques in cytological analysis of body fluids. *International Journal of Life Sciences, Biotechnology and Pharma Research, 14*(7), 996–1002.

Kim, K. T. (2022). Lumbar puncture: Considerations, procedure, and complications. *Encephalitis, 2*(4), 93–97. doi:10.47936/encephalitis.2022.00045.

Kitagawa, D., Kitano, T., Uchihara, Y., Ando, T., Nishikawa, H., Suzuki, R., Onaka, M., Kasamatsu, T., Shiraishi, N., Takemoto, K., Sekine, M., Suzuki, S., Suzuki, Y., Nakano, A., Nakano, R., Yano, H., Yoshida, S., Kawahara, M., Maeda, K., … Nakamura, F. (2023). Impact of multiplex polymerase chain reaction test in patients with meningitis or encephalitis. *Open Forum Infectious Diseases, 10*(12). doi:10.1093/ofid/ofad634.

Liesman, R. M., Strasburg, A. P., Heitman, A. K., Theel, E. S., Patel, R., & Binnicker, M. J. (2018). Evaluation of a commercial multiplex molecular panel for diagnosis of infectious meningitis and encephalitis. *Journal of Clinical Microbiology, 56*(4). http://jcm.asm.org/content/56/4/e01927-17.full.pdf+html. doi:10.1128/JCM.01927-17.

Liu, J., Jia, H., Yang, Y., Dai, W., Su, X., & Zhao, G. (2009). Cerebrospinal fluid cytology and clinical analysis of 34 cases with leptomeningeal carcinomatosis. *Journal of International Medical Research, 37*(6), 1913–1920. doi:10.1177/147323000903700629.

Lunding, J., Midgard, R., & Vedeler, C. A. (2000). Oligoclonal bands in cerebrospinal fluid: A comparative study of isoelectric focusing, agarose gel electrophoresis and IgG index. *Acta Neurologica Scandinavica, 102*(5), 322–325. doi:10.1034/j.1600-0404.2000.102005322.x.

Marcos, R., Santos, M., Marrinhas, C., Correia-Gomes, C., & Caniatti, M. (2016). Cytocentrifuge preparation in veterinary cytology: A quick, simple, and affordable manual method to concentrate low cellularity fluids. *Veterinary Clinical Pathology, 45*(4), 725–731. https://doi.org/10.1111/vcp.12423.

Marton, K. I., & Gean, A. D. (1986). The spinal tap: A new look at an old test. *Annals of Internal Medicine, 104*(6), 840–848. doi:10.7326/0003-4819-104-6-840.

McPherson, R. A., & Pincus, M. R. (2021). *Henry's clinical diagnosis and management by laboratory methods* (24th ed.). Elsevier.

Miftode, E. G., Dorneanu, O. S., Leca, D. A., Juganariu, G., Teodor, A., Hurmuzache, M., … Anton-Paduraru, D. T. (2015). Tuberculous Meningitis in Children and Adults: A 10-Year Retrospective Comparative Analysis. *PloS one, 10*(7), e0133477. https://doi.org/10.1371/journal.pone.0133477.

Mishra, O. P., Batra, P., Ali, Z., Anupurba, S., & Das, B. K. (2003). Cerebrospinal fluid lysozyme level for the diagnosis of tuberculous meningitis in children. *Journal of tropical pediatrics, 49*(1), 13–16. https://doi.org/10.1093/tropej/49.1.13.

Nørgaard-Pedersen, B., Lindholm, J., Albrechtsen, R., Arends, J., Diemer, N. H., & Riishede, J. (1978). Alpha-fetoprotein and human chorionic gonadotropin in a patient with a primary intracranial germ cell tumor. Cancer, 41(6), 2315–2320. https://doi.org/10.1002/1097-0142(197806)41:6⟨2315::AID-CNCR2820410633⟩3.0.CO;2-Q

Olsson, B., Lautner, R., Andreasson, U., Öhrfelt, A., Portelius, E., Bjerke, M., Hölttä, M., Rosén, C., Olsson, C., Strobel, G., Wu, E., Dakin, K., Petzold, M., Blennow, K., & Zetterberg, H. (2016). CSF and blood biomarkers for the diagnosis of Alzheimer's disease: A systematic review and meta-analysis. *The Lancet Neurology, 15*(7), 673–684. doi:10.1016/s1474-4422(16)00070-3.

Reiber, H., & Peter, J. B. (2001). Cerebrospinal fluid analysis: Disease-related data patterns and evaluation programs. *Journal of the Neurological Sciences, 184*(2), 101–122. doi:10.1016/S0022-510X(00)00501-3.

Perfect, J. R., Dismukes, W. E., Dromer, F., Goldman, D. L., Graybill, J. R., Hamill, R. J., ... Sorrell, T. C. (2010). Clinical practice guidelines for the management of cryptococcal disease: 2010 update by the Infectious Diseases Society of America. *Clinical Infectious Diseases, 50*(3), 291–322.

Reiber, H. (2001). Dynamics of brain-derived proteins in cerebrospinal fluid. *Clinica Chimica Acta, 310*(2), 173–186. doi:10.1016/s0009-8981(01)00573-3.

Rifai, N., Horvath, A. R., & Wittwer, C. T. (2018). Tietz textbook of clinical chemistry and molecular diagnostics. In St. Louis (Ed.), *Chapter 29: Body fluid analysis* (6th ed.). Elsevier.

Roccaforte, V, Daves, M, Proserpio, V, Sciarini, F, Sangiorgio, R, Costanzo, A, ... Bonato, C (2018). Evaluation of body fluid mode of Sysmex XN-9000 for white blood cell counts in cerebrospinal fluid. *J Lab Precis Med, 3*, 22.

Rock, R. B., Olin, M., Baker, C. A., Molitor, T. W., & Peterson, P. K. (2008). Central nervous system tuberculosis: Pathogenesis and clinical aspects. *Clinical Microbiology Reviews, 21*(2), 243–261. doi:10.1128/CMR.00042-07.

Ropper, A. H, Samuels, M. A, Klein, J. P, & Prasad, S. (2019). Adams and Victor's principles of neurology. *Chapter 2: Laboratory diagnosis of neurological disease; Chapter 43: Acute meningitis* (11th ed.). McGraw-Hill Education.

Ruotolo, R., Maffei, E., Sabbatino, F., Serio, B., Zeppa, P., & Caputo, A. (2023). Cytopathological differential diagnosis of malignant tumor cells in the cerebrospinal fluid: A retrospective analysis. *Diagnostic Cytopathology, 51*(12), 751–757. http://onlinelibrary.wiley.com/journal/10.1002/(ISSN)1097-0339. doi:10.1002/dc.25217.

Sanchez-Juan, P., Green, A., Ladogana, A., Cuadrado-Corrales, N., Sáanchez-Valle, R., Mitrováa, E., Stoeck, K., Sklaviadis, T., Kulczycki, J., Hess, K., Bodemer, M., Slivarichová, D., Saiz, A., Calero, M., Ingrosso, L., Knight, R., Janssens, A. C. J. W., van Duijn, C. M., & Zerr, I. (2006). CSF tests in the differential diagnosis of Creutzfeldt-Jakob disease. *Neurology, 67*(4), 637–643. doi:10.1212/01.wnl.0000230159.67128.00.

Seehusen, D. A., Reeves, M. M., & Fomin, D. A. (2003). Cerebrospinal fluid analysis. *American Family Physician, 68*(6), 1103–1108.

Siddiqui, I., Aleem, S., Kayani, N., & Baig, S. (2002). CSF oligoclonal bands in multiple sclerosis. *JPMA. The Journal of the Pakistan Medical Association, 52*(8), 351–353.

Smets, I., Verelst, S., Meyfroidt, G., Van Wijngaerden, E., Wilmer, A., van Loon, J., ... Dubois, B. (2020). Community-acquired bacterial meningitis in adults: Emergency department management protocol. *Acta Neurologica Belgica, 120*, 1033–1043. https://doi.org/10.1007/s13760-020-01428-w.

Spanos, A., Harrell, F. E., & Durack, D. T. (1989). Differential diagnosis of acute meningitis: An analysis of the predictive value of initial observations. *JAMA: The Journal of the American Medical Association, 262*(19), 2700–2707. doi:10.1001/jama.1989.03430190084036.

Stephani, C., Choi, A. H. K., & Moerer, O. (2021). Point-of-care detection of lactate in cerebrospinal fluid. *Intensive Care Medicine Experimental, 9*(1). icm-experimental.springeropen.com/articles. doi:10.1186/s40635-021-00385-9.

Temfack, E., Rim, J. J. B., Spijker, R., Loyse, A., Chiller, T., Pappas, P. G., Perfect, J., Sorell, T. C., Harrison, T. S., Cohen, J. F., & Lortholary, O. (2021). Cryptococcal antigen in serum and cerebrospinal fluid for detecting cryptococcal meningitis in adults living with human immunodeficiency virus: Systematic review and meta-analysis of diagnostic test accuracy studies. *Clinical Infectious Diseases, 72*(7), 1268–1278. http://cid.oxfordjournals.org/content/by/year. doi:10.1093/cid/ciaa1243.

Thwaites, G. E., Chau, T. T. H., & Farrar, J. J. (2004). Improving the bacteriological diagnosis of tuberculous meningitis. *Journal of Clinical Microbiology, 42*(1), 378–379. jcm.asm.org. doi:10.1128/JCM.42.1.378-379.2004.

Torres, S., & Naous, R. (2026). Cytopathologic diagnosis of cerebrospinal fluid metastasis from solid tumors: A 10-year institutional review. *Journal of the American Society of Cytopathology, 15*(1), 23–28. https://www.sciencedirect.com/science/journal/22132945. doi:10.1016/j.jasc.2025.07.009.

Tunkel, A. R., Hartman, B. J., Kaplan, S. L., Kaufman, B. A., Roos, K. L., Scheld, W. M., & Whitley, R. J. (2004). Practice guidelines for the management of bacterial meningitis. *Clinical Infectious Diseases, 39*(9), 1267–1284. https://doi.org/10.1086/425368.

Tumani, H., Petereit, H. F., Gerritzen, A., Gross, C. C., Huss, A., Isenmann, S., Jesse, S., Khalil, M., Lewczuk, P., Lewerenz, J., Leypoldt, F., Melzer, N., Meuth, S. G., Otto, M., Ruprecht, K., Sindern, E., Spreer, A., Stangel, M., Strik, H., … Zimmermann, T. (2020). S1 guidelines "lumbar puncture and cerebrospinal fluid analysis" (abridged and translated version). *Neurological Research and Practice, 2*(1). https://neurolrespract.biomedcentral.com/. doi:10.1186/s42466-020-0051-z.

Tumani, H., Teunissen, C., Süssmuth, S., Otto, M., Ludolph, A. C., & Brettschneider, J. (2008). Cerebrospinal fluid biomarkers of neurodegeneration in chronic neurological diseases. *Expert review of molecular diagnostics, 8*(4), 479–494. https://doi.org/10.1586/14737159.8.4.479.

Rimelen, V., Ahle, G., Pencreach, E., Zinniger, N., Debliquis, A., Zalmaï, L., Harzallah, I., Hurstel, R., Alamome, I., Lamy, F., Voirin, J., & Drénou, B. (2019). Tumor cell-free DNA detection in cerebrospinal fluid for primary central nervous system lymphoma. Acta Neuropathologica Communications, 7, Article 43. https://doi.org/10.1186/s40478-019-0692-8

Tunkel, A. R., Hartman, B. J., Kaplan, S. L., Kaufman, B. A., Roos, K. L., Scheld, W. M., & Whitley, R. J. (2004). Practice guidelines for the management of bacterial meningitis. *Clinical Infectious Diseases, 39*(9), 1267–1284. https://doi.org/10.1086/425368.

van de Beek, D., de Gans, J., Tunkel, A. R., & Wijdicks, E. F. M. (2006). Community-acquired bacterial meningitis in adults. *The New England Journal of Medicine, 354*(1), 44–53. https://doi.org/10.1056/NEJMra052116.

Vélez-Paez, J. L., Rueda-Barragán, F. E., Torres-Carpio, D. A., Guerrero, F. M., Romero, J. J., Vásconez-González, E., & Ortíz-Prado, E. (2022). Lactate as an important biomarker of cerebrospinal fluid in infectious diseases of the central nervous system: A comprehensive literature review. *Microbes, Infection and Chemotherapy, 2*, e1357.

Wang, Q., Wang, Y. F, Yang, Y., Kong, Y. X., & Peng, Y. (2023). The value of elevated cerebrospinal fluid lactate concentrations in post-neurosurgical bacterial meningitis. *BMC Neurology, 23*(1). doi:10.1186/s12883-023-03428-8.

Welinder-Olsson, C., Dotevall, L., Hogevik, H., Jungnelius, R., Trollfors, B., Wahl, M., & Larsson, P. (2007). Comparison of broad-range bacterial PCR and culture of cerebrospinal fluid for diagnosis of community-acquired bacterial meningitis. *Clinical Microbiology and Infection, 13*(9), 879–886. doi:10.1111/j.1469-0691.2007.01756.x.

Wick, M., Gross, C. C., Tumani, H., Wildemann, B., Stangel, M., & German Society of CSF Diagnostics and Clinical Neurochemistry (DGLN e.V.). (2021). Automated analysis of cerebrospinal fluid cells using commercially available blood cell analysis devices—A critical appraisal. Cells, 10(5), Article 1232. https://doi.org/10.3390/cells10051232

Wichmann, T. O., Damkier, H. H., & Pedersen, M. (2022). A brief overview of the cerebrospinal fluid system and its implications for brain and spinal cord diseases. *Frontiers in Human Neuroscience, 15*. https://www.frontiersin.org/journals/human-neuroscience. doi:10.3389/fnhum.2021.737217.

Yamada, S., & Mase, M. (2023). Cerebrospinal fluid production and absorption and ventricular enlargement mechanisms in hydrocephalus. *Neurologia Medico-Chirurgica, 63*(4), 141–151. doi:10.2176/jns-nmc.2022-0331.

Zimmermann, M., Ruprecht, K., Kainzinger, F., Heppner, F. L., & Weimann, A. (2011). Automated vs. manual cerebrospinal fluid cell counts: A work and cost analysis comparing the sysmex XE-5000 and the fuchs-rosenthal manual counting chamber. *International Journal of Laboratory Hematology, 33*(6), 629–637. doi:10.1111/j.1751-553x.2011.01339.x.

3

Cerebrospinal fluid analysis for metastatic malignancies

Sumaira Qayoom[a] and Namrata P. Awasthi[b]

[a]DEPARTMENT OF PATHOLOGY, KING GEORGE'S MEDICAL UNIVERSITY, LUCKNOW, UTTAR PRADESH, INDIA [b]DEPARTMENT OF PATHOLOGY, DR. RAM MANOHAR LOHIA INSTITUTE OF MEDICAL SCIENCES (DR. RMLIMS), LUCKNOW, UTTAR PRADESH, INDIA

3.1 Introduction

Leptomeningeal metastasis (LM), also known as "leptomeningeal carcinomatosis", is defined as cancer dissemination to the arachnoid mater, pia mater, and cerebrospinal fluid (CSF), and was first described in the 19th century (Hyun et al., 2016). Although once considered rare, due to the advancements in cancer treatment, the incidence has increased considerably in recent times. However, even most of these advanced treatment protocols fail to cross the blood-brain barrier (BBB), leading to their limited efficacy in CNS metastasis, making it a haven for cancer cells. LM is seen in 5%–8% of solid tumors and 5%–15% of hematologic malignancies, with breast cancer, lung cancer, and melanoma being the most common primaries. LM can occur generally through distinct mechanisms:

- *Hematogenous spread*: Cancer cells may cross the BBB directly from the bloodstream (Bönig et al., 2019).
- *Perineural and endoneural dissemination*: Cells may migrate along peripheral nerves, tracking into the CNS.
- *Direct extension*: Malignant cells can spread from preexisting parenchymal brain metastasis or bone vertebral metastasis that breaches the dural boundaries.

3.2 Pathogenesis of leptomeningeal metastasis

The pathogenesis of LM represents a complex adaptive process. Unlike brain parenchyma, CSF is acellular and low in protein, glucose, and cytokine content. The tumor cells must adapt to a mitogen-poor environment (Herrlinger et al., 2004a). For this, the tumor cells hijack host physiology, especially the complement component 3 (C3) and vascular endothelial growth factor (VEGF), which serve as critical mediators in subverting the blood-CSF barrier (B-CSF-B) and making CSF growth-supportive.

The central role of complement C3: The complement system is an important component of the innate immune system, but in LM, it is redirected by cancer cells to modify the B-CSF-B. The C3 involved in this axis is derived from cancer cells rather than the systemic circulation or

Advancing Body Fluid Diagnostics with Flow Cytometry: A Comprehensive Guide to Effusions and CSF.
DOI: https://doi.org/10.1016/B978-0-443-27426-8.00013-2

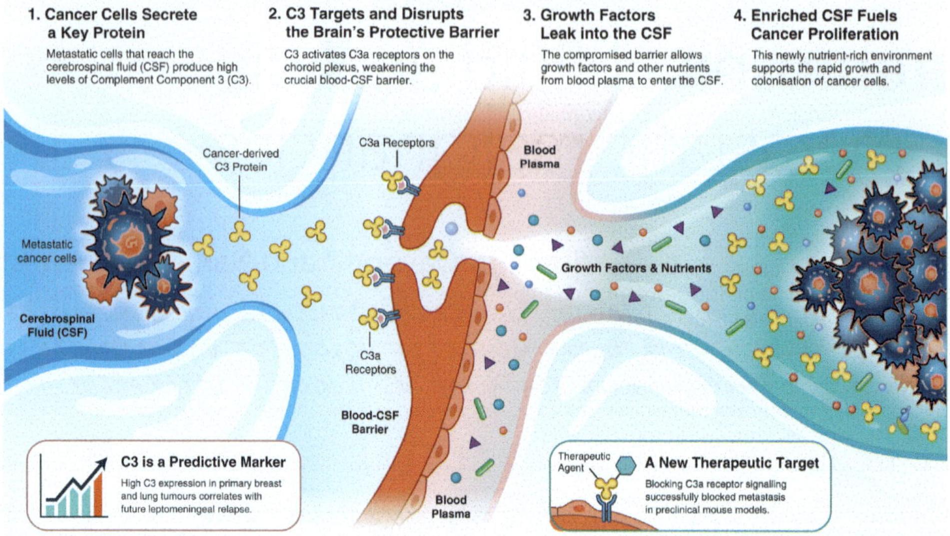

FIGURE 3.1 Schematic illustration of the central role of complement component C3 in the pathogenesis of leptomeningeal metastasis. Complement component 3 (C3) plays a critical role in the pathogenesis of leptomeningeal metastasis. C3 secreted by the metastatic cancer cells disrupts the blood-CSF barrier by activating C3a receptors. Growth factors leak into the CSF, promoting the survival and proliferation of metastatic cancer cells.

the host liver, and it does not stimulate the cancer cells directly but rather acts on the host's choroid plexus. This upregulation is a unique trait of cells capable of surviving in the CSF and is not necessarily found in brain metastasis. C3 is cleaved into C3a, an active anaphylatoxin, and binds to the C3a receptor (C3aR), a G-Protein-coupled receptor expressed on both the luminal and basolateral surfaces of the choroid plexus epithelium (Herrlinger et al., 2004a) (Fig. 3.1).

Activation of C3aR triggers the phosphorylation of myosin light-chain kinase (MLCK), which leads to the disorganization of tight junction proteins such as ZO-1 and claudin. By loosening these tight junctions, C3aR signaling reduces the trans-epithelial electrical resistance (TEER) and increases the permeability of the B-CSF-B. This "leaky gate" allows select plasma components, including growth factors like amphiregulin, to flood into the previously mitogen-poor CSF. Amphiregulin, an EGF-family ligand, then provides the necessary proliferative signals to the metastatic cells, enabling them to grow in suspension within the subarachnoid space (Herrlinger et al., 2004a).

The role of VEGF: While C3 primarily facilitates "opening the door" for nutrients, VEGF serves as a vital marker and mediator of the metabolic and angiogenic stress inherent to LM. VEGF is a potent inducer of angiogenesis and is often utilized by tumors to ensure survival under harsh conditions.

The leptomeningeal space is a challenging environment for metastatic cells. As these cells proliferate, they often enter a low-oxygen (hypoxic) state. VEGF is a well-documented vascular permeability factor that increases permeability, allowing the influx of plasma proteins and other systemic growth factors into the CSF, thereby enriching the mitogen-poor environment and supporting further tumor growth. As a primary inducer of angiogenesis, VEGF also facilitates the formation of new blood vessels (Boire et al., 2017).

Thus, the pathogenesis of LM involves a coordinated attack on CNS barriers, with complement C3 and VEGF playing a pivotal role. Complement C3 "initiates" the breach by targeting the choroid plexus to allow the initial nutrient entry, and VEGF "sustains" this breach as a vascular permeability and angiogenesis factor.

3.3 Role of cerebrospinal fluid analysis in leptomeningeal metastasis

CSF analysis is a critical diagnostic window to the CNS, serving as the gold standard for identifying and monitoring metastatic malignancy.

3.3.1 Diagnostic role of conventional cerebrospinal fluid cytology

The patients with LM present with headache, nausea, vomiting, gait instability, cranial nerve palsies, and seizures, and as these are nonspecific symptoms and overlap with the clinical manifestation of other infectious and inflammatory disorders, diagnosis of lumbar puncture (LP) needs high suspicion and a confirmatory test to establish the diagnosis. The diagnosis of LM rests on a "triple pillar" approach: Clinical neurological evaluation, neuroimaging, and CSF analysis. Magnetic resonance imaging (MRI) is the modality of choice. However, MRI sensitivity is approximately 60%–70% and is more likely to be positive in solid tumors than in hematologic malignancies due to the cells' adherence pattern. Despite the tremendous advancements in neuroimaging, CSF cytology remains the diagnostic gold standard for LM. A definitive diagnosis is established when malignant cells are morphologically identified in the fluid (Glantz et al., 1998). Once the cancer cells enter the CSF, they follow the natural flow of the fluid. These cells often exhibit "branched evolution," with a distinct molecular profile from the primary tumor, due to the unique microenvironment of the CNS (Kumthekar & Le Rhun, 2023). In patients with metastasis from solid primary tumors, cancer cells tend to adhere to the meninges, forming layers that are often identifiable on imaging. Contrary to this, in hematologic malignancies, malignant cells are more likely to remain free-floating within the fluid (Herrlinger et al., 2004a).

Malignant cells are identified by several hallmark characteristics: (Torzewski et al., 2008)

- Abnormal size with a high N; C ratio
- Conspicuous to prominent nucleoli
- Atypical mitosis and
- Intensive basophilia of the cytoplasm

The detection yield is often influenced by the tumor type. Solid tumors, such as breast or lung cancer, are more likely to adhere to the meninges, forming layers of tissue, whereas hematologic malignancies, such as leukemias and lymphomas, are more likely to remain freely floating in the CSF, often resulting in higher cell counts. Cytology is generally employed in three clinical scenarios: When a primary malignancy is already known, and dissemination is suspected; when clinical/radiological findings suggest neoplastic meningitis without a known primary; and as an incidental finding during a spinal tap performed for other reasons. The main limitation of CSF cytology is its low sensitivity, which ranges from 50% to 70% for initial sampling. To minimize false-negative results, various preanalytical variables should be taken care of as discussed later (Bae et al., 2013; Glantz et al., 1998).

3.3.1.1 Optimizing diagnostic accuracy: Preanalytical variables

- *Sample volume*: As the small sample volume increases the false-negative rates, it is recommended to withdraw a minimum of 10.5 mL of CSF for cytologic analysis to achieve an acceptable 3% error rate (Glantz et al., 1998).
- *Processing time*: As CSF is not growth conducive due to its low protein and sugar content, Samples **should ideally be processed within 2 h** to maintain cellular integrity. If delays are unavoidable, refrigeration or fixation in buffered formalin (1:1) may be used (Torzewski & Lackner, 2016).
- *Site of sampling*: The yield is significantly higher if the CSF is obtained from a site close to the symptomatic or radiographically demonstrated disease. For example, lumbar fluid is more diagnostic for spinal symptoms, while ventricular fluid (via an **Ommaya reservoir**) may be more accurate for cranial disease (Glantz et al., 1998).
- *Frequency of sampling*: A single negative tap does not exclude malignancy. Repeating the procedure once if the initial cytology is negative is essential, as two samplings achieve significantly higher detection rates (Bae et al., 2013).

3.3.2 Role of biochemical and immunological parameters

Routine CSF parameters provide supportive diagnostic data and possess prognostic significance.

- *CSF cell count*: While an elevated white blood cell count is common (found in 79%–92% of cases), up to 21% of patients with solid tumors and LM may have a normal CSF cell count. Therefore, cytology must be performed even if the cell count is normal (Djukic et al., 2017; Liu et al., 2009).
- *Protein and barrier function*: Elevated protein levels and B-CSF-B dysfunction (measured by the CSF/serum albumin quotient, QAlb) are found in the vast majority of cases. The rise in CSF protein levels is attributed either to increased protein synthesis by inflammatory or cancerous cells or to increased permeability of the BBB. Severe barrier dysfunction is often a therapy-independent predictor of poor survival.
- *Lactate and glucose*: Elevated CSF lactate ($\geq$3.5 mmol/L) is a common finding and serves as another therapy-independent predictor of poor prognosis. Low glucose levels (hypoglycorrhachia) are observed in 43%–46% of cases and may reflect tumor-cell metabolic consumption. Elevated CSF lactate levels may help differentiate infectious diseases, such as neuroborreliosis, which has a clinical picture and CSF cytology similar to those of lymphoid

malignancy, as elevated CSF LDH levels are seen in <10% of these patients (Djukic et al., 2012,2017).

- *Oligoclonal bands*: Interestingly, 26%–30% of LM patients exhibit oligoclonal bands restricted to the CSF, indicating a local intrathecal immune response, though this must be differentiated from autoimmune conditions like multiple sclerosis (Schipper et al., 1988; Weller et al., 1992).
- *VEGF*: VEGF is a potent inducer of angiogenesis and a survival factor for tumor cells under metabolic stress. High VEGF levels (>250 pg/mL) have an exceptional specificity (98.3%) for LM (Herrlinger et al., 2004a). Initial VEGF levels above 100 pg/mL are strong predictors of poor survival (median 2.4 months vs. not assessable in those with lower levels).
- *Complement C3*: C3 is upregulated specifically in cells capable of growing in the CSF. Cancer-cell-derived C3 acts on the choroid plexus epithelium to disrupt the B-CSF-B. This allows plasma growth factors, such as amphiregulin, to flood into the previously nutrient-poor CSF, supporting metastatic colonization. C3 levels in the CSF correlate with disease burden and clinical course, often rising during relapse.

3.3.3 Prognostic role of cerebrospinal fluid analysis in leptomeningeal metastasis

- Markedly elevated protein levels, an indirect marker of B-CSF-B dysfunction, predict survival independent of therapy (Chamberlain, 2010; Leal et al., 2011; Herrlinger et al., 2004b).
- Initial VEGF levels above 100pg/mL are associated with poor survival in LM (Herrlinger et al., 2004a).
- C3 levels in the CSF correlate with disease burden and clinical course, often rising during the relapse (Boire et al., 2017).

3.3.4 Monitoring role of cerebrospinal fluid analysis in leptomeningeal metastasis

CSF examination is an indispensable tool for evaluating therapeutic response and making critical management decisions.

- *Theragnostic decision node*: The identification of malignant cells often determines the initiation of intrathecal (IT) chemotherapy (e.g., methotrexate, cytarabine, or thiotepa). IT therapy is continued until two consecutive CSF samples are confirmed as "negative for malignant cells".
- *Tracking response*: Successful treatment typically leads to a marked reduction in biomarkers like VEGF and C3, as well as the clearance of malignant cells. Conversely, a clinical relapse can be suspected if these markers reappear.
- *Adjusting targeted therapy*: CSF analysis can identify various targetable molecular mutations, allowing clinicians to switch to CNS-penetrant targeted agents like osimertinib for lung cancer or tucatinib for HER2-positive breast cancer.

3.3.5 Leptomeningeal metastasis in hematolymphoid malignancies

LM represents one of the most clinically challenging complications in the management of hematolymphoid malignancies. The overall incidence of LM in HMs ranges from 5% to 15%, depending on the malignancy subtype and whether active prophylaxis is given (Domingues et al., 2021). LM in hematolymphoid malignancies is associated with an adverse prognosis. In adults, the relapse rate for leukemia can be as high as 75% after 1 year if active prevention is not done (Liu & Wang, 2023). Acute lymphoblastic leukemia (ALL) has a poorer prognosis once CNS involvement occurs compared to acute myeloid leukemia (AML), with an average survival of 6 months. Consequently, early detection and aggressive management are the primary determinants of patient morbidity and mortality (Momen et al., 2023). The diagnosis of CNS involvement in HMs has shifted from a simple morphology to an integrated approach comprising cytomorphological evaluation and multicolour flow cytometry.

3.3.5.1 Cerebrospinal fluid cytomorphology

The demonstration of malignant cells by cytomorphology is highly specific, but its sensitivity remains a significant limitation (Wu et al., 2022). The children's oncology group (COG) classification is widely used to categorize CNS status (Gossai et al., 2023; Muffly & Curran, 2019).

- CNS1: Absence of blasts on cytospin.
- CNS2: Presence of <5/μL WBCs with blasts present on cytospin.
- CNS3: Presence of ≥5 WBCs/μL with blasts, or the presence of clinical signs of CNS leukemia (e.g., facial nerve palsy).

Steinherz-Bleyer algorithm: The presence of blasts in CSF could be from the peripheral blood due to the traumatic LP. Applying the Steinherz-Bleyer Algorithm, if >10RBCs/ul are present, then the presence of the blasts should be attributed to the traumatic puncture rather than true CNS (Steinherz et al., 2019).

3.3.5.2 Flow cytometric immunophenotyping

The tumor cells are usually present in small numbers in CSF, leading to false-negative results. But with the advent of FCI, there has been a paradigm shift in diagnosis, as flow cytometry can identify small clonal populations, sometimes as low as 10 events out of 100,000 cells (Craig & Foon, 2008; Del Principe et al., 2014). Some studies have reported a sensitivity of as high as 95.3% (Buonocore et al., 2017). With such high sensitivity, FCS can detect occult leptomeningeal disease that may be missed on cytomorphology (Bromberg et al., 2007). Certain studies have shown a statistically significant decrease in survival and higher relapse rates in these patients.

3.3.5.3 Neuroimaging

MRI with contrast material detects leukemic or lymphomatous involvement in only 40%–60% of cases (Momen et al., 2023). Common findings include leptomeningeal enhancement or intraparenchymal mass lesions.

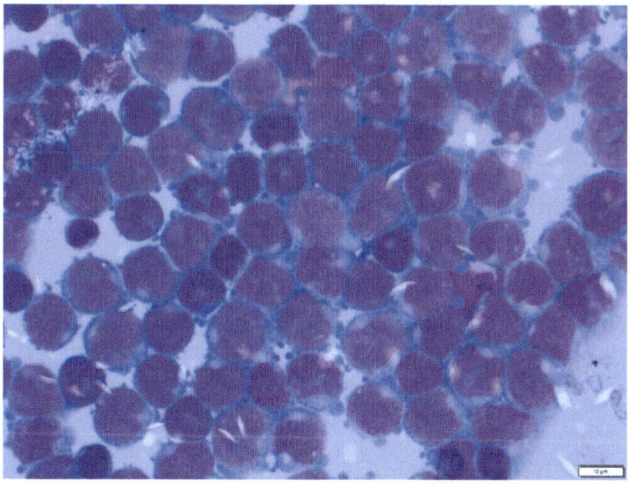

FIGURE 3.2 CSF cytology in acute lymphoblastic leukemia. CSF showing dispersed population of lymphoblasts (MGG; 100x). *Source: Image courtesy Dr Pariksha Gupta, PGIMER Chandigarh, India*

3.3.5.4 Specific hematolymphoid malignancies

3.3.5.4.1 Acute lymphoblastic leukemia

ALL is the most common HM to involve the CNS, appearing in 56.2% of cases in some series (Sruthi et al., 2025). T-ALL has a higher preponderance of developing CNS leukemia.

Cytomorphologic features in ALL (Fig. 3.2):

- Scattered or loosely clustered atypical lymphoid /blast cells
- Monomorphic small to medium-sized cells
- High N/C ratio
- Round to irregular nuclei
- Fine to slightly condensed chromatin
- Inconspicuous to occasional nucleoli
- Scant basophilic cytoplasm

3.3.5.4.2 Acute myeloid leukemia

In AML, CNS involvement is less common but remains significant, particularly in the AML-M3 (APML) and AML-M5 subtypes. In APML, smears show atypical promyelocytes with convoluted nuclei and granular cytoplasm.

Cytomorphologic features in APML:

- Medium to large atypical myeloid cells
- Folded, bilobed, or butterfly-shaped nuclei
- Abundant cytoplasm with azurophilic granules
- Auer rods may be present

Cytomorphologic features in AML-M5:

- Singly dispersed or loosely clustered cells
- Cells are medium to large blasts
- Irregular, folded, indented, or convoluted nuclei
- Fine chromatin with one or more conspicuous nucleoli
- Some cells may show cytoplasmic vacuolations

3.3.5.4.3 Nonhodgkin lymphoma

Diffuse large B-cell lymphoma (DLBCL) and Burkitt's lymphoma have a higher predisposition for leptomeningeal spread.

Cytomorphologic features of CSF involvement in DLBCL:

- Large atypical lymphoid cells
- High N: C ratio
- Irregular nuclear contours
- Vesicular chromatin with prominent nucleoli

Cytomorphologic features of CSF involvement in Burkitt lymphoma:

- Monomorphic medium-sized cells
- High N: C ratio
- Round nuclei with coarse to finely clumped chromatin
- Multiple small, distinct nucleoli
- Cytoplasmic vacuolation
- Apoptotic bodies may be seen in the background

3.3.5.4.4 Multiple myeloma

CNS invasion is rare but can present as intraparenchymal lesions or LM. Survival has historically been poor but is improving with novel therapies such as proteasome inhibitors (Diansari et al., 2017). The presence of plasma cells should raise the suspicion of CSF involvement in the appropriate clinical setting.

Cytomorphologic features:

- Singly lying or discohesive plasma cells
- Cells show an eccentric, round to oval nucleus
- Coarse clock-face or cartwheel chromatin
- Presence of Perinuclear Hof
- Presence of binucleate or multinucleate cells

Waldenström Macroglobulinemia (WM): CNS involvement here is known as Bing-Neel Syndrome, characterized by atypical lymphoid cells with plasmacytoid morphology in the CSF (Minnema et al., 2017).

3.3.5.5 Therapeutic relevance and monitoring

The involvement of the CSF directs the major changes in chemotherapy and radiation protocols.

- *CNS-directed therapy*: In CNS2 and CNS3 involvement cases, an intensified treatment regimen is required. Prophylactic treatment is essential to prevent the 75% relapse rate seen without intervention (Gossai et al., 2023).
- *Intrathecal (IT) therapy*: Common agents include methotrexate and cytarabine. Liposomal cytarabine has also been used to prolong the intervals between injections (Diansari et al., 2017).
- *Radiation therapy*: Prophylactic cranial radiotherapy is prevalent in T-ALL, despite risks of neurocognitive deficits and secondary malignancies. In some cases of multiple myeloma, proton beam craniospinal irradiation combined with immunomodulatory drugs (e.g., pomalidomide) has shown durable responses.
- *Genetic considerations*: Mutations such as those in the MTHFR gene are associated with increased toxicity from high-dose methotrexate and may even be associated with the occurrence of CNS lymphomas (Diansari et al., 2017).

To conclude, LM in hematolymphoid malignancies remains a daunting clinical entity and necessitates a vigilant diagnostic approach so that clinicians can better identify patients at risk and tailor CNS- directed therapies to improve survival and quality of life.

3.3.6 Leptomeningeal metastasis in solid malignancies

LM, also known as carcinomatous meningitis, represents mostly a late-stage complication of various epithelial malignancies. Historically regarded as a terminal event with a median survival of 2–4 months, LM is now increasingly recognized as a treatable but generally incurable manifestation of advanced solid malignancy. Its rising incidence reflects improved systemic control of extracranial disease, increased CNS surveillance with high-resolution MRI, and prolonged survival under targeted and immunotherapeutic regimens. Among solid tumors, breast cancer, lung cancer, and melanoma account for the majority of cases. However, virtually any epithelial malignancy may seed the leptomeninges.

3.3.6.1 Epidemiology and predisposing factors

LM is clinically detected in approximately 5%–10% of patients with solid cancers. However, true incidence may reach upto20% as suggested by certain autopsy series, as the condition is frequently underdiagnosed or asymptomatic.

Melanoma: LMD is seen in 5%–8% of patients with malignant melanoma. Autopsy data reveal leptomeningeal involvement in up to 20% cases (Smalley et al., 2024). Among solid tumors, melanoma has one of the highest overall risks of parenchymal brain invasion and LM. This high propensity to seed the leptomeninges may be due to melanocytes arising from neural crest cells and migrating during development to the skin, eyes, mucous membranes, and leptomeninges. This shared embryologic origin means melanocytic tumors retain neurotropic features that favor migration and survival in CNS tissues. Melanoma cells also acquire mutations and express adhesion molecules that mimic those found on vascular endothelial cells. These cells secrete matrix metalloproteinases (MMPs) that degrade the basement membrane and facilitate the invasion (Nguyen et al., 2023). The current diagnostic gold standard involves a combination of Gadolinium-enhanced MRI and CSF cytology. However, a single LP has a low sensitivity of only 44%–67% (Smalley et al., 2016; Nayak et al., 2013). Repeating the LP up to three times can increase the diagnostic yield to 80%–90%. Emerging "liquid biopsy" techniques, such as measuring circulating tumor DNA (ctDNA) via droplet digital PCR (ddPCR), have shown a stronger correlation with MRI findings than conventional cytology (Subirá et al., 2012).

Cytomorphologic features of metastatic melanoma in CSF:(Ballester et al., 2018)

- Moderate to high cellularity.
- Melanin pigment granules.
- Presence of melanophages (macrophages containing melanin pigment).
- Tumor cells are predominantly discohesive single cells or loose clusters.
- Individual tumor cells are usually large, epithelioid, spindle-shaped, and plasmacytoid. Occasionally, small, round, and bizarre multinucleated cells may be seen.
- Typical nuclear features include single or multiple prominent, centrally located eosinophilic nucleoli.
- Frequent mitosis and occasionally atypical mitotic figures.
- Tumor cells usually have moderate to abundant cytoplasm.
- Cytoplasmic melanin pigment, when present, is highly characteristic.
- In the amelanotic variant of melanoma, the absence of melanin pigment can pose a diagnostic challenge, and Immunohistochemistry markers S-100, HMB-45, Melan-A (Mart-1), and Sox-10 can help distinguish melanoma from other mimickers.

Breast cancer: Breast cancer, the most common epithelial malignancy responsible for LM, accounts for 27%–50% cases in large series (Murali et al., 2009). The incidence varies with the morphological type, grade, and molecular subtype of breast cancer. Lobular carcinoma has a unique propensity for leptomeningeal spread as compared to ductal carcinoma (Abouharb et al., 2014). Triple-negative breast cancer patients have the highest rates of LM and shorter duration between primary diagnosis and LM involvement (Niwińska et al., 2013). HER2Neu-positive breast cancer cells also have a predilection for the CNS. The higher incidence of LM in HER2Neu-positive breast cancer patients can also be attributed to the fact that Trastuzumab has poor penetration of the BBB (Bendell et al., 2003). Diffuse nonadherent LM is common, making these patients potential candidates for intrathecal therapy.

CSF cytomorphology in metastatic breast cancer:(Ballester et al., 2018; Singh et al., 2013)

- Moderate to high cellularity
- Tumor cells are arranged in cohesive clusters, three-dimensional aggregates, loose sheets, papillary fragments, and single scattered malignant cells. In lobular carcinoma, tumor cells may appear predominantly dispersed singly
- Tumor cells are medium to large in size and round to polygonal
- High N/C ratio with irregular nuclear membrane and conspicuous nucleoli
- Signet ring cells may sometimes be seen in lobular carcinoma

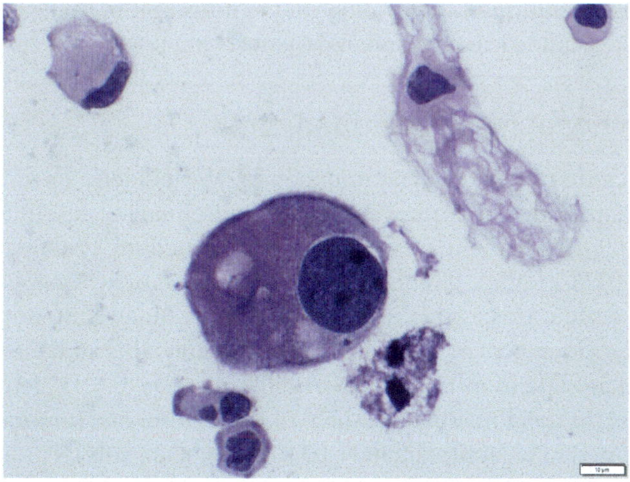

FIGURE 3.3 CSF cytology in metastatic carcinoma. CSF cytology showing large epithelial tumor cells with round to irregular nuclei, coarse chromatin, and abundant cytoplasm (Papanicolaou; 100x). *Source: Image courtesy Dr Parikshaa Gupta, PGIMER Chandigarh, India*

- Lobular carcinoma in CSF may mimic lymphoma, and also several conditions may mimic breast cancer, like reactive ependymomal or choroid plexus cells. Hence, sometimes immunocytochemistry is needed to make a correct diagnosis. Common markers used for cytokeratin (AE1/AE3), CK7, GATA-3, TRPS1, ER, PR, and CD45.

Lung cancer: Lung cancer is the second most common cause of LM and is seen in 9%–25% of NSCLC patients, depending on the histological type and molecular profile (Aroney et al., 1981). Adenocarcinoma is responsible for 84%–96% of lung cancer-related LM. LM occurs more frequently in EGFR-mutated (9.4%) compared to EGFR wild type (1.7%). ALK-rearranged NSCLC has a significant CNS tropism, with LM occurring in approximately 5% of patients.

Cytomorphologic features of metastatic lung cancer (Fig. 3.3):

- Cohesive epithelial cells or gland-forming clusters
- Mucin-producing cells
- Prominent nucleoli
- Necrotic background

Gastrointestinal tract malignancies: LM in gastrointestinal malignancies is an exceedingly rare and aggressive complication where cancer cells spread to the leptomeninges. In GI cancers, it most frequently occurs in gastric cancer (0.16% to 0.9%), followed by colorectal cancer (0.019% to 0.058%). The signet ring cell histological subtype is most strongly associated with leptomeningeal spread in both gastric and colorectal cancers due to its naturally decreased cell-cell adhesion and propensity for diffuse infiltration (Lucarini et al., 2024).

In summary, LM from epithelial malignancies- mainly breast, lung, and gastrointestinal cancers- represents a devastating and late-stage complication. Despite the emergence of highly sensitive liquid biopsy techniques, the cytological identification of malignant cells in the CSF remains the definitive diagnostic gold standard (Glantz et al., 1998).

3.3.7 Leptomeningeal metastasis in gliomas

LM was once considered a rare entity in neurooncology. The incidence of LM in glioma patients is increasingly recognized due to improvements in neuroimaging resolution, more frequent surveillance, and prolonged patient survival afforded by modern systemic therapies. The reported prevalence of LM in adult glioma patients typically ranges from 4% to 5% (De Groot et al., 2010). A large-scale retrospective review at Memorial Sloan Kettering Cancer (MSKCC) identified an overall incidence of 4.7% over 15 years. Notably, this study documented a rising trend in diagnosis, increasing from 1.8% between 2001 and 2005 to 6.9% between 2011 and 2016 (Andersen et al., 2019). In specific subtypes, such as oligodendroglial tumors (grades II and III), the frequency is reported to be approximately 3.9 (Roldán et al., 2011). For high-grade gliomas, particularly glioblastoma (WHO grade IV), the symptomatic incidence is similar, at roughly 3.8% to 4% [59](Dardis et al., 2014). However, autopsy studies suggest the true prevalence may be significantly higher—reaching 20% to 25%—indicating that many cases remain asymptomatic or undiagnosed during life (Vertosick & Selker, 1990).

3.3.7.1 Pathophysiology and mechanisms of spread

The leptomeningeal compartment is anatomically separated from the brain parenchyma by the glia limitans, a barrier formed by astrocytic foot processes, and from systemic circulation by the blood-CSF and BBBs. For a tumor arising within the CNS parenchyma or ventricular system to seed the leptomeninges, malignant cells must traverse one or more anatomical barriers (Freret & Boire, 2024).

Glioma cells migrate into the CSF pathways through several anatomical conduits.

- *CSF proximity*: Proximity to the CSF space is a primary driver of dissemination. In one series, 50% of patients with LM had tumors in direct contact with a lateral ventricle (Dardis et al., 2014).
- *Anatomical conduits*: Common putative origins include the cerebral cortex, superior temporal sulcus, transverse fissure, and central sulcus. The flow of CSF from these locations tends to carry malignant cells toward the cortex or caudally toward the brainstem and spinal cord (Dardis et al., 2014).
- *Subependymal spread*: Disease often spreads through the ventricles, leading to subependymal or ependymal enhancement. Evidence suggests that subependymal spread is clinically and prognostically equivalent to disseminated LM (Andersen et al., 2019).
- *Surgical factors*: There is ongoing debate regarding whether surgical interventions, such as **ventricular opening** or the use of cavitron ultrasonic surgical aspirators (CUSA), increase the risk of "spillage" and subsequent seeding. However, recent multivariate analyses suggest that preexisting tumor proximity to the CSF circulation is a more significant risk factor than the surgical opening of the ventricle itself (Freret & Boire, 2024).

3.3.7.2 Risk factors for leptomeningeal metastasis

Several clinicopathologic variables significantly influence the likelihood of a glioma patient developing LM:

- *Histological grade*: WHO grade IV gliomas (glioblastoma) carry the highest risk compared to lower-grade tumors (Yang et al., 2023).
- *Midline location*: Tumors residing in the thalamus, basal ganglia, or brainstem are significantly more likely to develop LM, with a hazard ratio (HR) of 2.624 compared to lateral tumors in the hemispheres (Yang et al., 2023).
- *Tumor recurrence*: In oligodendroglial tumors, the number of recurrences is the only factor statistically associated with LM development (Roldán et al., 2011).
- *PNET features*: Glioblastomas with primitive neuroectodermal tumor (PNET) features appear to have a higher tropism for the CSF and are overrepresented in LM cohorts (Andersen et al., 2019).
- *Age*: Some studies indicate that patients who develop LM are often younger than non-LM patients (median age 53 vs. 59), though this remains a point of investigation (Yang et al., 2023).

3.3.7.3 Diagnostic framework

A diagnosis of LM is established through a combination of clinical suspicion, neuroimaging, and CSF analysis.

- *Neuroimaging*: MRI of the brain and whole spinal cord with gadolinium contrast is the modality of choice. Two distinct enhancement patterns are often observed (Dardis et al., 2014; Liu et al., 2019):
- *Nodular pattern (Type B)*: Characterized by dot-shaped, focal enhancements
- *Linear pattern (Type A)*: Characterized by disseminated or diffuse enhancement along the sulci or gyri. Linear patterns are more frequently associated with midline tumors, whereas lateral tumors more commonly show nodular spread
- *CSF analysis*: While a positive CSF cytology is definitive, it is not essential for diagnosis if radiographic and clinical evidence are strong. In glioma patients, CSF cytology is positive in only about 24% to 30% of cases (Andersen et al., 2019). Biochemical markers often show elevated protein and, in some cases, high proportions of macrophages. Establishing "baseline" CSF results is recommended to monitor treatment response. Leptomeningeal dissemination from primary CNS tumors is variable based on histology and age group. It is most commonly seen in:
- Pediatric embryonal tumors (e.g., medulloblastoma and ATRT)
- Ependymal tumors
- Germ cell tumors
- Primary CNS lymphoma
- Gliomas, including glioblastoma (GBM) (Liu et al., 2019)

Early reports found CNS tumor LM to be relatively rare compared to systemic LM. However, cohort studies increasingly recognize that certain subtypes (especially medulloblastoma and ATRT) may manifest leptomeningeal spread at diagnosis or during progression (Boire et al., 2017). Liu et al. (2019) summarized that neuraxis metastases (a broader category including leptomeningeal and ventricular spread) arise from many CNS histologies, with medulloblastoma representing ~34% and other embryonal tumors contributing substantially (Engelhard & Corsten, 2005).

Medulloblastoma: Medulloblastoma is a highly malignant embryonal tumor predominantly affecting children. It is well recognized for its propensity to seed along the CSF pathways, resulting in drop metastases along the spinal cord and leptomeningeal dissemination at diagnosis or recurrence (Engelhard & Corsten, 2005). Studies have shown that 30%–40% of medulloblastoma patients may have leptomeningeal spread at the time of diagnosis, forming the basis of the metastasis (M) staging system used in pediatric neurooncology (Liu et al., 2019). On imaging, MRI typically reveals nodular enhancement on the brain and spinal cord surfaces with "sugarcoating" of leptomeninges and drop metastases along the spinal subarachnoid space. CSF cytology in medulloblastoma often yields clusters of small primitive cells with a high nuclear–cytoplasmic ratio and scant cytoplasm, reminiscent of the primary tumor morphology, but can be deceptively sparse. Repeated sampling improves sensitivity. Occasional rosette-like arrangements can also be seen. Early and timely detection of leptomeningeal spread in medulloblastoma is important as it impacts staging (M staging in pediatric neurooncology) and demands craniospinal irradiation.

Ependymomas: Ependymomas are close to CSF pathways, as these tumors arise from ependymal lining cells of the ventricles and spinal canal. Hence, have a tendency for subarachnoid dissemination, particularly in posterior fossa and spinal ependymomas (Liu et al., 2019). Leptomeningeal spread in ependymoma may be symptomatic or detected radiographically. Brain and spine MRI often reveal enhancing nodules along the leptomeninges. CSF cytology reveals small to medium uniform cells with round to oval nuclei and perivascular pseudorosettes.

Primary CNS lymphoma: Primary CNS lymphoma frequently invades the subarachnoid space due to its angiocentric growth pattern and predilection for perivascular infiltration. While leptomeningeal lesions are not universally seen, subarachnoid spread with positive CSF cytology is well-documented and often portends a need for aggressive therapy. Cytological examination typically reveals large atypical lymphoid cells with irregular nuclei and a high nucleocytoplasmic ratio. Flow cytometry enhances the detection and immunophenotyping of clonal populations.

High-grade gliomas: Glioblastoma rarely presents with overt LM at the time of diagnosis. However, it may develop during the disease progression. CSF cytology reveals pleomorphic glial cells with irregular nuclei and fibrillary cytoplasm. Detection is often difficult due to low cellularity. LM in GBM carries the most aggressive course, with median survival after diagnosis ranging from 3.2 to 5 months. If LM is present at the time of initial GBM diagnosis, the prognosis is even more dire, with an overall median survival of only 8.3 months (Andersen et al., 2019).

To summarize, LM in gliomas is no longer a rare or "calamitous" terminal event that mandates abandoning care. The integration of high-resolution MRI and careful CSF monitoring allows for earlier detection, often before the onset of debilitating symptoms (Table 3.1).

Table 3.1 Cytomorphologic features of malignant cells in cerebrospinal fluid.

Feature	Metastatic melanoma	Metastatic carcinoma	Lymphoma/leukemia	Glioma/other primary CNS tumor cells in CSF
Typical cellularity	Moderate to high	Moderate to high	Variable; may be scant or markedly cellular	Usually low to moderate, but variable
Cytologic pattern	Mostly **single dispersed cells** with occasional loose clusters	Cohesive clusters, sheets, acinar or papillary groups	Predominantly dispersed single cells	Single cells or small loose clusters
Cell size	Medium to large	Medium to large	Often variable, usually medium-sized to large	Variable
Cell shape	Epithelioid, plasmacytoid, spindle, and bizarre giant-cell forms	Polygonal or columnar epithelial-type cells	Round to irregular lymphoid cells	Round, oval, or irregular astrocytic/glial-type cells
Nuclear features	Marked pleomorphism, irregular nuclear contours, and prominent macronucleoli	Moderate to marked nuclear atypia, irregular membranes, conspicuous nucleoli	Relatively monomorphic atypical lymphoid cells, coarse chromatin, and a high N:C ratio	Variable nuclear atypia depending on grade and type
Cytoplasmic features	Moderate to abundant cytoplasm; may contain coarse brown melanin pigment	Moderate cytoplasm: vacuolation or mucin may be present	Scant to moderate cytoplasm	Delicate cytoplasm; may show fibrillary processed
Background	Melanin pigment, melanophages, and tumor debris	Necrosis, mucin, tumor diathesis	Clean or proteinaceous; lymphoglandular bodies	Proteinaceous background; occasional glial fibrillary material
Most useful morphologic clue	Pleomorphic large cells with **prominent nucleoli and pigment**	**Cohesive epithelial groups** with glandular/papillary tendency	**Dispersed monomorphic atypical lymphoid population**	Known CNS primary + cells with glial-type morphology/fibrillary
Useful ancillary tests	S100, SOX10, HMB45, and Melan-A	Cytokeratin, EMA, and organ-specific markers	CD45, CD20/CD3, and TdT **flow cytometry**	GFAP, OLIG2, and clinicoradiologic correlation

3.4 Diagnostic pitfalls in cerebrospinal fluid analysis

The diagnostic accuracy of CSF analysis in the context of LM is important for clinical practice. While identifying malignant cells remains the gold standard, the biological mimickers and technical pitfalls can lead to misdiagnosis. Clinicians must be aware of conditions in which reactive inflammatory states, accidental tissue contamination, and the inherent spatial variability of the CSF compartment can mimic or suggest malignancy. Both false-negative and false-positive interpretations may lead to undertreatment or inappropriate escalation of therapy.

1. *Cellular mimickers: The "nonmalignant" intruders.* One of the most significant pitfalls in cytomorphology is the presence of nonneoplastic cells mimicking malignant cells.
   - *Reactive lymphocytes*: In patients with leukemia, lymphoma, or CNS infections, reactive lymphocytes (primarily of T-cell origin) are difficult to distinguish from malignant lymphoid cells. These cells may appear large with "blastic" features, leading to "doubtful" morphology in up to 40% of samples (Bommer et al., 2011).
   - *Chondrocytes*: If the LP needle is inserted too far anteriorly, contaminating the CSF with chondrocytes from the intervertebral disc. These cartilage cells often appear in large clusters or scattered singly and can mimic chordomas or metastatic tumors like renal cell carcinoma (Pal et al., 2019).
   - *Other accidental contaminants*: The route of the needle may also introduce bone marrow cells, squamous cells from the skin, ependymal cells, or choroidal cells. The presence of hematopoietic elements from the vertebral body can particularly mimic a hematologic malignancy in the CSF (Kruskall et al., 1983; Takeda et al., 1981).
2. *Clinical and biochemical mimicry*: The clinical and biochemical profile of LM frequently overlaps with **infectious meningitis**, creating a profound diagnostic pitfall.
   - *The "infectious" mask*: At times, LM can perfectly mimic bacterial meningitis. The patient presented with fever, headache, and high CSF cell counts (409×10^6/L) with 95% neutrophil predominance, along with severe hypoglycorrhachia (low glucose) and high protein. These findings typically trigger empirical antibiotic or antituberculosis therapy, which may delay the correct palliative approach for weeks while the patient's condition deteriorates (Chun-Yin et al., 2019).
   - *Variable pleocytosis*: While high neutrophils suggest bacterial infection, and lymphocytes suggest viral or TB causes, LM can manifest with any leukocyte population, including monocytic predominance. Furthermore, immunocompromised states or prior antibiotic use can further obscure these typical patterns (Djukic et al., 2022).
3. *Spatial pitfalls*: The rostro-caudal gradient: A critical and often overlooked pitfall is the spatial inhomogeneity of the CSF space. Routine parameters are often interpreted as if the CSF were a well-mixed, homogeneous compartment, but it is actually highly convoluted and exhibits oscillatory flow (Wellmer et al., 2001).
   - *Ventricular versus lumbar discrepancy*: Concentrations of blood-derived compounds, such as proteins and cells, typically increase from the ventricles to the lumbar sac. For example, the protein content in ventricular CSF is often 2.5 times lower than in lumbar fluid (Wellmer et al., 2001).

- *Sampling error*: A patient may have "typical" findings of bacterial meningitis or LM in the lumbar CSF that are absent in the ventricular CSF, or vice-versa. In some cases, pathogens or malignant cells may grow in only one part of the compartment. Consequently, a normal ventricular tap cannot definitively rule out a basal or spinal malignancy (Bommer et al., 2011).
- *Lactate as a "stable" indicator*: Unlike cells and proteins, CSF lactate (with a cut-off of ≥3.5 mmol/L) distributes more homogeneously throughout the space. It is often the most reliable parameter to differentiate bacterial/fungal processes from viral ones, though it remains nonspecific and can be elevated in severe noninfectious cerebral suffering, including intracranial hemorrhage or malignant diseases (Wellmer et al., 2001).

4. *Temporal pitfalls and the role of treatment*: The timing of an LP relative to the disease course can create "false-negative" or "false-positive" impressions.
   - *The "false-negative" early phase*: Early in the course of bacterial meningitis (or potentially rapid-onset malignancy), the causative agent can be present while leukocyte counts and protein remain normal (Troendle & Pettigrew, 2019).
   - *Slow resolution*: Conversely, pleocytosis can persist for weeks or even months after a pathogen has been successfully eliminated, leading to the mistaken impression of therapeutic failure (Nau et al., 1990).
   - *The antibody Index paradox*: After successful treatment, serum antibody levels may drop faster than CSF levels, leading to a paradoxical rise in the antibody index (AI) that does not reflect acute infection (Reiber, 2021).
5. *Technical and analytical pitfalls*: The process of obtaining and analyzing the fluid itself introduces several confounding variables.
   - *Blood contamination (traumatic tap)*: This is the most common pitfall, and blood in the CSF affects leukocyte counts, protein measurements, and the estimation of intrathecal immunoglobulin (Ig) synthesis (Ward, 1980).
   - *Failure of reibergrams*: Even a low blood contamination (0.1%) can simulate an **intrathecal IgM synthesis** on a Reiber-Felgenhauer nomogram. Because IgM is a large molecule, its presence in CSF is highly sensitive to accidental blood contamination. Beyond 0.1% contamination (roughly >5000 erythrocytes/μL), these nomograms should generally not be used for diagnosis (Djukic et al., 2022).
   - *Preanalytical degeneration*: CSF is a "hostile" environment for cells once removed from the body due to low protein and sugar content. Delay in processing can lead to either leukocyte death or degenerative changes, which obscure the actual morphology of the cells (Djukic et al., 2022).

3.4.1 Overcoming pitfalls: Integrated diagnostics

To mitigate these pitfalls, an integrated approach is essential. Overcoming the mimickers and pitfalls of CSF analysis requires a meticulous adherence to technical protocols. Clinicians must remember that "normal" values at one site (ventricular) do not preclude pathology at another (lumbar), and that the presence of "blastic/blastoid" cells may sometimes reflect a reactive immune response rather than a fatal relapse.

- *Repeat sampling*: If initial results are discordant with the clinical picture, the LP should be repeated. This usually increases the yield for cytology.
- *Flow cytometry (FC)*: In cases where reactive lymphoid cells mimic malignant cells, FC provides an objective and qualitative analysis. In one study, FC improved sensitivity from 73% to 96% compared to morphology alone. It is highly useful for identifying occult disease in patients with "atypical" or "suspicious" cytomorphology (Bommer et al., 2011) (Table 3.2).

3.5 Emerging diagnostic techniques in cerebrospinal fluid analysis

The diagnostic landscape for LM is currently undergoing a paradigm shift. Historically, clinicians have relied on the "triple pillar" of neurological examination, neuroimaging, and CSF cytology. However, these traditional tools are often described as "insensitive," as they often fail to detect disease until symptoms are severe and irreversible. Even though CSF cytology is highly specific, it often has a sensitivity below 50% on the first tap, requiring repeated procedures and leading to a significant delay in treatment.

To bridge this "diagnostic gap," a spectrum of emerging techniques, ranging from liquid biopsies targeting circulating cells and DNA to high-sensitivity proteomic biomarkers and nanotechnology, is transforming the CSF from a simple diagnostic fluid into a molecular mirror of the central nervous system.

1. *The rise of liquid biopsy: Circulating tumor cells* (CTCs)

One of the most significant recent advancements in CSF diagnostic testing is the detection and enumeration of CTCs. While cytology relies on a pathologist's visual identification of intact cells, which has certain limitations, CTC assays use automated, quantitative methods to "catch" malignant cells based on specific surface markers (Kessel et al., 2025).

Mechanism and sensitivity: Most CTC assays for epithelial tumors (like breast and lung cancer) target the epithelial cell adhesion molecule (EpCAM). Studies show that EpCAM-based FC has a sensitivity of 78%–100%, significantly higher than standard (Bussel et al., 2020).

Power of quantification: CTC assays allow clinicians to measure the burden of the disease as CTC assays are quantitative, unlike cytology, which is binary (positive or negative). Certain studies have reported a count of ≥61 CTCs per 3 mL of CSF as a critical threshold where the risk of mortality doubles (Diaz et al., 2022).

Commercial innovations: Technologies like "CellSearch" and "CNSide" are now being adapted for CSF. CNSide uses a multimarker panel (EpCAM, cytokeratin, and nonepithelial markers) to detect cells that might have undergone "epithelial-mesenchymal transition," which would make them unidentifiable for EpCAM-only tests (Wooster et al., 2022).

2. Genetic profiling through cell-free DNA (cfDNA)

Perhaps the most rapidly advancing area of research has emerged in cfDNA and circulating tumor DNA (ctDNA). When tumor cells die or undergo turnover, they shed DNA fragments into the CSF.

Table 3.2 Diagnostic pitfalls and mimickers in CSF cytology.

Category	Mimicker/pitfall	Key cytologic features	How it mimics LM	Distinguishing clues	Ancillary tests
Sampling error	Low tumor burden	Rare and atypical cells	False-negative	Repeat LP; ≥10 mL volume	Repeat cytology
	Compartmentalization	Negative lumbar CSF	Missed disease	MRI spine; repeat sampling	Imaging correlation
	Delayed processing	Degenerated cells	Misread as macrophages	Smudged nuclei, cytoplasmic blebs	Rapid fixation
Reactive changes	Viral meningitis	Activated lymphocytes and plasmacytoid forms	Mimics lymphoma	Polymorphous population	Flow cytometry
	TB/fungal meningitis	High protein, lymphocytes	Mimics carcinomatous meningitis	Clinical infection signs	Microbiology
	Immune therapy-related meningitis	Atypical lymphocytes	Mimics malignant cells	Treatment history	Flow cytometry
Inflammatory cells	Activated macrophages	Large, vacuolated cytoplasm	Mimics adenocarcinoma	Bean-shaped nucleus; no tight clusters	CD68 IHC
	Hemosiderin macrophages	Pigment-laden	Mimics melanoma	Iron stain positive	Prussian blue
Benign epithelial cells	Ependymal cells	Cohesive clusters; uniform nuclei	Mimics carcinoma	Cilia; smooth nuclear contours	GFAP negative, CK negative
	Choroid plexus cells	Papillary fragments	Mimics papillary carcinoma	Clinical shunt history	–
Contamination	Peripheral blood	Atypical circulating tumor cells	False-positive	Compare the peripheral smear	Peripheral blood review

The CSF advantage: As the BBB restricts DNA transport, genomic testing of plasma (blood) is often insensitive for CNS tumors. By contrast, ctDNA in the CSF is highly enriched and undiluted by the "sea" of normal DNA found in the blood. This allows detection of mutations at much higher variant allelic frequencies (VAFs).

Targeted versus broad detection:

- *Droplet digital PCR (ddPCR)*: This is an ultrasensitive technique for identifying single, known mutations. In EGFR-mutant lung cancer, ddPCR of CSF cfDNA achieved an accuracy of 94.1%, identifying malignant involvement in cases where MRI and cytology were both negative (Swinkels et al., 2000; Suryavanshi et al., 2020)
- *Next-generation sequencing (NGS)*: NGS can identify "actionable" mutations—like ALK, ROS1, or MET fusions—that may be present in the CNS metastasis but not in the primary tumor due to "branched evolution" (Wilcox et al., 2022). This gives clinicians an option to choose a targeted therapy.

Monitoring resistance: Consecutive CSF DNA tests can detect resistance mutations (like EGFR T790M) that emerge during treatment, allowing oncologists to change to CNS-penetrant therapies like osimertinib before the patient's symptoms worsen (Hickman et al., 2023). These mutations may not be present in the primary tumor due to the "branched evolution."

3. *Soluble biomarkers and proteomic panels*

While cells and DNA provide genetic data, the proteomic milieu of the CSF reveals the tumor's metabolic "footprint".

Tumor marker panels: Recently, a panel of three markers—CEA, CA125, and CA19-9—has been validated as a highly accurate tool for diagnosing LM in lung cancer (Huang et al., 2025). Carcinoembryonic antigen (CEA) is a high-molecular-weight glycoprotein that does not easily cross the BBB. Therefore, an elevated CSF CEA level (especially when the serum-to-CSF ratio is less than 60:1) is highly specific for leptomeningeal infiltration (Kang et al., 2010).

Economic viability: While ctDNA tests offer high sensitivity, their cost remains a significant limitation compared to conventional tumor marker assays. A study showed that a CSF tumor marker panel costs only 0.76% of a single hospitalization bill, whereas ctDNA testing can account for over 60% of the cost. This makes soluble markers a more viable tool for frequent response assessment (Huang et al., 2025).

4. *Advanced molecular techniques: mNGS and CNV*

In addition to simple DNA sequencing, researchers are utilizing metagenomic next-generation sequencing (mNGS) and copy number variation (CNV) analysis to differentiate malignant cells from their mimickers.

- *mNGS:* Originally, this tool was designed to detect infectious pathogens (like bacteria or viruses); mNGS captures all DNA in a sample. Since most sequences in mNGS are human, clinicians can use this "background" data to look for signs of cancer (Zhang et al., 2024).
- *CNV analysis:* This technique looks for large-scale deletions or amplifications of DNA fragments (typically >10 Mb in length). As the CNV analysis of CSF has demonstrated

Table 3.3 Comparison of various techniques used in CSF examination.

Technique	Sensitivity	Specificity	Cost	Best use
Cytology	Moderate	High	Low	First-line
Tumor markers	High (panel)	High	Very low	Monitoring
ctDNA	High	Very high	High	Precision diagnosis
mNGS-CNV	Moderate	Very high	High	Complementary
Flow cytometry	High (hematologic)	High	Moderate	Lymphoma

CSF, Cerebrospinal fluid.

100% specificity, this test can be used as a confirmatory test when cytology is equivocal. It is particularly effective for tumors with complex genetic alterations (Zhang et al., 2024).

5. *In vivo detection and nanotechnology*

Some of the most futuristic "emerging" techniques involve moving beyond the laboratory and examining the patient's CNS in real time.

- *Quantum dots:* These are tiny semiconductor particles with unique luminescent properties. They can be conjugated with antibodies (like anti-HER2) to label cancer cells in the CSF with precision (Weston et al., 2011).
- *Magnetic iron oxide nanoparticles (MIONs):* MIONs can be used as targeted contrast agents for MRI. When conjugated to specific tumor markers, they specifically alter the T2 signal of cancer cells, potentially allowing MRIs to "see" microscopic disease that is currently invisible (Table 3.3).

Emerging techniques in CSF examination are transforming LMD from a "hidden" disease into a measurable one. While cytology remains the legal gold standard, the integration of CTC enumeration, ctDNA sequencing, and tumor marker panels allows for earlier diagnosis, more accurate monitoring, and the selection of targeted therapies that were previously impossible (Funovics et al., 2004).

AI disclosure

During the preparation of this work, the author(s) used Notebook LLM and Chat GPT to summarize the articles and check language. After using this tool/service, the author(s) reviewed and edited the content as needed and take(s) full responsibility for the content of the publication.

References

Abouharb, S., Ensor, J., Loghin, M. E., Katz, R., Moulder, S. L., Esteva, F. J., Smith, B., Valero, V., Hortobagyi, G. N., & Melhem-Bertrandt, A. (2014). Leptomeningeal disease and breast cancer: The importance of tumor subtype. *Breast Cancer Research and Treatment, 146*(3), 477–486. www.wkap.nl/journalhome.htm/0167-6806. doi:10.1007/s10549-014-3054-z.

Andersen, B. M., Miranda, C., Hatzoglou, V., Deangelis, L. M., & Miller, A. M. (2019). Leptomeningeal metastases in glioma: The memorial sloan kettering cancer center experience. *Neurology, 92*(21), e2483. http://www.neurology.org. doi:10.1212/WNL.0000000000007529.

Aroney, R. S., Dalley, D. N., Chan, W. K., Bell, D. R., & Levi, J. A. (1981). Meningeal carcinomatosis in small cell carcinoma of the lung. *The American Journal of Medicine, 71*(1), 26–32. doi:10.1016/0002-9343(81)90254-0.

Bae, Y. S., Cheong, J. W., Chang, W. S., Kim, S., Oh, E. J., & Kim, S. H. (2013). Diagnostic accuracy of cerebrospinal fluid (CSF) cytology in metastatic tumors: An analysis of consecutive CSF samples. *Korean Journal of Pathology, 47*(6), 563–568. http://koreanjpathol.org/upload/journal/KJPathol%2047-563.pdf. doi:10.4132/KoreanJPathol.2013.47.6.563.

Ballester, L. Y., Glitza Oliva, I. C., Douse, D. Y., Chen, M. M., Lan, C., Haydu, L. E., Huse, J. T., Roy-Chowdhuri, S., Luthra, R., Wistuba, I. I., & Davies, M. A. (2018). Evaluating circulating tumor DNA from the cerebrospinal fluid of patients with melanoma and leptomeningeal disease. *Journal of Neuropathology and Experimental Neurology, 77*(7), 628–635. http://journals.lww.com/jneuropath. doi:10.1093/jnen/nly046.

Bendell, J. C., Domchek, S. M., Burstein, H. J., Harris, L., Younger, J., Kuter, I., Bunnell, C., Rue, M., Gelman, R., & Winer, E. (2003). Central nervous system metastases in women who receive trastuzumab-based therapy for metastatic breast carcinoma. *Cancer, 97*(12), 2972–2977. doi:10.1002/cncr.11436.

Boire, A., Zou, Y., Shieh, J., Macalinao, D. G., Pentsova, E., & Massagué, J. (2017). Complement component 3 adapts the cerebrospinal fluid for leptomeningeal metastasis. *Cell, 168*(6), 1101. doi:10.1016/j.cell.2017.02.025.

Bommer, M., Nagy, A., Schöpflin, C., Pauls, S., Ringhoffer, M., & Schmid, M. (2011). Cerebrospinal fluid pleocytosis : Pitfalls and benefits of combined analysis using cytomorphology and flow cytometry. *Cancer Cytopathology, 119*(1), 20–26. doi:10.1002/cncy.20127.

Bönig, L., Möhn, N., Ahlbrecht, J., Wurster, U., Raab, P., Puppe, W., Sühs, K.-W., Stangel, M., Skripuletz, T., & Schwenkenbecher, P. (2019). Leptomeningeal metastasis: The role of cerebrospinal fluid diagnostics. *Frontiers in Neurology, 10*. doi:10.3389/fneur.2019.00839.

Bromberg, J. E. C., Breems, D. A., Kraan, J., Bikker, G., van der Holt, B., Smitt, P. S., van den Bent, M. J., van't Veer, M., & Gratama, J. W. (2007). CSF flow cytometry greatly improves diagnostic accuracy in CNS hematologic malignancies. *Neurology, 68*(20), 1674–1679. doi:10.1212/01.wnl.0000261909.28915.83.

Buonocore, D. J., McCash, S. I., Roshal, M., Mazzella, A. L., & Lin, O. (2017). Cerebrospinal fluid from patients with hematopoietic neoplasms: The need for an integrated approach. *Journal of the American Society of Cytopathology, 6*(5), 198–204. http://www.elsevier.com/journals/journal-of-the-american-society-of-cytopathology/2213-2945. doi:10.1016/j.jasc.2017.06.004.

Bussel, M. T. J. V., Pluim, D. I, Kerklaan, B. M., Bol, M., Sikorska, K., Linders, D. éT. C., Broek, D. V. D., Beijnen, J. H., Schellens, J. H. M., & Brandsma, D. I. (2020). Circulating epithelial tumor cell analysis in CSF in patients with leptomeningeal metastases. *Neurology, 94*(5). doi:10.1212/WNL.0000000000008751.

Chamberlain, M. C. (2010). Leptomeningeal metastasis. *Current Opinion in Oncology, 22*(6), 627–635. doi:10.1097/CCO.0b013e32833de986.

Chun-Yin, L. W., Ian, L. Y. H., Hilary, K. H-C., & Shek-Kwan, C. R (2019). Pitfalls in cerebrospinal fluid analysis: A case report of carcinomatous meningitis mimicking infective causes. *Journal of the Formosan Medical Association, 118*(5), 953–954. doi:10.1016/j.jfma.2019.02.004.

Craig, F. E., & Foon, K. A. (2008). Flow cytometric immunophenotyping for hematologic neoplasms. *Blood, 111*(8), 3941–3967. ttp://bloodjournal.hematologylibrary.org/cgi/reprint/111/8/3941. doi:10.1182/blood-2007-11-120535.

Dardis, C., Milton, K., Ashby, L., & Shapiro, W. (2014). Leptomeningeal metastases in high-grade adult glioma: Development, diagnosis, management, and outcomes in a series of 34 patients. *Frontiers in Neurology, 5*. doi:10.3389/fneur.2014.00220.

De Groot, J. F., Fuller, G., Kumar, A. J., Piao, Y., Eterovic, K., Ji, Y., & Conrad, C. A. (2010). Tumor invasion after treatment of glioblastoma with bevacizumab: Radiographic and pathologic correlation in humans and mice. *Neuro-Oncology, 12*(3), 233–242. doi:10.1093/neuonc/nop027.

Del Principe, M. I., Buccisano, F., Cefalo, M., Maurillo, L., Di Caprio, L., Di Piazza, F., Sarlo, C., De Angelis, G., Irno Consalvo, M., Fraboni, D., De Santis, G., Ditto, C., Postorino, M., Sconocchia, G., Del Poeta, G., Amadori, S.,

& Venditti, A. (2014). High sensitivity of flow cytometry improves detection of occult leptomeningeal disease in acute lymphoblastic leukemia and lymphoblastic lymphoma. *Annals of Hematology, 93*(9), 1509–1513. link.springer.de/link/service/journals/00277/index.htm. doi:10.1007/s00277-014-2080-6.

Diansari, Y., Aninditha, T., & Andriani, R. (2017). P14.04 Characteristics of cerebrospinal fluid in hematologic malignancy. *Neuro-Oncology, 19*(suppl_3), iii102. doi:10.1093/neuonc/nox036.391.

Diaz, M., Singh, P., Kotchetkov, I. S., Skakodub, A., Meng, A., Tamer, C., Young, R. J., Reiner, A. S., Panageas, K. S., Ramanathan, L. V., & Pentsova, E. (2022). Quantitative assessment of circulating tumor cells in cerebrospinal fluid as a clinical tool to predict survival in leptomeningeal metastases. *Journal of Neuro-Oncology, 157*(1), 81–90. doi:10.1007/s11060-022-03949-1.

Djukic, M., Lange, P., Erbguth, F., & Nau, R. (2022). Spatial and temporal variation of routine parameters: Pitfalls in the cerebrospinal fluid analysis in central nervous system infections. *Journal of Neuroinflammation, 19*(1). doi:10.1186/s12974-022-02538-3.

Djukic, M., Schmidt-Samoa, C., Lange, P., Spreer, A., Neubieser, K., Eiffert, H., Nau, R., & Schmidt, H. (2012). Cerebrospinal fluid findings in adults with acute Lyme neuroborreliosis. *Journal of Neurology, 259*(4), 630–636. doi:10.1007/s00415-011-6221-8.

Djukic, M., Trimmel, R., Nagel, I., Spreer, A., Lange, P., Stadelmann, C., & Nau, R. (2017). Cerebrospinal fluid abnormalities in meningeosis neoplastica: A retrospective 12-year analysis. *Fluids and Barriers of the CNS, 14*(1). doi:10.1186/s12987-017-0057-2.

Domingues, R. B., de Moura Leite, F. B. V., & Senne, C. (2021). Cerebrospinal fluid findings in patients with hematologic neoplasms and meningeal infiltration. *Acta Neurologica Belgica, 121*(6), 1543–1546. http://www.springerlink.com/content/2240-2993. doi:10.1007/s13760-020-01397-0.

Engelhard, H. H., & Corsten, L. A. (2005). Leptomeningeal metastasis of primary central nervous system (CNS) neoplasms. *Cancer treatment and research, 125*, 71–85. doi:10.1007/0-387-24199-x_5.

Freret, M. E., & Boire, A. (2024). The anatomic basis of leptomeningeal metastasis. *Journal of Experimental Medicine, 221*(4). doi:10.1084/jem.20212121.

Funovics, M. A., Kapeller, B., Hoeller, C., Su, H. S., Kunstfeld, R., Puig, S., & Macfelda, K. (2004). MR imaging of the her2/neu and 9.2.27 tumor antigens using immunospecific contrast agents. *Magnetic Resonance Imaging, 22*(6), 843–850. doi:10.1016/j.mri.2004.01.050.

Glantz, M. J., Cole, B. F., Glantz, L. K., Cobb, J., Mills, P., Lekos, A., Walters, B. C., & Recht, L. D. (1998). Cerebrospinal fluid cytology in patients with cancer: Minimizing false- negative results. *Cancer, 82*(4), 733–739. doi:10.1002/(SICI)1097-0142(19980215)82:4⟨733::AID-CNCR17⟩3.0.CO;2-Z.

Gossai, N. P., Devidas, M., Chen, Z., Wood, B. L., Zweidler-McKay, P. A., Rabin, K. R., Loh, M. L., Raetz, E. A., Winick, N. J., Burke, M. J., Carroll, A. J., Esiashvili, N., Heerema, N. A., Carroll, W. L., Hunger, S. P., Dunsmore, K. P., Winter, S. S., & Teachey, D. T. (2023). Central nervous system status is prognostic in T-cell acute lymphoblastic leukemia: A children's oncology group report. *Blood, 141*(15), 1802–1811. https://www.journals.elsevier.com/blood. doi:10.1182/blood.2022010653.

Herrlinger, U., Wiendl, H., Renninger, M., Förschler, H., Dichgans, J., & Weller, M. (2004a). Vascular endothelial growth factor (VEGF) in leptomeningeal metastasis: Diagnostic and prognostic value. *British Journal of Cancer, 91*(2), 219–224. doi:10.1038/sj.bjc.6601953.

Herrlinger, U., Förschler, H., Küker, W., Meyermann, R., Bamberg, M., Dichgans, J., & Weller, M. (2004b). Leptomeningeal metastasis: Survival and prognostic factors in 155 patients. *Journal of the Neurological Sciences, 223*(2), 167–178. doi:10.1016/j.jns.2004.05.008.

Hickman, R. A., Miller, A. M., & Arcila, M. E. (2023). Cerebrospinal fluid: A unique source of circulating tumor DNA with broad clinical applications. *Translational Oncology, 33*. doi:10.1016/j.tranon.2023.101688.

Huang, K., He, H., Jin, W., Liu, X., Weng, Y., Yan, D., Zhao, Y., Yang, H., Zhan, R., Xu, Q., & Chen, Q. (2025). Tumor marker test in cerebrospinal fluid for leptomeningeal metastasis diagnosis and response assessment in non-small-cell lung cancer. *European Journal of Neurology, 32*(6). doi:10.1111/ene.70245.

Hyun, J. W., Jeong, I. H., Joung, A. R., Cho, H. J., Kim, S. H., & Kim, H. J. (2016). Leptomeningeal metastasis: Clinical experience of 519 cases. *European Journal of Cancer, 56*, 107–114. http://www.journals.elsevier.com/european-journal-of-cancer/. doi:10.1016/j.ejca.2015.12.021.

Kang, S. J, Kim, K. S, Ha, Y. S, Huh, S. oY., Lee, J. iH., Kim, J. K., & Kim, M. J. (2010). Diagnostic value of cerebrospinal fluid level of carcinoembryonic antigen in patients with leptomeningeal carcinomatous metastasis. *Journal of Clinical Neurology (Korea), 6*(1), 33–37. doi:10.3988/jcn.2010.6.1.33.

Kessel, E., Woerdman, N., Pluim, D., Kerklaan, B. M., Broek, D., Huitema, A. D. R., & Brandsma, D. (2025). Added prognostic value of circulating tumor cell numbers in CSF of patients with leptomeningeal metastasis from epithelial tumors. *European Journal of Cancer, 220*. doi:10.1016/j.ejca.2025.115377.

Kruskall, M. S., Carter, S. R., & Ritz, L. P. (1983). Contamination of cerebrospinal fluid by vertebral bone-marrow cells during lumbar puncture. *New England Journal of Medicine, 308*(12), 697–700. doi:10.1056/NEJM198303243081206.

Kumthekar, P., & Le Rhun, E. (2023). Brain metastases and leptomeningeal disease. *Continuum, 29*(6), 1727–1751. doi:10.1212/con.0000000000001354.

Le Zhang, K. F., Ren, H., Fan, S., Wang, J., & Guan, H. (2024). Comparison of the diagnostic significance of cerebrospinal fluid metagenomic next-generation sequencing copy number variation analysis and cytology in leptomeningeal malignancy. *BMC Neurology, 24*(1). doi:10.1186/s12883-024-03655-7.

Leal, T., Chang, J. E., Mehta, M., & Robins, H. I. (2011). Leptomeningeal metastasis: Challenges in diagnosis and treatment. *Current Cancer Therapy Reviews, 7*(4), 319–327. doi:10.2174/157339411797642597.

Liu, H., Zhang, J., Liu, Y., Sun, Y., Li, C., Gu, C., Wang, H., Zhang, H., Yu, C., & Zhang, M. (2019). Neuraxis metastases of primary central nervous system tumors: A review of clinicopathological and radiographic characters of 198 cases in a single center. *Cancer Management and Research, 11*, 9829–9841. https://www.dovepress.com/getfile.php?fileID=54145. doi:10.2147/CMAR.S217672.

Liu, J., Jia, H., Yang, Y., Dai, W., Su, X., & Zhao, G. (2009). Cerebrospinal fluid cytology and clinical analysis of 34 cases with leptomeningeal carcinomatosis. *Journal of International Medical Research, 37*(6), 1913–1920. doi:10.1177/147323000903700629.

Liu, S., & Wang, Y. (2023). Diagnosis and management of adult central nervous system leukemia. *Blood Science, 5*(3), 141–149. https://journals.lww.com/bls/pages/aboutthejournal.aspx. doi:10.1097/BS9.0000000000000162.

Lucarini, A., Arrivi, G., Liotta, E., Causi, F. S. L., Cicco, L. D., Mazzuca, F., Osti, M. F, Balducci, G., & Mercantini, P. (2024). Leptomeningeal carcinomatosis in early gastric cancer: A case report and literature review. *Healthcare, 12*(12), 1184. doi:10.3390/healthcare12121184.

Minnema, M. C., Kimby, E., D'Sa, S., Fornecker, L. M., Poulain, S., Snijders, T. J., Kastritis, E., Kremer, S., Fitsiori, A., Simon, L., Davi, F., Lunn, M., Castillo, J. J., Patterson, C. J., Le Garff-Tavernier, M., Costopoulos, M., Leblond, V., Kersten, M. J., Dimopoulos, M. A., & Treon, S. P. (2017). Guideline for the diagnosis, treatment and response criteria for Bing-Neel syndrome. *Haematologica, 102*(1), 43–51. http://www.haematologica.org/content/haematol/102/1/43.full.pdf. doi:10.3324/haematol.2016.147728.

Momen, N., Tario, J., Fu, K., & Qian, Y.-W. (2023). Correction: Initial and follow-up evaluations on cerebrospinal fluid involvement by hematologic malignancy. *Journal of Hematopathology, 16*(3), 141–142. doi:10.1007/s12308-023-00553-8.

Muffly, L., & Curran, E. (2019). Pediatric-inspired protocols in adult acute lymphoblastic leukemia: Are the results bearing fruit? *Hematology, 2019*(1), 17–23. doi:10.1182/hematology.2019000009.

Murali, R., Loughman, N. T., McKenzie, P. R., Watson, G. F., Thompson, J. F., & Scolyer, R. A. (2009). Cytological features of melanoma in exfoliative fluid specimens. *Journal of Clinical Pathology, 62*(7), 638–643. http://jcp.bmj.com/cgi/reprint/62/7/638. doi:10.1136/jcp.2009.065599.

Nau, Schuchardt, & Prange (1990). Comments on listeriosis of the CNS zur listeriose des zentralnervensystems. *Fortschritte der Neurologie Psychiatrie, 58*(11), 408–422. doi:10.1055/s-2007-1001204.

Nayak, L., Fleisher, M., Gonzalez-Espinoza, R., Lin, O., Panageas, K., Reiner, A., Liu, C.-M., DeAngelis, L. M., & Omuro, A. (2013). Rare cell capture technology for the diagnosis of leptomeningeal metastasis in solid tumors. *Neurology, 80*(17), 1598–1605. doi:10.1212/wnl.0b013e31828f183f.

Nguyen, A., Nguyen, A., Dada, O. T., Desai, P. D., Ricci, J. C., Godbole, N. B., Pierre, K., & Lucke-Wold, B. (2023). Leptomeningeal metastasis: A review of the pathophysiology, diagnostic methodology, and therapeutic landscape. *Current Oncology, 30*(6), 5906–5931. https://www.mdpi.com/journal/curroncol. doi:10.3390/curroncol30060442.

Niwińska, A., Rudnicka, H., & Murawska, M. (2013). Breast cancer leptomeningeal metastasis: Propensity of breast cancer subtypes for leptomeninges and the analysis of factors influencing survival. *Medical Oncology, 30*(1). doi:10.1007/s12032-012-0408-4.

Pal, S., Nautiyal, N., Kaira, K., & Bharosay, V. (2019). Cytological features of cutaneous granular cell tumor: A brief report. *Journal of Cytology, 36*(4), 217. doi:10.4103/joc.joc_31_19.

Reiber, H. (2021). Non-linear ventriculo – Lumbar protein gradients validate the diffusion-flow model for the blood-CSF barrier. *Clinica Chimica Acta, 513*, 64–67. doi:10.1016/j.cca.2020.12.002.

Roldán, G., Chan, J., Eliasziw, M., Cairncross, J. G., & Forsyth, P. A. (2011). Leptomeningeal disease in oligodendroglial tumors: A population-based study. *Journal of Neuro-Oncology, 104*(3), 811–815. doi:10.1007/s11060-011-0551-3.

Schipper, H. I., Bardosi, A., Jacobi, C., & Felgenhauer, K. (1988). Meningeal carcinomatosis: Origin of local IgG production in the CSF. *Neurology, 38*(3), 413–416. doi:10.1212/wnl.38.3.413.

Singh, G., Mathur, S. R., Iyer, V. K., & Jain, D. (2013). Cytopathology of neoplastic meningitis: A series of 66 cases from a tertiary care center. *CytoJournal, 10*(1). doi:10.4103/1742-6413.114212.

Smalley, I., Boire, A., Brastianos, P., Kluger, H. M., Hernando-Monge, E., Forsyth, P. A., Ahmed, K. A., Smalley, K. S. M., Ferguson, S., Davies, M. A., & Glitza Oliva, I. C. (2024). Leptomeningeal disease in melanoma: An update on the developments in pathophysiology and clinical care. *Pigment Cell and Melanoma Research, 37*(1), 51–67. http://onlinelibrary.wiley.com/journal/10.1111/(ISSN)1755-148X. doi:10.1111/pcmr.13116.

Smalley, K. S. M., Fedorenko, I. V., Kenchappa, R. S., Sahebjam, S., & Forsyth, P. A. (2016). Managing leptomeningeal melanoma metastases in the era of immune and targeted therapy. *International Journal of Cancer, 139*(6), 1195–1201. doi:10.1002/ijc.30147.

Sruthi, S., Sangam, P. S., Pattar, U. N., Priyanka, P., & Kori, N. P. (2025). Central nervous system involvement in haematolymphoid malignancies – Role of cerebrospinal fluid cytology in the diagnosis. *Journal of Diagnostic and Academic Pathology, 2*(2), 96–100. doi:10.4103/jdap.jdap_18_25.

Steinherz, P. G., Seibel, N. L., Sather, H., Ji, L., Xu, X., Devidas, M., & Gaynon, P. S. (2019). Treatment of higher risk acute lymphoblastic leukemia in young people (CCG-1961), long-term follow-up: A report from the children's oncology group. *Leukemia, 33*(9), 2144–2154. http://www.nature.com/leu/index.html. doi:10.1038/s41375-019-0422-z.

Subirá, D., Serrano, C., Castañón, S., Gonzalo, R., Illán, J., Pardo, J., Martínez-García, M. ía, Millastre, E., Aparisi, F., Navarro, M., Dómine, M., Gil-Bazo, I., Segura, P. Pé, Gil, M., & Bruna, J. (2012). Role of flow cytometry immunophenotyping in the diagnosis of leptomeningeal carcinomatosis. *Neuro-Oncology, 14*(1), 43–52. doi:10.1093/neuonc/nor172.

Suryavanshi, M., Jaipuria, J., Panigrahi, M. K., Goyal, N., Singal, R., Mehta, A., Batra, U., & Doval, V. T. (2020). CSF cell-free DNA EGFR testing using DdPCR holds promise over conventional modalities for diagnosing leptomeningeal involvement in patients with non-small cell lung cancer. *Lung Cancer, 148*, 33–39. doi:10.1016/j.lungcan.2020.07.034.

Swinkels, D. W., Kok, J. B., Hanselaar, A., Lamers, K., & Boerman, R. H. (2000). Early detection of leptomeningeal metastasis by PCR examination of tumor-derived K-ras DNA in cerebrospinal fluid. *Clinical chemistry, 46*(1), 132–133.

Takeda, M., King, D. E., Choi, H. Y., Gomi, K., & Lang, W. R. (1981). Diagnostic pitfalls in cerebrospinal fluid cytology. *Acta Cytologica, 25*(3), 245–250.

Torzewski, M., Lackner, K. J., Bohl, J., & Sommer, C. (2008). *Integrated cytology of cerebrospinal fluid integrated cytology of cerebrospinal fluid* (pp. 1–93). Springer Berlin Heidelberg. http://www.springerlink.com/openurl.asp?genre=book&isbn=978-3-540-75884-6. doi:10.1007/978-3-540-75885-3.

Torzewski, M., & Lackner, K. J. (2016). Cerebrospinal fluid cytology: A highly diagnostic method for the detection of diseases of the central nervous system. *LaboratoriumsMedizin, 40*(s1). doi:10.1515/labmed-2016-0044.

Troendle, M., & Pettigrew, A. (2019). A systematic review of cases of meningitis in the absence of cerebrospinal fluid pleocytosis on lumbar puncture. *BMC Infectious Diseases, 19*(1). doi:10.1186/s12879-019-4204-z.

Vertosick, F. T., & Selker, R. G. (1990). Brain stem and spinal metastases of supratentorial glioblastoma multiforme: A clinical series. *Neurosurgery, 27*(4), 516–522. doi:10.1227/00006123-199010000-00002.

Ward. (1980). Cerebrospinal fluid data. 1. Interpretation in intracranial hemorrhage and meningitis. *Postgraduate Medicine, 68*(4), 181–188. doi:10.1080/00325481.1980.11715571.

Weller, M., Stevens, A., Sommer, N., Schabet, M., & Wiethölter, H. (1992). Humoral CSF parameters in the differential diagnosis of hematologic CNS neoplasia. *Acta Neurologica Scandinavica, 86*(2), 129–133. doi:10.1111/j.1600-0404.1992.tb05053.x.

Wellmer, A., Prange, J., Gerber, J., Zysk, G., Lange, P., Michel, U., Eiffert, H., & Nau, N. (2001). D- and L-lactate in rabbit and human bacterial meningitis. *Scandinavian Journal of Infectious Diseases, 33*(12), 909–913. doi:10.1080/00365540110076732.

Weston, C. L., Glantz, M. J., & Connor, J. R. (2011). Detection of cancer cells in the cerebrospinal fluid: Current methods and future directions. *Fluids and Barriers of the CNS, 8*(1). doi:10.1186/2045-8118-8-14.

Wilcox, J. A., Li, M. J, & Boire, A. A. (2022). Leptomeningeal metastases: New opportunities in the modern era. *Neurotherapeutics, 19*(6), 1782–1798. doi:10.1007/s13311-022-01261-4.

Wooster, M., McGuinness, J. E., Fenn, K. M., Veena, M., Singh, L. E., Franks, S., Lee, D., Cieremans, A. B., Lassman, D. L., Hershman, K. D., Crew, M. K., Accordino, M. S., Trivedi, F., Iwamoto, M. R., Welch, A., Haggiagi, R. D., Schultz, L., Huynh, E., Sales, D., Fisher, J., Mayer, A., Kreisl, T., & Kalinsky, K. (2022). Diagnosis of leptomeningeal metastasis in women with breast cancer through identification of tumor cells in cerebrospinal fluid using the CNSide™ assay. *Clinical Breast Cancer, 22*(4). doi:10.1016/j.clbc.2021.11.002.

Wu, S. Y., Short, N. J., Nasr, L., Dabaja, B. S., & Fang, P. Q. (2022). Central nervous system prophylaxis and treatment in acute leukemias. *Current Treatment Options in Oncology, 23*(12), 1829–1844. https://www.springer.com/journal/11864. doi:10.1007/s11864-022-01032-5.

Yang, J., Kwon, J. W., Shin, S. H., Yoo, H., Wang, K. C., Lee, S. H., & Gwak, H. S. (2023). Leptomeningeal metastasis in gliomas: Clinical characteristics and risk factors. *Journal of Korean Neurosurgical Society, 66*(4), 465–475. https://www.jkns.or.kr/upload/pdf/jkns-2022-0166.pdf. doi:10.3340/jkns.2022.0166.

4

Challenges and limitations in ascertaining a "gold standard" technique for malignant effusions and cerebrospinal fluid

Parikshaa Gupta

DEPARTMENT OF CYTOLOGY AND GYNECOLOGIC PATHOLOGY, POSTGRADUATE INSTITUTE OF MEDICAL EDUCATION AND RESEARCH (PGIMER), CHANDIGARH, INDIA

4.1 Introduction

Malignant effusions involving pleural, peritoneal, and pericardial spaces generally signify advanced disease. These are associated with a variety of malignancies and can develop due to direct tumor spread, lymphatic obstruction, or increased vascular permeability.[1] Cerebrospinal fluid (CSF) analysis is central to the diagnosis of infectious, inflammatory, neoplastic, hemorrhagic, and neurodegenerative disorders of the central nervous system (CNS). Cytological examination of these fluids remains a primary diagnostic procedure due to its minimally invasive nature, cost-effectiveness, negligible complications, high accuracy, and rapid turnaround.[2] However, despite the widespread reliance on fluid cytology, a universally accepted "gold standard" for confirming malignancy in body fluids remains elusive. This chapter explores the challenges and limitations of existing diagnostic methods and highlights the evolving landscape of ancillary techniques that are often employed as gold standards in confirming the diagnosis of malignant effusions and CSF. As CSF has a more dynamic nature than body cavity effusions, it will be discussed as a separate section.

4.2 Understanding the concept of "gold standard"

In diagnostic medicine, a "gold standard" refers to the best available method against which all the newer tests or procedures are evaluated or measured. It is considered the most accurate and

Advancing Body Fluid Diagnostics with Flow Cytometry: A Comprehensive Guide to Effusions and CSF.
DOI: https://doi.org/10.1016/B978-0-443-27426-8.00005-3

Table 4.1 Advantages and limitations of cytology in the diagnosis of malignant effusions.

Advantages of effusion cytology	Limitations of effusion cytology
Minimally invasive	Variable sensitivity; false negatives
Cost-effective and rapid	Difficulty in distinguishing reactive atypia from malignant cells
High specificity	Limited utility in samples with low-cellularity or cellular degeneration
Potential for ancillary testing (immunocytochemistry, flow cytometry, and molecular assays)	Lack of tissue architecture

reliable test for confirming a particular disease or condition. In the context of malignant effusions, a gold standard technique would ideally have high sensitivity, specificity, reproducibility, and consistency across various clinical settings.

4.3 Need for a gold standard

- Benchmarking: To evaluate the performance of new diagnostic tests.
- Clinical decision-making: Establishing an accurate diagnosis is critical for deciding the appropriate treatment strategy and optimal patient management.
- Research validity: A gold standard is essential for validating novel biomarkers and techniques before their clinical application.

Despite its theoretical clarity, designating/identifying a gold standard test in malignant effusions is complex due to the variability in sample quality, tumor biology, and available infrastructure.

4.4 Effusion cytology as the gold standard in malignant effusions

Effusion cytology plays a pivotal role in identifying malignancy, determining tumor type and origin, and even guiding therapy. Cytomorphologic examination of the malignant cells can help in determining the specific tumor type seen in effusions, which is essential for appropriate patient management. For example, effusion cytology helps in identifying characteristic cytomorphologic features that are useful in differentiating mesotheliomas from metastatic adenocarcinoma. Besides its traditional diagnostic role, cytology has also proven useful in providing prognostic insights and guiding therapeutic decisions in oncology.

While it offers several advantages, including safety, speed, and cost-effectiveness, it is plagued by limitations such as variable sensitivity, cellular degeneration, overlapping cellular morphologies, interobserver variability, and lack of architectural context (Table 4.1).

4.5 Correlation with clinical and radiologic details

Effusion cytology alone often cannot confirm malignancy with high confidence. Clinical history, examination, and radiologic findings are indispensable and provide the vital context that enhances the diagnostic accuracy of effusion cytology.

4.5.1 Clinical details

Clinical information contextualizes cytologic findings and helps formulate a working list of differential diagnoses even before the microscopic examination. Pertinent details include age, gender, presenting symptoms, past or current malignancies, treatment history, presence of fever, weight loss, or other systemic manifestations. Additionally, information regarding the patient's known comorbidities (e.g., cirrhosis, renal failure, and autoimmune disease) and geographic and epidemiologic context (e.g., for TB-endemic areas) is also important. All these details can guide the cytomorphologic interpretation, help reduce diagnostic ambiguity, and prioritize further investigations.

For instance, in a patient with known breast cancer, the detection of atypical cells in pleural fluid raises a strong suspicion of metastasis. Without the relevant clinical history, these cells might be considered reactive or suspicious for a malignancy, but not diagnostic. With the known history, suspicion for metastatic disease is heightened, prompting the use of a specific immunocytochemical panel for breast markers (e.g., GATA3, mammaglobin, and GCDFP-15), which helps confirm the metastasis.

4.5.2 Imaging details

Radiologic modalities including X-rays, ultrasonography, CT scans, and MRI play a crucial role in the comprehensive evaluation of patients with effusions. Imaging allows for the detection of minimal effusions, which are often not discernible clinically. With the aid of radiology, simultaneous estimation of the severity of effusion in terms of fluid volume is also possible in addition to guiding safe aspiration in real-time. Furthermore, the imaging modalities can help detect other abnormalities including underlying masses, organomegaly, peritoneal thickening, nodules, or lymphadenopathy. Evaluation of tumor spread patterns, disease monitoring, as well as assessment of therapeutic response is also possible.

For instance, atypical cells in a pleural effusion from a young, nonsmoking female with a lung mass on a CT scan are highly suggestive of metastatic lung adenocarcinoma. Subsequent immunocytochemistry (TTF1) on the cell block section can help confirm the diagnosis. Similarly, the presence of papillary clusters of atypical cells and occasional psammoma bodies in an adult female patient with bilateral complex ovarian masses is highly indicative of metastatic ovarian carcinoma. Immunocytochemistry (ICC) for WT-1, PAX8, and p53 can help confirm the diagnosis of high-grade serous carcinoma. Cytology alone may not have suggested this diagnosis without imaging correlation. In patients with lymphocyte-rich effusions, radiologic evidence of hepatosplenomegaly or generalized lymphadenopathy favors a diagnosis of lymphoma over a benign reactive effusion.

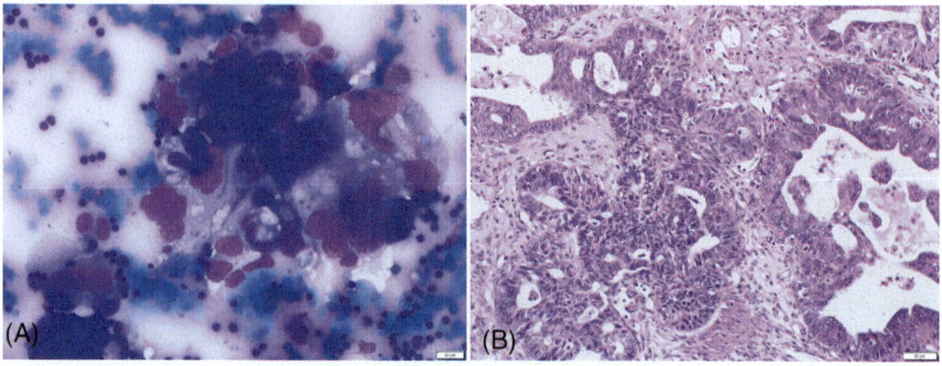

FIGURE 4.1 Histopathology as the gold standard in malignant effusions. (A) Conventional sediment smears showing clusters of highly pleomorphic tumor cells with large irregular nuclei, coarse chromatin, prominent nucleoli, and abundant vacuolated cytoplasm, metastatic adenocarcinoma in the effusion sample (MGG; 40x). (B) Corresponding histopathology section from the adnexal mass confirming a high-grade serous carcinoma (Hematoxylin & Eosin; 40x).

Similarly, in the case of ascites in an adolescent with a retroperitoneal mass, the young age and presence of a radiologically confirmed mass are useful diagnostic clues for the presence of a malignant neoplasm. The list of differentials in such a case can be further refined with the availability of relevant biochemical tests; for instance, if the serum alpha fetoprotein is elevated in this adolescent patient, the differentials can generally be narrowed down to metastasis from a yolk sac tumor or hepatoblastoma, and the diagnosis can be confirmed by use of specific ICC markers on cell block sections.

4.6 Histopathology as the gold standard in malignant effusions

While effusion cytology serves as the widely used first-line diagnostic tool for the evaluation of malignant effusions, it has some inherent limitations, as listed earlier in Table 4.1. In some cases, cytology may raise suspicion of malignancy but fail to definitively classify or subtype the underlying neoplasm. In such cases, histopathology is often considered the cornerstone for establishing a definitive diagnosis, mainly due to its superior ability to assess tissue architecture and to provide an adequate sample for performing immunohistochemistry and molecular testing.

Histopathology comfortably allows for a more comprehensive immunohistochemical workup, which is sometimes essential for tumor subtyping and differentiating primary from metastatic disease. For example, in cases of metastatic adenocarcinoma, markers such as TTF-1, CK7, CK20, PAX-8, WT-1, CDX2, and SATB2 can help identify the primary tumor origin in correlation with the clinico-radiologic details (Fig. 4.1). Furthermore, histopathologic examination permits the integration of molecular diagnostics including next-generation sequencing (NGS), which are vital in deciding targeted therapies.

Table 4.2 Advantages and limitations of histopathologic examination as a gold standard for the diagnosis of malignant effusions.

Advantages of histopathologic examination	Limitations of histopathologic examination
Allows evaluation of the architectural patterns (e.g., glandular and papillary)	More invasive than cytology and relatively higher complication rate
High diagnostic accuracy in solid tumors	Generally not required in all cases and especially ones where benign outcome is favored
Allows subtyping and grading of neoplasms	Often not feasible in all patients
Provides sufficient samples for immunohistochemical and molecular testing for diagnostic, prognostic, and therapeutic purposes	Requires specialized skills and sophisticated infrastructure

Although histopathologic analysis is more invasive than cytology, especially when obtained via thoracoscopy or laparoscopy, its diagnostic superiority often justifies or overrides the procedural risk, more so in cases with inconclusive or equivocal cytological interpretation (Table 4.2). In clinical scenarios where the treatment decisions hinge on precise tumor typing, grading, and molecular characterization, histopathologic confirmation becomes indispensable.

4.7 Flow cytometric immunophenotyping as the gold standard in malignant effusions

Flow cytometric immunophenotyping (FCI) is invaluable for diagnosing hematologic or lymphoreticular malignancies (LRM) involving effusions. Involvement of body cavity effusions by LRM is reportedly rare, with an incidence ranging from 3% to 30%. Of all the LRMs, nonHodgkin lymphomas, especially the T-lymphoblastic lymphomas, are the most common ones to involve effusions. It is essential to accurately diagnose such cases, as these are associated with a poorer prognosis and hence require more aggressive management. Diagnosing these LRMs based on cytomorphology alone can often be challenging and may lead to misinterpretations, which can adversely affect patient outcomes.

FCI on the effusion fluids can help confirm the diagnoses in cytologically suspicious cases. Different fluorescence-labeled antibodies in various combinations can be used; the most commonly employed antibodies include CD45, CD19, CD20, CD2, CD3, CD4, CD5, CD7, CD8, CD10, CD1a, CD11b, CD13, CD14, CD16, CD23, CD33, CD30, CD34, CD38, CD56, CD64, CD117, CD138, HLA-DR, TdT, κ, λ light chains, TCRα/β, and TCRγ/δ. With the help of these antibodies, FCI can identify specific cell populations, evaluate clonality, as well as provide comprehensive lineage specific information. Furthermore, FCI also allows accurate subtyping of various LRMs in the cytologic samples (Table 4.3) (Fig. 4.2).

While FCI is most widely applied in diagnosing and characterizing lymphoid and myeloid neoplasms, recent advances have extended its utility to epithelial malignancies. In particular, FCI can be used to evaluate epithelial markers such as EpCAM (CD326) or cytokeratins (e.g., CK7 and CK20) for detecting the presence of malignant epithelial cells and circulating tumor cells in

Table 4.3 Advantages and limitations of flow cytometric immunophenotyping as a gold standard for the diagnosis of malignant effusions.

Advantages of flow cytometric immunophenotyping	Limitations of flow cytometric immunophenotyping
High sensitivity (detects <1% abnormal cells)	Limited utility in solid tumors
Can reliably distinguish malignant from reactive lymphocytes	Requires viable single-cell suspensions
Enables disease monitoring	Often not feasible in all patients
Allows precise subtyping of many lymphoreticular malignancies	Requires specialized skills and sophisticated instruments
High diagnostic accuracy in lymphoreticular malignancies	Compromised results in degenerated and necrotic samples

effusions. This approach is particularly helpful in cases with equivocal cytologic interpretation, where the cell block is unavailable or inadequate, and a prompt diagnosis is needed. However, the interpretation is challenging and requires strict standardization. Although some studies have evaluated the role of FCI in diagnosing epithelial malignant cells in effusions and recommend its use, it is generally used as a complementary modality in select cases.

Furthermore, flow cytometry can also be used to identify and characterize cancer stem cells (CSCs) in effusion samples. These cells often express unique stemness markers such as CD44, CD133, and ALDH1, and thus can be detected by using fluorescently labeled specific antibodies. The presence of CSCs in patients with malignant effusions is associated with higher tumor aggressiveness, metastasis, and therapeutic resistance. Detecting CSCs in effusions by FCI, thus, provides useful prognostic information and can help guide therapeutic strategies.

4.8 Cell block immunocytochemistry as the gold standard in malignant effusions

To overcome the limitations of conventional cytology, cell blocks are now routinely prepared from the effusion samples at a majority of cytopathology centers worldwide. The cell blocks act as a bridge between cytology and histopathology by providing the much-needed architectural details, thereby improving the diagnostic accuracy. A variety of methods can be employed for cell block preparation, depending on the available resources. The effusion sample is centrifuged and the sedimented cells are ultimately processed into paraffin blocks for histological and immunohistochemical evaluation, thereby enhancing the cytomorphologic diagnosis.

With the use of specific immunocytochemical antibodies, CB-ICC can help confirm the type of malignancy (epithelial, mesenchymal, germ cell, or lymphomatous), as well as identify the site of the primary tumor in the majority of the cases. Based on the clinical, radiologic, and cytomorphologic features, appropriate antibody panels can be employed to confirm as well as characterize the malignant effusions. CB-ICC can reliably distinguish metastatic adenocarcinomas from malignant mesotheliomas in effusions. Similarly, the specific origin of the tumor cells

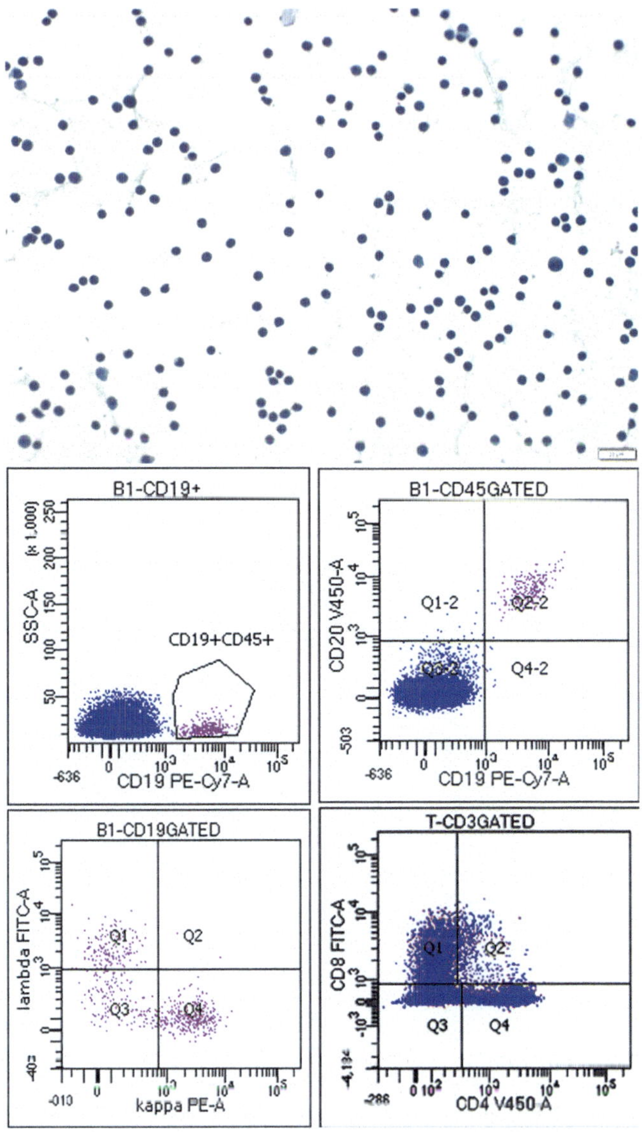

FIGURE 4.2 Flow cytometric immunophenotyping as the gold standard in malignant effusions. Liquid-based cytology preparation showing scattered monomorphic small lymphoid cells (Papanicolaou; 40x); flow cytometric immunophenotyping revealed reactive B and T-lymphoid cells, indicating a reactive effusion.

(e.g., ovarian, breast, prostate, lung, colorectal, neuroendocrine, etc.) can be accurately detected using specific ICC markers (Table 4.4).

ICC for different prognostic biomarkers (e.g., PD-L1 in lung cancer) can also be performed, thereby aiding in better patient management. Recent advancements have also permitted the detection of certain genetic abnormalities (e.g., *HER2* amplification, *TP53* mutations, and

Table 4.4 Advantages and limitations of cell block immunocytochemistry as a gold standard for the diagnosis of malignant effusions.

Advantages of cell block immunocytochemistry	Limitations of cell block immunocytochemistry
Allows for the architectural evaluation	Dependent on sufficient cell yield and processing quality
Can help distinguish malignant from benign cells	Expensive
Allows use of multiple markers on serial sections	Limited by the availability of sufficient material
Allows precise subtyping of most malignancies	Requires specialized skills and sophisticated instruments
Facilitates immunochemical characterization using specific antibody panels for prognostication as well as therapeutic decision-making	Compromised results in degenerated and necrotic samples

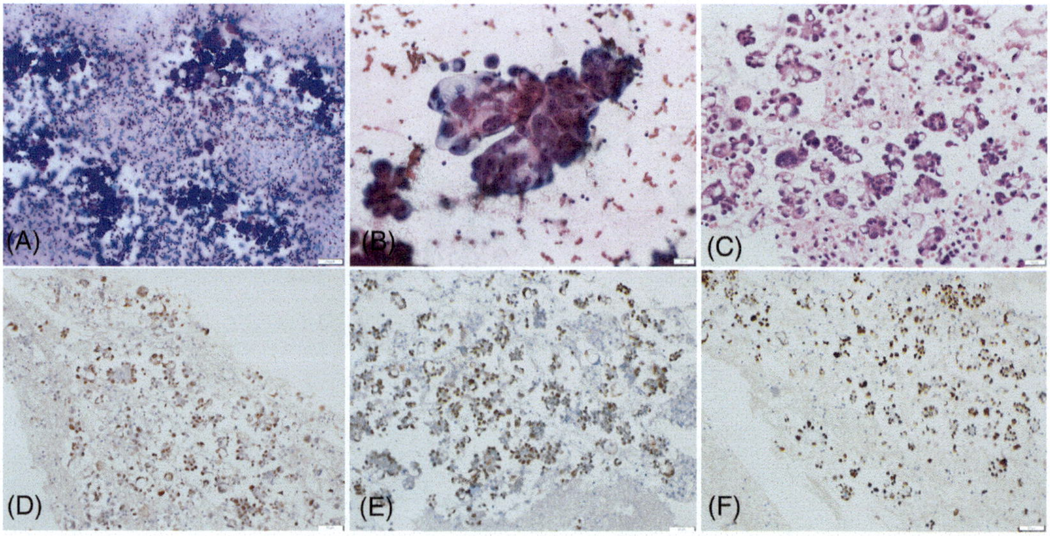

FIGURE 4.3 Cell block immunocytochemistry as gold standard in malignant effusions. (A–B) Conventional sediment smears showing many clusters of tumor cells in a case of metastatic adenocarcinoma in the effusion sample (A: MGG; 40x) (B: Papanicolaou; 40x). (C) Cell block section is highly cellular and shows similar tumor cells (Hematoxylin & Eosin; 40x). (D–F) Immunocytochemistry showing tumor cells to be positive for PAX-8 (D; 20x), WT-1 (E; 20x), and mutant expression for p53 (F; 20x).

Anaplastic lymphoma kinase (ALK) rearrangements) in effusion cell blocks by using their surrogate immunocytochemical markers (Fig. 4.3). This provides significant prognostic information and guides the use of targeted therapies.

Furthermore, CB-ICC also allows for the identification of CSCs in effusions by employing immunostaining for markers like CD44, CD24, and ALDH1. The detection of these cells provides useful insights into tumor heterogeneity, tumor prognosis, treatment response, and recurrence risk.

Table 4.5 Advantages and limitations of molecular testing as a gold standard for the diagnosis of malignant effusions.

Advantages of molecular testing	Limitations of molecular testing
Identifies actionable mutations	Dependent on viable tumor cell yield
Can help distinguish malignant from benign cells	Expensive
Guides targeted therapy	Limited accessibility
Applicable even on low-cellularity samples	Requires specialized skills and sophisticated instruments

4.9 Molecular testing as the gold standard in malignant effusions

Molecular diagnostics provide high sensitivity and specificity for detecting specific gene mutations, amplifications, or rearrangements that are essential for personalized treatment. Unlike the conventional cytomorphologic evaluation, which has limited sensitivity, molecular testing has, in a way, redefined diagnostic accuracy and clinical utility in oncology. It can be employed as a gold standard in certain clinical contexts, particularly when cytological results are inconclusive or when precise molecular characterization is essential for deciding targeted therapies. Molecular testing in effusions is being used for a variety of malignancies, including lung, breast, ovarian, colorectal, and hematologic malignancies. Table 4.5 highlights the advantages and limitations of molecular testing in malignant effusions.

4.9.1 Molecular testing techniques used in effusions

- NGS: Allows comprehensive profiling of multiple genes from a small volume of effusion fluid.
- Real-time polymerase chain reaction (PCR): Allows rapid detection of known mutations (e.g., *EGFR* mutations in lung adenocarcinoma).
- Fluorescence in situ hybridization (FISH): Used for detecting gene rearrangements, deletions, and amplifications (e.g., *ALK*, *ROS1*, and *HER2*).
- Digital PCR (dPCR): Highly sensitive for detecting rare mutations in low-cellularity samples.
- Methylation analysis and miRNA profiling: Emerging tools for further tumor characterization, especially in cases with equivocal cytology.

Molecular profiling may also help identify typical gene signatures associated with cancer stem cell phenotypes, thereby improving the understanding of tumor biology and predicting treatment resistance.

4.10 Toward a multidisciplinary gold standard in malignant effusions: Integration of multiple modalities

No single test can be universally applied as the gold standard in all malignant effusions, reflecting the real-world complexity of effusion evaluation. While cytology offers significant diagnostic insights, it has its limitations. To enhance diagnostic accuracy, it is essential to integrate clinical details, radiologic imaging, and various ancillary techniques like flow cytometry, cell block immunocytochemistry (CB-ICC), histopathology, and molecular testing. Establishing an accurate effusion diagnosis benefit from a multimodal approach, wherein each modality contributes its unique strengths and compensates for the prevailing limitations. This integrative, patient-centered multimodality approach ensures accurate diagnosis and guides optimal treatment strategies.

4.11 Gold standard in cerebrospinal fluid analysis

The concept of a single "gold standard" in CSF diagnostics is tricky, particularly because most CNS malignancies can be diagnosed using composite clinical, radiological, and laboratory criteria rather than a single confirmatory test. Furthermore, there are several biological and pathophysiological factors that limit the use of a single gold standard. CSF composition is dynamic and varies with disease stage, immune response, blood-brain barrier integrity, and treatment. Normal CSF has very low-cellularity and protein concentration. Even minor contamination significantly alters its parameters. There are higher chances of collection-related variability as the lumbar puncture technique influences the cell counts and protein concentration.[26] The inherent paucicellular nature of CSF limits the sensitivity of cytological techniques.[27] Processing delays cause rapid cellular degeneration, further affecting the test yields.

Various analytical techniques that can be employed as a gold standard for CSF analysis include: cytology, biochemical markers, FCI, ICC, and molecular tests. As highlighted earlier, a true gold standard should demonstrate near-perfect sensitivity and specificity, high reproducibility, minimal operator dependency, and feasibility in diverse healthcare settings. In CSF diagnostics, most of the techniques fall short in one or more of these domains.

4.12 Cerebrospinal fluid cytology as a gold standard in cerebrospinal fluid analysis

CSF cytological examination is based on direct visualization of various cells in cytospin preparations. Traditionally, a positive cytology result has been regarded as the reference standard for the diagnosis of malignant involvement of the CNS, particularly in leptomeningeal metastasis and hematolymphoid malignancies.

Although it is widely available, reliable, and inexpensive, diagnostic CSF cytology relies heavily on collection of adequate CSF volume (preferably 5–10 mL), prompt processing (within 1–4 h) with specialized cell concentration techniques (cytospin), proper staining, and accurate

Table 4.6 Strengths and limitations of CSF cytology as a gold standard.

Strengths	Limitations
High specificity	Variable sensitivity; false negatives common due to intermittent tumor shedding, and low-cellularity
Direct cytomorphologic assessment; helpful in determining cell types	Rapid cellular degeneration
Cost-effective	High operator dependency; morphological overlap between reactive and malignant cells
Wide availability	Limited prognostic value

CSF, cerebrospinal fluid.

cytomorphologic interpretation.[27] Furthermore, it has a limited role in diagnosing specific infections and inflammatory diseases.

Clinical contexts where CSF cytology can serve as a *gold standard*:

Leptomeningeal metastasis from solid tumors: In suspected leptomeningeal carcinomatosis, cytology remains the diagnostic reference standard. A positive finding confirms disease, even when neuroimaging is inconclusive.[28]

Hematolymphoid malignancies: In acute leukemia and high-grade lymphoma, cytology establishes CNS involvement, especially in resource-limited settings without advanced molecular diagnostics and flow cytometry.[29]

Monitoring disease response: Serial CSF cytology has been used to monitor treatment response in patients being administered intrathecal chemotherapy.[30]

The strengths and limitations of CSF cytology as a gold standard have been highlighted in Table 4.6.

Therefore, although CSF cytology remains a foundational technique, it is best described as a context-specific reference standard, rather than a universal gold standard in CSF analysis. For the same reason, currently CSF evaluation in suspected malignancy patients often employs a composite diagnostic approach combining radiology, cytology, flow cytometry, ICC, and molecular assays.

4.13 Biochemical markers as a gold standard in cerebrospinal fluid analysis

Biochemical analysis of CSF constitutes one of the most commonly performed tests for CSF evaluation. Different biochemical parameters such as glucose, protein, lactate, chloride, adenosine deaminase, and oligoclonal bands (OCBs) provide critical supportive evidence in diagnosing various infectious, inflammatory, demyelinating, and metabolic disorders of the CNS.[31] Although widely used and standardized, biochemical markers are rarely considered independent gold standards. Instead, they function as essential adjunctive tools to help support a clinical diagnosis. Commonly assessed biochemical parameters and their uses have been listed in Table 4.7.

Table 4.7 Commonly assessed CSF biochemical parameters and their uses.

Biochemical parameters	Utility
Glucose and CSF-to-serum glucose ratio	Differentiating bacterial versus viral meningitis
Total protein	Differentiating bacterial versus viral meningitis
Lactate	Differentiating bacterial versus viral meningitis
Adenosine deaminase	Diagnosis of tubercular meningitis
Immunoglobulin levels and oligoclonal bands (OCBs)	Diagnosis of multiple sclerosis
Albumin quotient	Indicator of blood-brain barrier permeability

CSF, cerebrospinal fluid.

Table 4.8 Strengths and limitations of CSF biochemical markers as a gold standard.

Strengths	Limitations
High objectivity; automated and standardized techniques	Lack of specificity
Quantitative results	Affected by multiple systemic factors
Cost-effective and widely available	Variable cut-off values across laboratories
Rapid turnaround time	Limited role in early disease

CSF, cerebrospinal fluid.

Clinical contexts where biochemical markers function as a *gold standard*:

Multiple sclerosis (MS) and demyelinating disorders: Detection of OCBs in CSF, particularly when absent in serum, is strongly associated with MS.[32]

Differentiating bacterial and viral meningitis: Elevated lactate supports bacterial etiology. Low glucose (hypoglycorrhachia) is typical in bacterial and tuberculous meningitis.[31]

Tubercular meningitis: Adenosine deaminase (ADA) levels help support the diagnosis of tubercular meningitis.[31]

Table 4.8 highlights the strengths and limitations of biochemical tests in CSF analysis. Biochemical markers in CSF analysis are an indispensable component of diagnostic tests for CNS diseases due to their high objectivity, rapid availability, and standardized methodology. However, their limited disease specificity and overlap across multiple pathological conditions prevent them from being employed as universal gold standards. Therefore, CSF biochemical markers best serve as critical adjunctive tools within a composite diagnostic framework rather than independent gold standards in CSF analysis.

4.14 Immunocytochemistry as gold standard in cerebrospinal fluid analysis

ICC is an advanced diagnostic technique used primarily for the detection and characterization of malignant cells in CSF. It involves the application of antigen-specific monoclonal or polyclonal antibodies to cytological preparations (cytospin or cell blocks) to identify lineage-specific,

Table 4.9 Strengths and limitations of CSF immunocytochemistry as a gold standard.

Strengths	Limitations
High sensitivity	Dependent on CSF cellularity
High specificity	Interpretation is operator dependent
Confirmation of the morphologic diagnosis	Expensive; skilled manpower
Accurate tumor subtyping	Limited role in infections
Identification of targets for personalized therapy	Limited role in inflammatory diseases

CSF, cerebrospinal fluid.

tumor-specific, or pathogen-associated markers. While conventional CSF cytology remains the reference standard for diagnosing leptomeningeal malignancies, ICC improves this diagnostic sensitivity and specificity.[33]

Clinical contexts where CSF ICC can serve as *gold standard:*

- Leptomeningeal metastasis: Conventional cytology is considered the gold standard for diagnosing leptomeningeal metastasis. However, sensitivity is limited. Specific ICC markers can help confirm the presence of malignant cells in CSF.[33]
- Subtyping metastatic malignancies: Lineage-specific ICC markers can help in accurately subtyping the malignant cells.

Differentiating CNS lymphoma from reactive lymphomatous meningitis: In CSF, morphology alone may not reliably distinguish reactive lymphocytosis from neoplastic lymphoid cells. ICC helps provide immunophenotypic confirmation.[34]

Table 4.9 highlights the strengths and limitations of CSF ICC as a gold standard.

CSF ICC plays a crucial role in the evaluation of malignant involvement of the CNS. It improves specificity, diagnostic precision, and aids in tumor classification. However, its dependence on cytologic interpretation, high cost, and technical limitations prevent it from being used as a universal gold standard in CSF analysis. It is best conceptualized as a specialized confirmatory modality in neurooncology rather than a diagnostic benchmark.

4.15 Flow cytometry as a gold standard in cerebrospinal fluid analysis

Flow cytometry (FCM) detects abnormal immunophenotypic cell populations in CSF using fluorescent-labeled antibodies. It is especially useful in diagnosing hematolymphoid malignancies.[35]

Clinical contexts where CSF flow cytometry can serve as a *gold standard:*[35,36]

- Suspected CNS lymphomas: FCM enables detection of monoclonal B-cell populations.
- Leukemic CNS involvement: FCM helps in accurate characterization of leukemic blasts.

Table 4.10 Strengths and limitations of flow cytometry as a gold standard in CSF analysis.

Strengths	Limitations
Higher sensitivity than cytology	Requires viable cells
Confirmation of the cytologic diagnosis; exact characterization of lymphoreticular malignancies	Dependent on CSF cellularity
Objective and quantitative results	No morphologic assessment
Detects minimal disease	Expensive; skilled manpower
Rapid turnaround time	Limited role in other malignancies and infections

CSF, cerebrospinal fluid.

- Differentiating primary CNS lymphoma from reactive lymphomatous meningitis: In CSF, morphology alone may not reliably distinguish reactive lymphocytosis from neoplastic lymphoid cells. FCM helps provide immunophenotypic confirmation.
- Monitoring minimal residual disease: FCM helps in early detection of occult leptomeningeal disease.[37]

Table 4.10 highlights the strengths and limitations of flow cytometry in CSF analysis.

Flow cytometry plays a significant role in diagnosis of hematologic malignancies involving CSF; however, is not universally applicable across all CSF disorders and hence, is best used as an adjunctive technique in CSF analysis rather than a standalone gold standard.

4.16 Molecular techniques as a gold standard in cerebrospinal fluid analysis

Molecular diagnostic techniques have transformed CSF analysis over the past two decades. Methods such as PCR, multiplex nucleic acid amplification panels, and NGS enable rapid and sensitive detection of infectious agents and tumor-derived genetic material in CSF.[38] NGS and metagenomic sequencing approaches enable unbiased, high-throughput detection of nucleic acids from pathogens or host-derived genetic material. Metagenomic next-generation sequencing (mNGS) is particularly valuable in diagnosing CNS infections of unclear etiology.[39]

Clinical contexts where CSF molecular tests can serve as a *gold standard*:[38-40]

Viral infections: PCR for viruses including herpes simplex virus, enteroviruses, varicella-zoster virus, cytomegalovirus, and JC virus in suspected cases is now the diagnostic gold standard owing to its high sensitivity and specificity.

Partially treated bacterial meningitis: PCR and NGS have a much higher sensitivity for detecting the pathogenic bacteria in the CSF in patients who have already received antibiotics.

Tubercular meningitis: Molecular tests such as nucleic acid amplification assays (Xpert MTB/RIF) enable rapid detection of *Mycobacterium tuberculosis* compared to conventional cultures.

Table 4.11 Strengths and limitations of molecular tests as a gold standard in CSF analysis.

Strengths	Limitations
High analytical sensitivity; outperform conventional cultures	Detect nucleic acids rather than viable organisms or intact cells
Can detect organisms despite antibiotic exposure	Risk of contamination
Rapid turnaround times; results generally available within hours	Variable standardization
Broad coverage; useful for detecting viral, fungal, tubercular, culture-negative bacteria, parasites, and rare pathogens	High cost and skilled manpower
Identification of specific therapeutic targets and prognostication	Limited accessibility in low-resource settings

CSF, cerebrospinal fluid.

Fungal infections and rare pathogens: Molecular methods can easily detect microorganisms that are slow-growing, difficult to culture, and present in low numbers in the CSF.
Culture-negative infections: Molecular methods have broad coverage and can be used to detect microorganisms that do not grow on culture media.
CNS malignancies: Detection of circulating tumor DNA (ctDNA) in CSF is being explored as a liquid biopsy approach for diagnosing primary CNS tumors, leptomeningeal metastasis, and mutation profiling.

Table 4.11 highlights the strengths and limitations of various molecular tests in CSF analysis.

Molecular diagnostic techniques have revolutionized CSF analysis and, in certain infections, particularly viral encephalitis, have completely replaced traditional methods as the diagnostic gold standard. However, their inability to confirm viability, variable standardization, high cost, and limited accessibility prevent them from serving as a universal gold standard across all CNS disorders. In current practice, molecular techniques serve as a potent component of a composite diagnostic strategy, rather than a standalone universal benchmark.

4.17 Composite gold standard in cerebrospinal fluid analysis: The practical approach

As discussed, a variety of techniques are available for analysis of CSF; however, all of these have one or more limitations. Furthermore, global disparities in healthcare infrastructure prevent universal adoption of any single technique. Thus, rather than a single gold standard, a composite diagnostic framework provides the highest accuracy. Hence, modern CSF diagnostics essentially rely on multidimensional integration of clinical details, neuroimaging, CSF cytology, ICC, flow cytometry, biochemistry, and molecular diagnostics.

4.18 Conclusion

Cytology remains foundational for detection of malignancies in both body cavity fluids and CSF, but is insufficient alone for definitive diagnosis in challenging cases. Concurrent or subsequent histopathologic examination, biochemical tests, ICC, FCI, and molecular diagnostics serve as essential complements, each with the potential to act as a gold standard depending on the clinical context. By integrating multiple modalities, the limitations of each technique are mitigated, and the diagnostic utility is refined to maximize accuracy and ensure better patient management. To be precise, in the current era of personalized medicine, the "gold standard" is no longer a singular entity, but it should be interpreted as a coordinated strategy leveraging the strengths of multiple diagnostic tools.

References

Bashour, S. I., Mankidy, B. J., & Lazarus, D. R. (2022). Update on the diagnosis and management of malignant pleural effusions. *Respiratory Medicine, 196*. http://www.harcourt-international.com/journals/rmed/. doi:10.1016/j.rmed.2022.106802.

Bobillo, S., Crespo, M., Escudero, L., Mayor, R., Raheja, P., Carpio, C., Rubio-Perez, C., Tazón-Vega, B., Palacio, C., Carabia, J., Jiménez, I., Nieto, J. C., Montoro, J., Martínez-Ricarte, F., Castellvi, J., Simó, M., Puigdefàbregas, L., Abrisqueta, P., & Bosch, F. (2021). Cell free circulating tumor DNA in cerebrospinal fluid detects and monitors central nervous system involvement of B-cell lymphomas. *Haematologica, 106*(2), 513–521. doi:10.3324/haematol.2019.241208.

Bönig, L., Möhn, N., Ahlbrecht, J., Wurster, U., Raab, P., Puppe, W., Sühs, K. W., Stangel, M., Skripuletz, T., & Schwenkenbecher, P. (2019). Leptomeningeal metastasis: The role of cerebrospinal fluid diagnostics. *Frontiers in Neurology, 10*. http://www.frontiersin.org/Neurology. doi:10.3389/fneur.2019.00839.

Bromberg, J. E. C., Breems, D. A., Kraan, J., Bikker, G., Van Der Holt, B., Smitt, P. S., Van Den Bent, M. J., Van't Veer, M., & Gratama, J. W. (2007). CSF flow cytometry greatly improves diagnostic accuracy in CNS hematologic malignancies. *Neurology, 68*(20), 1674–1679. http://www.neurology.org. doi:10.1212/01.wnl.0000261909.28915.83.

Candeloro, R., Ferri, C., Laudisi, M., Baldi, E., Pugliatti, M., & Castellazzi, M. (2025). The diagnostic utility of oligoclonal bands in multiple sclerosis: A time-course analysis. *Biomedicines, 13*(2), 440. doi:10.3390/biomedicines13020440.

Chu, X., Tian, W., Ning, J., Xiao, G., Zhou, Y., Wang, Z., Zhai, Z., Tanzhu, G., Yang, J., & Zhou, R. (2024). Cancer stem cells: Advances in knowledge and implications for cancer therapy. *Signal Transduction and Targeted Therapy, 9*(1). doi:10.1038/s41392-024-01851-y.

Das, D. K. (2006). Serous effusions in malignant lymphomas: A review. *Diagnostic Cytopathology, 34*(5), 335–347. doi:10.1002/dc.20432.

Davidson, B. (2021). Molecular testing on serous effusions. *Diagn Cytopathol, 49*(5), 640–646. doi:10.1002/dc.24392.

Deak, D., Gorcea-Andronic, N., Sas, V., Teodorescu, P., Constantinescu, C., Iluta, S., Pasca, S., Hotea, I., Turcas, C., Moisoiu, V., Zimta, A.-A., Galdean, S., Steinheber, J., Rus, I., Rauch, S., Richlitzki, C., Munteanu, R., Jurj, A., & Petrushev, B. (2021). A narrative review of central nervous system involvement in acute leukemias. *Annals of Translational Medicine, 9*(1), 68. doi:10.21037/atm-20-3140.

Dermawan, J. K. T., & Policarpio-Nicolas, M. L. (2020). Malignancies in pleural, peritoneal, and pericardial effusions. *Archives of Pathology and Laboratory Medicine, 144*(9), 1086–1091. https://allen.silverchair-cdn.com/allen/content_public/journal/aplm/144/9/10.5858_arpa.2019-0429-oa/4/i1543-2165-144-9-1086.pdf?. doi:10.5858/arpa.2019-0429-OA.

Dong, H. P., Holth, A., Berner, A., Davidson, B., & Risberg, B. (2007). Flow cytometric immunphenotyping of epithelial cancer cells in effusions - Technical considerations and pitfalls. *Cytometry Part B - Clinical Cytometry, 72*(5), 332–343. doi:10.1002/cyto.b.20172.

Gomes, H. R. (2022). Cerebrospinal fluid analysis: Current diagnostic methods in central nervous system infectious diseases. *Arquivos de Neuro-Psiquiatria, 80*, 290–295. https://www.scielo.br/j/anp/a/Q86xqQVvQPCHQ8L54xYzNyC/abstract/?lang=en. doi:10.1590/0004-282X-ANP-2022-S114.

Gostomczyk, K., Łukaszewska, E., Borowczak, J. ę, Bator, A., Zdrenka, M., Bodnar, M., & Szylberg, Ł (2024). Flow cytometry in the detection of circulating tumor cells in neoplastic effusions. *Clinica Chimica Acta, 552*, 117651. doi:10.1016/j.cca.2023.117651.

Grigoriadou, G. I., Esagian, S. M., Ryu, H. S., & Nikas, I. P. (2020). Molecular profiling of malignant pleural effusions with next generation sequencing (NGS): Evidence that supports its role in cancer management. *Journal of Personalized Medicine, 10*(4), 1–18. https://www.mdpi.com/2075-4426/10/4/206/pdf. doi:10.3390/jpm10040206.

Gupta, P., Gupta, N., Bal, A., Rastogi, P., Prakash, G., Malhotra, P., Dey, P., Srinivasan, R., & Das, A. (2023). Cytomorphological characterisation of angioimmunoblastic T cell lymphoma: A case-control study. *Journal of Clinical Pathology, 76*(5), 320–326. http://jcp.bmj.com/. doi:10.1136/jclinpath-2021-207887.

Gupta, P., Pandey, T., Gautam, U., Rajwanshi, A., Srinivasan, R., Gupta, N., Rohilla, M., Varma, N., & Dey, P. (2021). Lymphoreticular malignancies in serous effusions: Cytomorphologic, flow cytometric and immunocytochemical analysis. *Diagnostic Cytopathology, 49*(5), 647–656. http://onlinelibrary.wiley.com/journal/10.1002/(ISSN)1097-0339. doi:10.1002/dc.24729.

Hagmeyer, L., Schäfer, S., Engels, M., Pietzke-Calcagnile, A., Treml, M., Herkenrath, S.-D., Heldwein, M., Hekmat, K., Matthes, S., Scheel, A., Wolf, J., Büttner, R., & Randerath, W. (2021). High sensitivity of PD-L1 analysis from pleural effusion in nonsmall cell lung cancer. *ERJ Open Research, 7*(1), 00787–02020. doi:10.1183/23120541.00787-2020.

Hakala, S., Hämäläinen, A., Sandelin, S., Giannareas, N., & Närvä, E. (2025). Detection of cancer stem cells from patient samples. *Cells, 14*(2). https://www.mdpi.com/journal/cells. doi:10.3390/cells14020148.

Harding, W. C., Halawa, A. R., Aiche, M. M., Zafar, B., Ali, H. J. R., Bashoura, L., & Faiz, S. A. (2025). Pleural effusion: Shedding light on pleural disease beyond infection and malignancy. *Medicina (Lithuania), 61*(3). https://www.mdpi.com/journal/medicina. doi:10.3390/medicina61030443.

Kassirian, S., Hinton, S. N., Cuninghame, S., Chaudhary, R., Iansavitchene, A., Amjadi, K., Dhaliwal, I., Zeman-Pocrnich, C., & Mitchell, M. A. (2023). Diagnostic sensitivity of pleural fluid cytology in malignant pleural effusions: Systematic review and meta-analysis. *Thorax, 78*(1), 32–40. http://thorax.bmj.com/. doi:10.1136/thoraxjnl-2021-217959.

Kelbich, P., Hrach, K., Spicka, J., Vachata, P., Radovnicky, T., Hanuljakova, E., & Krejsek, J. (2022). Basic analysis of the cerebrospinal fluid: An important framework for laboratory diagnostics of the impairment of the central nervous system. *Current Issues in Molecular Biology, 44*(8), 3666–3680. https://www.mdpi.com/journal/cimb. doi:10.3390/cimb44080251.

Levine, G. J., & Cook, J. R. (2020). Cerebrospinal fluid and central nervous system cytology. *Cowell and Tyler's diagnostic cytology and hematology of the dog and cat* (pp. 210–228). doi:10.1016/B978-0-323-53314-0.00014-6.

Lin, Y. Y., Lin, Y. S., & Liang, C. W. (2024). Heterogeneity of cancer stem cell-related marker expression is associated with three dimensional structures in malignant pleural effusion produced by lung adenocarcinoma. *Cytopathology, 35*(1), 105–112. http://onlinelibrary.wiley.com/journal/10.1111/(ISSN)1365-2303. doi:10.1111/cyt.13321.

Olie, S. E, Andersen, C. Ø, van de Beek, D., & Brouwer, M. C. (2024). Molecular diagnostics in cerebrospinal fluid for the diagnosis of central nervous system infections. *Clin Microbiol Rev, 37*(4), e0002124. doi:10.1128/cmr.00021-24.

Pairman, L., Beckert, L. E. L., Dagger, M., & Maze, M. J. (2022). Evaluation of pleural fluid cytology for the diagnosis of malignant pleural effusion: A retrospective cohort study. *Internal Medicine Journal, 52*(7), 1154–1159. http://onlinelibrary.wiley.com/journal/10.1111/(ISSN)1445-5994. doi:10.1111/imj.15725.

Sahu, S., Gupta, P., & Dey, P. (2021). Molecular testing on serous effusion: An update. *Cytojournal, 18*, 35. doi:10.25259/cytojournal_55_2020.

Sahu, S., Gupta, P., Susheilia, S., Gautam, U., & Dey, P. (2021). Application of multicolour flow cytometry in the detection of metastatic carcinoma in serous effusions: Special emphasis in atypical cytology. *Cytopathology, 32*(2), 169–179. http://onlinelibrary.wiley.com/journal/10.1111/(ISSN)1365-2303. doi:10.1111/cyt.12922.

Scott, H., & Phil, R. (2024). What you need to know about: The management of malignant pleural effusion. *British Journal of Hospital Medicine, 85*(12). http://www.magonlinelibrary.com/loi/hmed. doi:10.12968/hmed.2024.0311.

Sener, U., Kumthekar, P., & Boire, A. (2021). Advances in the diagnosis, evaluation, and management of leptomeningeal disease. *Neuro-Oncology Advances, 3*, v86–v95. academic.oup.com/noa. doi:10.1093/noajnl/vdab108.

Shidham, V. B. (2022). Metastatic carcinoma in effusions. *CytoJournal, 19*, 1–31. https://cytojournal.com/metastatic-carcinoma-in-effusions/. doi:10.25259/CMAS_02_09_2021.

Shidham, V. B., & Janikowski, B. (2022). Immunocytochemistry of effusions: Processing and commonly used immunomarkers. *CytoJournal, 19*. https://cytojournal.com/immunocytochemistry-of-effusions-processing-and-commonly-used-immunomarkers/. doi:10.25259/CMAS_02_15_2021.

Shidham, V. B., & Layfield, L. J. (2021). Approach to diagnostic cytopathology of serous effusions. *CytoJournal, 18*(1). https://cytojournal.com/approach-to-diagnostic-cytopathology-of-serous-effusions/. doi:10.25259/CMAS_02_03_2021.

Shidham, V. B., & Layfield, L. J. (2022). Immunocytochemistry of effusion fluids: Introduction to SCIP approach. *CytoJournal, 19*. https://cytojournal.com/immunocytochemistry-of-effusion-fluids-introduction-to-scip-approach/. doi:10.25259/CMAS_02_05_2021.

Tawfik, B., Brown, L., Fuda, F., Chen, W., Niu, S., Sailors, J., & Sadeghi, N. (2018). Utility and proposed algorithm of CSF flow cytometry in hematologic malignancies. *Annals of Hematology, 97*(9), 1707–1716. link.springer.de/link/service/journals/00277/index.htm. doi:10.1007/s00277-018-3336-3.

Tsai, T. H., Wu, S. G., Chang, Y. L., Wu, C. T., Tsai, M. F., Wei, P. F., Yang, C. H., Yu, C. J., Yang, P. C., & Shih, J. Y. (2012). Effusion immunocytochemistry as an alternative approach for the selection of first-line targeted therapy in advanced lung adenocarcinoma. *Journal of Thoracic Oncology, 7*(6), 993–1000. https://www.journals.elsevier.com/journal-of-thoracic-oncology/. doi:10.1097/JTO.0b013e31824cc46b.

Tsamis, K. I., Sakkas, H., Giannakis, A., Ryu, H. S., Gartzonika, C., & Nikas, I. P. (2021). Evaluating infectious, neoplastic, immunological, and degenerative diseases of the central nervous system with cerebrospinal fluid-based next-generation sequencing. *Molecular Diagnosis and Therapy, 25*(2), 207–229. http://rd.springer.com/journal/40291. doi:10.1007/s40291-021-00513-x.

van Linder, B. M. H., Brandsma, D., Verhoeven, E., Jongejan, L., Monkhorst, K., de Langen, A. J., van den Broek, D., Pluim, D., van Wezel, T., van Kessel, E., & Boelens, M. C. (2025). Improving the diagnosis of leptomeningeal metastases by molecular profiling of cell-free DNA from cerebrospinal fluid. *European Journal of Cancer, 229*. https://www.sciencedirect.com/science/journal/09598049. doi:10.1016/j.ejca.2025.115783.

Vidriales, M. B., San-Miguel, J. F., Orfao, A., Coustan-Smith, E., & Campana, D. (2003). Minimal residual disease monitoring by flow cytometry. *Best Practice and Research: Clinical Haematology, 16*(4), 599–612. http://www.elsevier.com/inca/publications/store/6/2/3/0/0/3/index.htt. doi:10.1016/S1521-6926(03)00067-7.

Weston, C. L., Glantz, M. J., & Connor, J. R. (2011). Detection of cancer cells in the cerebrospinal fluid: Current methods and future directions. *Fluids and Barriers of the CNS, 8*(1). http://www.fluidsbarrierscns.com/content/8/1/14. doi:10.1186/2045-8118-8-14.

Yadav, K., Cree, I., Field, A., Vielh, P., & Mehrotra, R. (2022). Importance of cytopathologic diagnosis in early cancer diagnosis in resource-constrained countries. *JCO Global Oncology, 8*(8). doi:10.1200/go.21.00337.

5

Flow cytometry: Basic principles and techniques

Khaliqur Rahman and Mona Singh

DEPARTMENT OF HEMATOLOGY, SANJAY GANDHI POST GRADUATE INSTITUTE OF MEDICAL SCIENCES, LUCKNOW, UTTAR PRADESH, INDIA

5.1 Introduction

Flow cytometry is a highly sophisticated and precise technique that analyses single cells at a time when made to pass through single or multiple lasers. It is a powerful tool with multidisciplinary applications and has now become indispensable for the phenotypic characterization of cells.

As the name suggests, it is a technique that measures *(metry)* the characters of a cell *(cytos)* when they are flowing *(flow)* in a fluid stream in front of lasers. Flow cytometry can measure the optical and fluorescence characteristics of a cell or particle after optimum processing, while in a suspended state (Adan et al., 2017). Optical (light) scatter obtained is measured in two directions: Forward scatter (FSC) and side scatter (SSC), based on size and cellular complexity, respectively. Fluorescence achieved by staining with fluorescent dyes or fluorescently conjugated antibodies is also measured similarly (McKinnon, 2018). These light signals are then converted into digital electronic signals, which are represented as different kinds of plots. Interpretation of data is done via traditional two-parameter dot plot gating method (widely applied) or, furthermore, advanced cluster data analysis algorithms (McKinnon, 2018).

5.2 History

About 77 years of technical innovation and joint efforts of numerous physicists, biophysicists, biologists, and engineers have led this endeavor to the present-day world. Original observation of cells queued while being suspended in a fluid sheath to facilitate counting was first described by Moldavan (1934).

Coons et al. (1942) used an anthracene-conjugated antibody (UV excitable fluorochrome) for the detection of *Streptococcus pneumonia*. A decade later, the first fluorescein isothiocyanate (FITC) conjugated antibody was described by Coons and Kaplan (1950). Coulter (1945) employed the concept of conductance variation to measure cell size and event count. A few years later, Coulter also measured cell volume suspended in liquid medium using the technique of electrical

Advancing Body Fluid Diagnostics with Flow Cytometry: A Comprehensive Guide to Effusions and CSF.
DOI: https://doi.org/10.1016/B978-0-443-27426-8.00007-7

impedance variation (Picot et al., 2012). Eventually, Coulter's cell counters evolved into modern-day cell analyzers for blood cell counting (Brecher et al., 1956).

In 1974, the first flow cytometry-based differential counter was launched for commercial purposes by the Hemalog-D (Technicon, Tarrytown, NY, USA), which provided a differential count for the white cells. Simultaneously, Becton-Dickinson (now known as BD Biosciences, San Jose, CA, USA) developed a second version of fluorescence-activated cell sorting (FACS) based cell sorter, originally developed by Bonner et al. (1972). Two-parameter-sorted-1 (TPS-1) was the first cell sorter developed by Coulter Electronics, which measured FSC and single fluorescence (Picot et al., 2012). Toward the end of the decade, scientists at NIH developed FACS IV, the first dual-laser cytometer, marketed by Becton-Dickinson (Steinkamp et al., 1973). The revolutionary advancements over the years and decades added to the complexities of the analyzers, enabling faster data analysis, widespread clinical applications, and research.

In modern-day practice, approximately 17 or more groups manufacture cytometers and cell sorters, each having commendable capacity for sample processing and analysis.

5.3 Principle

The fundamental principle is based on light scattering and fluorescence emission. Usually, the source of light is a laser beam that strikes the moving particle or cell. The amount of light scattered is based on the structure and morphology of the cell, whereas fluorescence emitted is determined by the amount of fluorescent-tagged antibodies to the cell or its component (Macey, 2007).

Two types of cytometers have been described: Nonsorting and sorting. As the name suggests, nonsorting ones can perform light scattering and fluorescence emission, while the sorting ones have an additional ability of sorting the particles or cells. For example, FACS are cytometers with the property of sorting out the fluorescently labeled cells from a mixed population (Wilkerson, 2012).

The main compartments of a flow cytometer are fluidics, optics, an electronic network, and a computer. Optics further has two subcomponents: Excitation and collection. The purpose of fluidics is to suspend the cells and guide them to the light source in a linear fashion. The excitation optics focus a laser beam onto a single-cell at a time; the emitted fluorescence or light scatter from the cell is gathered by the collection optics and is transmitted to the electronic network. The signals are then converted into digital data and are fed to a computer system, wherein they can be analyzed using various software (Shapiro & Telford, 2009; Wilkerson, 2012).

5.4 Components of a flow cytometer

5.4.1 Fluidic system

This component plays the role of a transportation medium through which the cells move through the instrument. It includes two compartments: Sheath fluid and pressurized lines. Commonly used sheath fluid is phosphate-buffered saline, which is injected into the flow chamber through pressurized lines. Cells suspended in the sample tube are also injected via a pressurized air-line. The sample stream flows centrally surrounded by the sheath fluid stream, known as a coaxial flow, achieved based on the pressure difference between the sheath and sample stream (Macey, 2007; Wilkerson, 2012). Since the sample pressure is higher than the sheath fluid, cells align in a single-file fashion while they pass through the laser beam, which ensures uniform illumination of each cell. This phenomenon is known as hydrodynamic focusing. The rate of flow can be controlled depending upon the analysis-based requirement (Adan et al., 2017) and is done by changing the pressure difference between the two previously mentioned streams. It must be ensured that the fluidics system is free of air bubbles and debris.

5.4.2 Optical system

The optics of a flow cytometer are composed of lasers, lenses (excitation optics), and collection optics. A laser produces a narrow, high-intensity beam of light, often in a single wavelength (color). In flow cytometry, lasers are crucial as they excite fluorescent dyes or proteins attached to cells, enabling the detection and analysis of various cell characteristics such as size, granularity, and protein expression. Several lasers are available for flow cytometry, depending on the type of fluorochromes being excited. A few examples are the argon laser (488 nm, blue), the helium-neon laser (633 nm, Red), and the violet laser (405 nm, Violet). Many instruments have an additional ultraviolet laser to excite UV-sensitive fluorochromes or the red diode for the far red (630 nm) range fluorochromes (Macey, 2007; Wilkerson, 2012). Older flow cytometers used gaseous state flow cytometers. However, the modern instruments use diode-based solid-state lasers, which are more compact, energy-efficient, and require less maintenance.

The path of these lasers is fitted with different kinds of lenses and prisms, which shape and focus the laser beam.

As the photons strike a cell, they deflect around the edges, known as light scattering. Two forms of scatter are identified, FSC and SSC. FSC is due to diffraction along the same axis as that of the laser beam and hence is proportional to cell surface area or size. The deflection of light is typically between 0 and 20 degrees. SSC, on the other hand, measures reflected and refracted light at 90 degrees to that of the laser beam, and is therefore proportional to the internal complexity of the cell. A combined measurement of FSC and SSC can help differentiate cells in a mixed population (Reggeti & Bienzle, 2011). Similarly, the fluorescence emitted by cells is typically detected at a 90-degree angle to the incident laser light. This minimizes interference from the laser's scattered light and allows the detector to capture the fluorescent light only.

Collection optics is composed of fiber-optic cables, which bring the scattered light from the cell-laser interaction, passing it through optical mirrors and filters to separate various wavelengths, and finally to the optical detectors (Reggeti & Bienzle, 2011). These detectors are made specific by placing appropriate filters, which could be long pass, short pass, and band pass types. The long pass transmits wavelengths of light equal to or longer than the specified range, whereas the short pass filters allow wavelengths of light equal to or shorter than the specified range. The bandpass filters allow a narrow range of wavelengths to pass through (Reggeti & Bienzle, 2011).

5.4.3 Signal detection and processing—Electronics

The specific fluorescent signals, as dissected out using different kinds of previously mentioned mirrors and filters, strike the detectors, which convert these light signals into electronic signals.

These photodetectors may be photodiodes (PDs), photomultipliers (PMTs), or avalanche photodiodes (APDs). PMT multiplies the photonic signals many-fold and thus has a greater sensitivity compared to PD. It is used to detect the weaker signals generated by SSC and fluorescence, while the PD is used to detect the stronger signals of FSC (Reggeti & Bienzle, 2011). APDs are relatively newer detectors used in flow cytometers. It also amplifies the signals with external amplification, but as compared to PMT, these have higher efficiency and are much smaller in size, making it easier to manufacture multiparametric instruments with smaller footprints.

Two types of amplifiers, linear and logarithmic, are responsible for the conversion of light signals into analog signals; furthermore, these are converted into digital signals by converters for computer-based processing. This digital data, using various available software, shall be displayed as scatter plots or histograms. Linear amplification is used if a limited dynamic range is required (for example, in DNA analysis) or for the FSC/SSC. Whereas, Logarithmic amplification is used for a broader dynamic range, e.g., in surface/cytoplasmic marker expression (Snow, 2004).

5.5 Fluorescence and fluorochrome

Any fluorescent compound has a fixed range of wavelengths at which it absorbs light energy. This causes the electrons to jump into their higher orbits (excited state); as they return to the ground state, excess energy is emitted as photons. This conversion of energy is fluorescence (Ormerod, 2008). The spectrum at which this compound is excited is known as the "excitation spectrum," while the spectrum at which it emits the fluorescence is known as the "emission spectrum." As a rule, the emission wavelength will always be longer than the excitation wavelength (Ormerod, 2008). This difference in the wavelength is the "Stokes shift," which is the determining feature of how strong the fluorochrome is. The higher the difference, the better the color separation using optical filters. Each fluorochrome has an optimal condition at which there is maximal absorbance and maximal emission wavelengths. A positive signal detection against a negative background makes this technique sensitive. In modern-day technology, multiparametric analysis of cells is made possible due to the simultaneous detection of up to 14 or more fluorochromes at different wavelengths (Ormerod, 2008).

In FCM-based immunophenotyping, the fluorochrome is like a reporter molecule of an antigen-antibody reaction, similar to the horse radish peroxidase (HRP) of immunohistochemistry (IHC) or a radioisotope of radioimmunoassay (RIA).

Different kinds of fluorescent particles are available in markets, which are being summarized in the following section:

Small organic molecules: These molecules have a consistent emission spectrum but with a small Stokes shift of approximately 50–100 nm (McKinnon, 2018). Some of these include fluorescein, Alexa Fluor 488, Texas Red, Alexa Fluor 647, and Cy5, which are commonly used for antibody conjugation. They are well known for their stability and resistance to photobleaching.

Phycobiliproteins: These are large proteins, for example, phycoerythrin (PE) weighing 240,000D, which are obtained from cyanobacteria, dinoflagellates, and algae. These exhibit a large Stokes shift and exceptional stability, making them widely applicable in quantitative flow cytometry. However, they are prone to photobleaching upon repeated exposure to excitation sources. Notable examples include allophycocyanin (APC) and peridinin chlorophyll protein (PerCP)

Quantum dots: These are semiconductor nanocrystals with narrow range emission spectra. UV or violet lasers optimally excite quantum dots, but excitation is suboptimal if multiple lasers are employed. These are difficult to conjugate and compensate for, and hence are now largely replaced in multiparametric analysis.

Polymer dyes: These are composed of polymer chains that collect light signals. Based on the length of polymer chains and molecular subunits, absorption and emission of light at specific wavelengths are possible. Hence, they work well in a multilaser excitation system. Examples of these include brilliant violet (BV), brilliant ultraviolet (BUV), and brilliant blue (BB).

Tandem dyes: These dyes chemically bind to phycobiliproteins or polymer dyes or small organic fluorochromes (Cy3, Cy5, and Cy7), creating a dye that works on the concept of fluorescence energy transfer (FRET) (McKinnon, 2018), thereby enhancing the availability of fluorochromes. For instance, in Texas Red and PE coupling, the emission range of PE is 585 nm, which excites Texas Red (maximum excitation-589 nm) via FRET, allowing the PE-Texas Red combination to be excited by a single laser. These have large Stoke shift values and hence are extremely bright, although stability is an issue.

Viability dyes: The viability of a cell is assessed using either the exclusion of dyes (Propidium iodide, DAPI) or by dye binding to intracellular components, indicating the integrity of the cell membrane. Several amine-binding dyes are available, such as live/dead reagents (ThermoFisher), Zombie dyes (Biolegend), or Fixable Viability dyes (BD Biosciences).

Quantum efficiency/yield: It is a measure of the brightness of the fluorophore and is determined by the efficiency of a fluorophore to convert the absorbed photon into the fluorescent signal. It is essentially the number of photons emitted to that of no of photons absorbed. Its value ranges from 0 to 1, where a value of "0" means no photon is emitted and "1" means that every photon absorbed is emitted. The higher the QE, the brighter the fluorophore is, and the higher the median fluorescent intensity (MFI).

Stain index (SI): It is a measure of the resolution of any fluorophore and the effectiveness of a fluorochrome in distinguishing positive and negative signals. It is measured as [(MFI of positive–MFI of negative)/2SD of negative population], where MFI is median fluorescence intensity, and SD is Standard deviation. Unlike just using MFI, SI factors in background noise, making it a more reliable metric for fluorochrome performance. It is to be noted that the SI is not only dependent on the fluorophore, but also on the antibody conjugate. Hence, the CD8-BV421 may have a different SI as compared to the CD20-BV421.

A good fluorochrome should have a

- high quantum efficiency, which is equivalent to brightness (no of photons emitted/no of photons absorbed), (Table 5.1)
- should have a high SI (Table 5.1)
- short lifetime, the decay rate is in nanoseconds
- easily detected by the machine and fit in your pane (excited by available lasers and emission captured by PMTs)

Table 5.1 Features of some of the common primary dyes.

Fluorophore	Excitation max	Emission max	Quantum efficiency	Stain index
Fluorescein isothiocyanate (FITC)	488	519	0.92	Low to moderate
Phycoerythrin (PE)	488	578	0.82	High
Peridinin chlorophyll protein (PerCP)	488	675	0.98	Moderate
Allophycocyanin (APC)	650	654	0.68	Moderate
BV421	405	421	0.66	High

- nontoxic
- do not require any special processing

5.6 Cell preparation techniques

The first step for a flow cytometric analysis is to ensure a single-cell suspension of the sample. The peripheral blood/bone marrow/body fluids are inherently in the fluid state and hence are the most common sample types. But, fine needle aspirate material or the fresh tissue from a biopsy specimen needs to be mechanically or enzymatically dissociated to make a single-cell suspension.

The next step depends on the type of cell that needs to be evaluated. For the leukocyte immunophenotyping, one of the most common uses of flow cytometry, we need to get rid of red blood cells and enrich the leukocytes. This can be attained by using Ficoll-Hypaque/the density gradient method, which enriches the mononuclear cells and removes the RBCs and granulocytes, or the RBC lysis protocols, which enrich all the leukocytes.

RBCs are lysed using an ammonium chloride-based lysing agent, which in turn can be used with or without fixatives. Based on the sequence of steps, the protocol can be named as stain-lyse-wash or the lyse-wash-stain method. Routinely, 100–200 μL of a whole blood sample is used, which has a final yield of ~0.5–1 million cells per tube. These many events are sufficient for the routine immunophenotyping analysis. However, for a rare event analysis and for measurable residual disease analysis, we need to acquire and analyze a higher number of events to a tune of ~2–5 million cells per tube. In these circumstances, we need a higher volume of whole blood/bone marrow sample, where RBCs are lysed first, and the WBCs are concentrated by spinning them down to small volumes, which is subsequently stained. This protocol of taking 0.5 mL or more of a sample and lysing RBCs for the preparation of cell pellets is known as bulk lysis.

The staining for the kappa and lambda needs special consideration during the cell processing. The plasma present in the sample normally has kappa/lambda, present either in intact immunoglobulins or in the free form. Even a small amount of plasma will have sufficient kappa/lambda light chains that can bind to all the antikappa/antilambda antibodies added during the staining, and the B cells will appear as negative. Hence, it is mandatory to remove all the plasma before adding antikappa/antilambda by washing it multiple times. Usually, three to four prior washes are required for the optimal antikappa/antilambda staining.

Body fluids, as such, may not need RBC lysing agents. But if they are hemorrhagic, we need to get rid of red cells. The whole sample (in case of CSF) or a portion of other body fluids (pleural/ascitic fluid) is centrifuged to form the cell pellet, which is washed after removing the supernatant and finally stained with the antibody cocktails. Stepwise cell preparation technique for the blood and bone marrow is mentioned as annexures toward the end of this chapter.

5.7 Antibodies used in flow cytometry

Almost all the antibodies used in flow cytometry are the primary fluorophore-labeled antibodies. Use of unlabeled primary antibodies is rare. These antibodies are mostly of the IgG subclass and are produced either through the hybridoma technique or using recombinant DNA technology. These antibodies are specific for the antigenic epitope against which they are developed. This specificity is enhanced by using a particular clone of antibodies. All these clones work similarly, but there is always a possibility that one clone may work better than the others. It is mandatory to read the literature of any commercial antibody before using it, in order to know its reactivity and expression pattern.

Proteolytically, the antibody can be cleaved into three fragments–two fragment antigen-binding (Fab) regions and a fragment crystallizable (Fc) region. The Fab includes the variable ends of an antibody that specifically bind to the antigenic epitopes, while the Fc determines the antibody's biological effector functions, such as antibody-dependent cytotoxicity and complement-dependent cytotoxicity. These Fc fragments bind to monocytes/B cells, etc. The desirable binding in flow cytometry is the Fab binding (targeted for specific epitope/antigen), while the Fc binding is the nondesirable binding and may lead to false signals. When the stain-lyse-wash protocol is used, there is the presence of some amount of plasma in the sample. The immunoglobulins present in this serum normally saturate the Fc binding sites on the monocytes/B cells, and the added specific antibodies do not find a place to bind to the Fc binding sites. But when the lyse stain wash protocol is used, the plasma is washed off during the initial washing steps. Now, the added specific antibodies may find a place on the Fc binding sites and may lead to nondesirable signals. Hence, there is a need to use some Fc blocking reagents, like Human BD Fc block reagents (BD Biosciences), to stop these nondesirable signals. The newer recombinant DNA technology can make the truncated antibodies with only the Fab component, or they may have mutated Fc components that do not have these nondesirable bindings and are free from background signals.

The selection of antibodies and their appropriate fluorochrome is the cornerstone of panel designing. This is based primarily on the laboratory's need and availability. Antibodies may have different names based on the manufacturer or the discovery. To avoid confusion, these have been given a unique cluster designation (CD) number like CD1, CD2, CD3, and so on. Currently, there are more than 370 CD markers that are directed against different mammalian epitopes. It is important to read the technical details when using an antibody, as the target epitope may differ in different clones. For example, CD3 has three commonly used clones: UCHT1, SK7, and HIT3. All bind to the epsilon portion of TCR. But the clone UCHT1 binds to both the extracellular and intracellular portions, while the HIT3 binds only to the extracellular component. Similarly, CD16 (Leu-11b) (clone GO22) is prepared in mice immunized with granulocytes. But CD16 (Leu-11c) (clone B73.1) is developed in mice immunized with NK lymphocytes. Hence, for an MDS panel, one should choose CD16 (Leu-11b) while for NK cell analysis, one should choose CD16 (Leu-11c).

Antibodies used in FCM are usually monoclonal antibodies, as these are more specific for one particular antigenic epitope. However, in certain situations, like antikappa and antilambda for the clonality assessment of B cells and CD38 for the gating of plasma cells for the multiple myeloma measurable residual disease analysis, the use of polyclonal/multiepitopic antibodies is better.

Antibodies validated for IHC/western blot (WB)/immunofluorescence (IF) may not work well with flow cytometry. This is due to the steric properties. In flow cytometry, the antibodies used are for the antigenic epitopes, which are in their native states with a multidimensional configuration, while other methods like IHC/WB/IF use the processed cells and hence the epitopes are in a linear configuration.

5.7.1 Panel designing

Designing a successful multicolor flow cytometry panel is an essential component for accurate and reliable data in flow cytometry. It is a complex process, requiring careful selection and optimization of fluorochromes, antibodies, and controls to ensure sensitivity, minimize spectral overlap, and allow for precise discrimination of cellular populations. The first and foremost step is to define the experimental goals, such as what exactly we want to see, which populations we will target, and what markers and types of expression (surface or intracellular) are expected. It is better to prioritize the most important markers so that high-performing fluorochromes may be assigned to them. The choice of fluorochrome should match the instrument configuration. In experiments with smaller panels (<6 fluorochromes), it is easier to assign the bright fluorochrome with a good SI across all the channels. But for larger panels, we may have to compromise on a few. As a general rule, a bright fluorochrome is assigned to the antibody that is either dimly expressed, seen in rarer populations, or is crucial for the characterization and discrimination of cells. It is advisable to avoid pairing fluorochromes with overlapping emission spectra as this increases the likelihood of spectral spillover and complicates data interpretation. It is also advisable to avoid using highly similar fluorochrome pairs on markers expected to coexpress on the same cells.

After initial panel design, testing is necessary to confirm performance. This step often involves titrating antibodies to find the optimal concentration, testing for nonspecific binding, and validating compensation settings. Adjusting antibody concentrations can minimize background noise, while optimizing compensation ensures that spillover correction is accurate without loss of data integrity. Optimal compensations and use of isotype and fluorescence minus one (FMO) controls are useful in determining gating boundaries for dim markers, especially in complex panels. Biological controls, such as unstimulated and stimulated cells, can also help validate assay specificity and sensitivity.

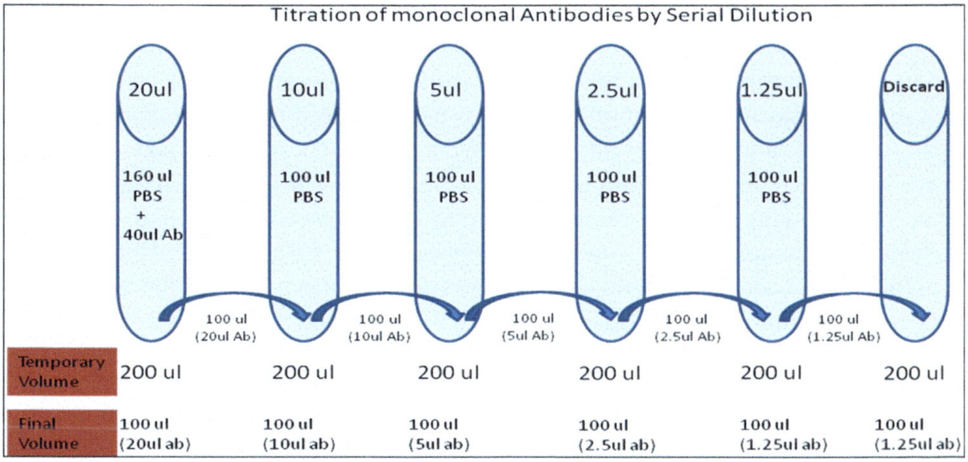

FIGURE 5.1 Titration experiment. Methodology of serial dilution for the titration of monoclonal antibody.

A well-designed flow cytometry panel is key to obtaining accurate, reproducible data, especially in high-dimensional, multicolor experiments. By carefully selecting fluorochromes, minimizing spectral overlap, using proper controls, and optimizing antibody concentrations, one can ensure optimum functioning of these panels and correct results.

5.7.2 Titration of antibodies

The antibodies supplied nowadays mention the recommended volume of antibodies to be used per test. Normally, 1 μg of antibody is enough to stain 10^8 cells with 10^4 antigens/cell. At typical cell concentrations of 10^6 cells, there is still a 10–100-fold excess of antibodies, and variations in cell number will not measurably affect the staining intensity. Nonsaturating stains lead to Volume-dependent and time-dependent staining intensities. Titration is usually needed for optimization of signal, minimization of background, and in resource-limited labs, mainly to conserve reagents. As the concentration of antibody is increased, the MFI of the cells for that particular fluorochrome increases. On further increasing the concentration, there is saturation of the positive signals, while the negative signal shows increasing spread. Hence, the separation between the positive and negative signals becomes poorer. Thus, the optimal concentration will be the one where the MFI of the positive population (signal) is well separated from the MFI of the negative population (noise), i.e., the best signal-to-noise (S/N) ratio (Maecker et al., 2004; Maecker & Trotter, 2009). This is achieved by using a serial dilution of antibodies in different tubes and acquiring them individually to see the positive signal and negative signals, and the S/N ratio (Fig. 5.1). The concentration at which this is best is taken as the final concentration.

5.7.3 Cocktail preparation

For any clinical laboratory that does flow cytometric immunophenotyping on a daily/routine basis, it is always advisable to prepare an antibody cocktail rather than using the antibodies individually in each tube. The cocktails are stable for weeks to months, if stored carefully. This avoids pipetting errors, saves time, saves money, and also excludes the possibility of missing one or more antibodies during staining. Based on the workload, a cocktail of 10–20 tests is prepared at one time point. The amount of antibody used is usually based on the titration experiments. Phosphate-buffered saline (PBS) is added to these antibody cocktails to make a volume that can be easily pipetted and is uniform throughout the panels. For example, if the total volume of antibodies for making 20 test cocktails adds up to 784 μL, we will add 216 μL of PBS so that the total volume becomes 1000 μL. Now, for staining each tube, we add 50 μL of this cocktail. The date of preparation of cocktails should be mentioned on the vial. A typical antibody cocktail used for B tube in B measurable residual disease (MRD)/B ALL is shown in Table 5.2.

5.8 Instrument setup

Before acquiring and analyzing the samples on a flow cytometer, it is of utmost importance to ensure that the instrument is optimally performing. This is required to ensure consistent analysis of samples. The main components of instrument setup are the alignment of the laser and detectors, setting up the laser delay, optimizing instrument detector settings, stabilizing the fluids, optimizing PMT voltages, spectral compensation, etc. The ultimate goal is the correct discrimination between different populations.

The photonic power of a laser needs to be directed and focused onto a very small spot. This spot can be spherical or elliptical. Given a stream velocity of 20 m/s, a beam spot of 20 μm, and a 15-μm cell, the cell is illuminated for only 0.015 microseconds. To maximize the likelihood that sufficient fluorescence events are produced by a labeled cell, and to best measure that fluorescence, it is necessary to bombard that cell with as many photons as possible. Hence, the laser should be aligned in such a way that it hits the cell optimally. Similarly, the detectors should also be placed correctly in the path of light so that an optimal detection takes place. The flow cytometer usually has a series of lasers that are spatially separated and aligned in such a way that they are focused on the sample stream and hit the cells at different time points (differing in milliseconds). The laser delay is set to ensure that the signals are coming from the same cell and not from different cells. The central laser (usually blue) is taken as a reference, and the delay for the laser that hits the cell earlier than the reference laser is in "minus/−" and the laser delay for the laser that hits the cells after the reference laser is in "plus/+" (Fig. 5.2).

5.8.1 Photomultipliers voltage setup

This voltage functions similarly to an audio-visual system amplifier. As we increase the amplification, the voice becomes louder, but after a certain point, unwanted background noise starts to appear. Similarly, as PMT voltages are increased, the positive signal also increases. However, beyond an optimal point, the negative signal may show false positivity due to signal spread. Therefore, the optimal baseline voltage is the one that maximizes the separation between negative and positive signals (the S/N ratio). The ultimate goal of PMT voltage optimization is to distinguish positive from negative populations and to keep all populations of interest on the scatter plot without making them off-scale.

Table 5.2 An example of antibody cocktail preparation along with their source, catalog number, and titrated volume. Phosphate-Buffered Saline is added to complete the volume so that a similar volume of different antibody cocktails may be added subsequently.

–	BV605	V500	BV421	FITC	PE	PerCP	PeCy7	APC	APCR700	APCH7
Antibody	CD20	CD45	CD123	CD86	CD73	CD34	CD19	CD10	CD304	CD38
Clone	2H7	HI30	9F5	2331	AD2	8G12	SJ25C1	HI10A	Neuropilin	HB7
Source	BD	BD	BD	BD	BD	BD	BD	BC	BD	BD
Cat no	563783	560777	562517	555657	550257	347222	557835	IM3699	566038	656646
Titrated volume	2 µL	1.5 µL	2 µL	2 µL	2 µL	8 µL	1.5 µL	2 µL	2.5 µL	2 µL
Final volumes	Total Ab per test = 25.5 µL		Total Ab for 20 test = 510 µL		Volume of PBS added = 490 µL		–	Final Volume of antibody cocktail per test is 50 µL		

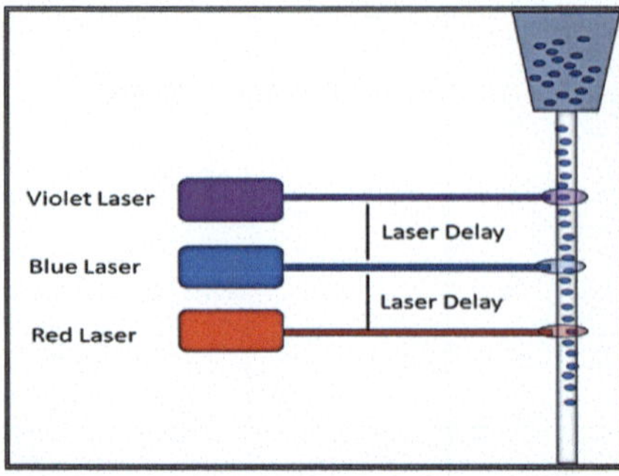

Laser	Position	Delay	
		Trigger on FSC	Trigger on Fluorescence
Blue	Second	0.00	0.00
Red	Third	35.34	35.36
Violet	First	-34.04	-34.07

FIGURE 5.2 Orientation of lasers in a typical flow cytometer and the laser delays. The central laser is taken as a reference, and the delay for the laser that hits the cell earlier and later than the reference laser is in "minus/−" and "plus/+" respectively.

In the modern-day cytometers, the laser alignment, detector alignment, laser delay, area scaling, PMT voltages, etc., are set up by using the setup beads. These beads are the polystyrene beads with different sizes and fluorescent intensities. The bead lots have a target MFI value, and during calibration, the equipment changes its properties to reach those target values, and this is known as baseline quality control (QC) or the characterization QC. Subsequently, it is required to run the same beads on a daily basis, during which the equipment detects the changes in PMT voltages (delta PMT) to achieve the same target values and the deviations from the target values, on a daily basis (CV). This result is known as performance QC/daily QC. These baseline PMT voltages are only the starting points and may not be optimal for every experiment. Some adjustments are needed based on the cell type, staining protocols, etc. The optimization requires a fully stained sample with a complete range of signals (negative, dim, and positive). Some tweaking in terms of lowering of PMT if the brightest signal goes off-scale, or the background is high, is permissible. However, it is advisable to avoid the PMT voltage changes and rather work on reducing the antibody titer, increasing the number of washes, and using Fc blocking reagents. The final PMT voltages for an experiment are then saved with a unique name, and a compensation experiment is carried out with the same voltage settings.

5.8.2 Spectral overlap and compensation setup

The conventional fluorophores that we use in flow cytometry do not have a sharp emission peak and are detected over a broad range of the spectrum. The detector is designed in a way that it detects a narrow band of the fluorochrome only and not the whole spectrum. This narrow band of spectrum is usually the maximum excitation profile of that particular fluorophore. There is always an overlap of this emitted spectrum between different fluorophores. This is termed as spectral overlap, and the mathematical way to correct it is known as compensation.

This is exemplified in the Fig. 5.3, where three fluorochromes (FL 1, FL 2, and FL 3) have been used, and there are three detector channels. Although the detector captures the specific and maximum emission spectrum from a particular fluorochrome, there is some amount of spectral emission that goes to different detector channels (the shaded area). As a result, the detector

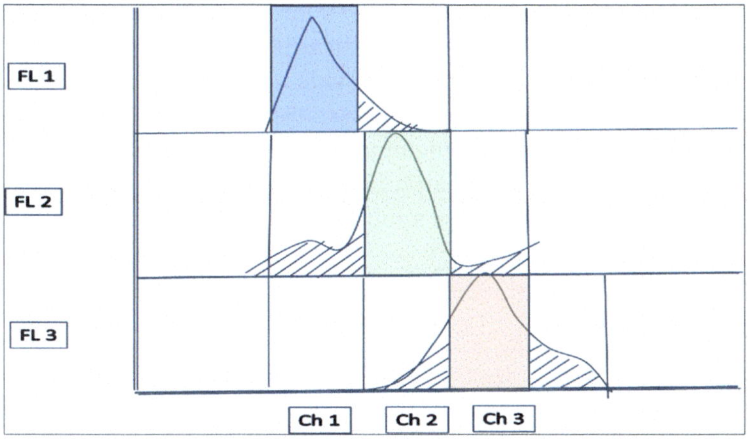

FIGURE 5.3 Spectral overlap. A rough sketch showing the spectral spillover of one fluorophore into another.

↓ Into- From →	FITC	PE	PerCP-Cy5.5	PE-Cy7	APC	APC-R700	APC-H7	BV421	BV510	BV605
FITC	100.00	0.66	0.04	0.01	0.03	2.04	0.10	0.12	0.57	0.00
PE	69.54	100.00	0.11	0.34	0.15	13.08	0.25	0.26	0.51	6.44
PerCP-Cy5.5	9.83	16.27	100.00	0.65	0.53	6.84	0.54	0.34	0.10	13.07
PE-Cy7	2.71	3.50	17.00	100.00	0.17	44.24	6.61	0.00	0.06	3.27
APC	4.00	0.04	10.58	0.00	100.00	100.20	6.82	0.99	0.01	0.98
APC-R700	0.41	0.00	3.01	0.00	4.80	100.00	0.92	0.00	0.01	0.06
APC-H7	2.00	0.01	0.95	2.42	2.69	53.18	100.00	1.00	0.01	0.03
BV421	0.00	0.36	0.00	0.00	0.41	0.01	0.02	100.00	2.27	10.00
BV510	15.16	0.46	0.00	0.00	0.21	19.12	0.39	60.94	100.00	1.46
BV605	0.83	2.96	0.05	0.03	0.04	0.04	0.14	1.70	6.79	100.00

FIGURE 5.4 Compensation matrix. Screenshot showing the compensation matrix with the spillover values (SOVs).

channel 2 (Ch 2) not only gets signals from the FL 2, but also a small amount of spectral emission from FL1 and FL3. To determine the actual signal from FL2 in Ch 2, it is necessary to subtract the proportion of signals coming from FL1 and FL3. This can be accomplished by running single-stained tubes, allowing the equipment to recognize that the expected signal should be in that specific channel, and any signal detected in other channels is undesirable and should be subtracted.

Compensation is done using the commercial beads or through cell-based methods. For cell-based compensation, a single-stained tube for each fluorochrome tagged to an antibody against a marker that is generally brightly expressed in a fraction of normal cells, along with some negative cells, is used. CD4 is one such marker that has positive signals (CD4+ T cells), negative signals (CD8+ T cells and B Lymphocytes), as well as dim positive monocytes. All individual stained tubes are run singly, and then a command is given to the software to calculate compensation. The software calculates the spillover values (SOVs) and generates the compensation matrix (Fig. 5.4).

Some basic rules should be followed for the compensation setup. These include: the fluorescence spectrum (% Spillover) of the compensation control reagent should be identical to the reagent (fluorochrome) to be used in the experiment. Hence, the compensation of FITC may not be similar to that of AF488, although both these fluorochromes are detected in the first blue channel. The positive population should be as bright as possible, as it ensures that the values in the spillover channel are above background and provide greater accuracy for calculating low (but real) spillovers. Sufficient events are needed to get a statistically accurate SOV. The complexity of compensation increases with the increasing number of fluorochromes and different antibody panels. The same 12 color compensation for a "B" tube may not be optimum for the "M" tube or "T" Tube. Hence, it is advisable to have a tube-specific and reagent-specific compensation matrix.

5.9 Analysis

Several methods have been developed and are commercially available. The utmost necessity is to simplify analysis. The main principle of data analysis is to selectively study the population of interest. This method is known as "gating." Several such regions can be defined on the same graph. Most commonly, we use FSC and SSC plots to gate. Different components based on their cell structure segregate out on the graph. Various kinds of gates are used in flow cytometry. These are used sequentially to refine the population of interest. These are mentioned in the following section:

- *Threshold or trigger gate*: Its role is to remove the background/Electronic noise. Usually, the FSC/SSC is taken as the threshold. If the size of the particle is too small, like bacteria, exosomes, etc, a fluorescent parameter is used as a threshold. The software excludes all the signals that are lower than the set threshold.
- *Flow stability or time gate:* Helps to capture events once the flow stream has stabilized and eliminates effects of clogging, back-pressure, sample exhaustion, air bubble aspiration, and other instrument issues.
- *Singlet gating:* FSC area versus FSC height. This helps in excluding the doublets.
- *Live gate:* FSC versus SSC. Helps in removing the debris, which usually has a very low FSC and higher SSC.
- *Subset gating*: Based on one or two markers that define a population, like CD45 vs SSC for leukocytes, CD19 vs SSC for B cells, CD138 vs CD38 for plasma cells, etc.
- *Dump gating*: When the objective is to exclude many different kinds of cells, then the antibodies specific to these unwanted cells are put together in one/same fluorochrome, and all the cells expressing those are excluded.

Fig. 5.5 shows various kinds of gates used in flow cytometric analysis. After a successful gating, the next job is to analyze the expression pattern of different antigens on that population of interest. This is done in terms of positivity or negativity and also by the degree of positivity (dim, moderate, or bright). Different kinds of positive and negative controls are used: Unstained negative control, isotype negative control, FMO controls, and the internal biological controls, which are present in the sample itself. A false positivity should also be kept in mind, which can be due to nondesirable Fc binding, increased background, suboptimal PMT voltages, or poor

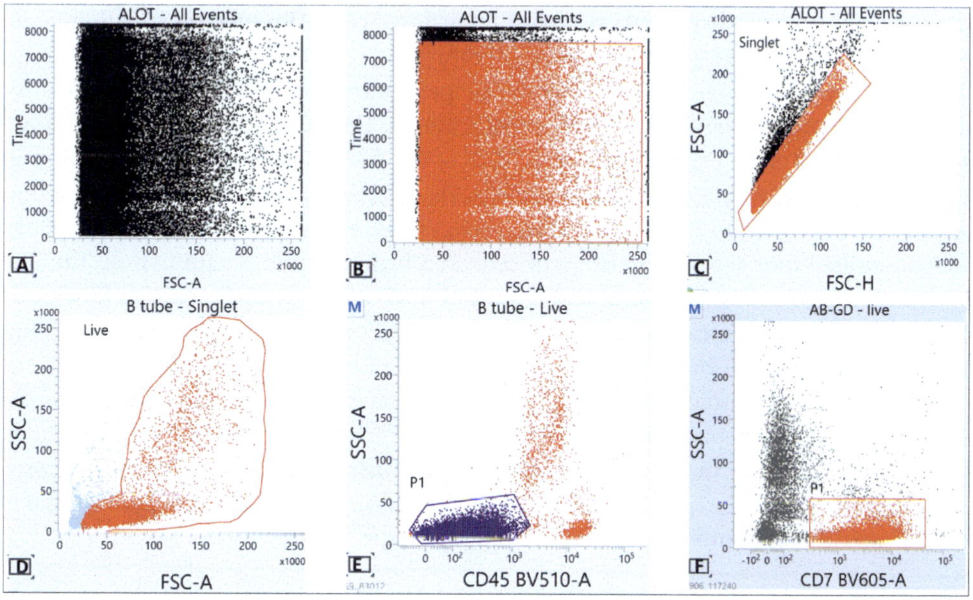

FIGURE 5.5 Scatter plots showing various gating strategies. (A) Ungated, (B) Time gate, (C) Singlet gate, (D) Live gate, (E and F) Subset gating.

compensations. The degree of positivity is usually expressed by comparing the cell of interest with the internal positive or negative controls. A simple MFI-based expression can be compared with the negative control population in the tube. These can be mentioned as dim, moderate, or bright if the difference in MFI of the positive and negative population is <1 log, 1–2 log, or >2 log, respectively.

5.9.1 Data representation

Different types of plots that are made to represent the data include histograms, scatter plots, contour plots, density plots, etc. Widely used are the Univariate histograms when a single parameter has to be visualized (Lugli et al., 2010). The expression pattern of a single marker can be assessed by graphs and expressed as a percentage of cells above a certain threshold; alternatively, the calculation of the mean (median) of fluorescence intensity shall be used. Overlay of histograms is made possible by most of the software, allowing robust comparison of the parameters of interest.

While analyzing two parameters, dot plots have been the most popular graphs. Every dot represents individual cells. Monochromatic dot plots have their limitations, such as a difficult distinction between cells with the same amount of fluorescence, especially in rare event analysis. Density gradient plots, such as contour or color dot plots, help overcome this issue (Lugli et al., 2009; Herzenberg et al., 2006). Pseudocolor dot plots are used in the FlowJo software, wherein different colors are used as per cell density, thus, different cell populations can be identified.

Routinely, flow data is displayed on a two-dimensional plot. Although it can rather be imagined in a multidimensional space wherein each dimension corresponds to a single fluorescent parameter or scatter. To make this feasible, principal component analysis (PCA) was used to reduce the dimensions; it generates a new set of variables (known as principal components) as linear combinations of the original variables. Most of the datasets are captured by the dominant principal component, which then acts as an axis. Use of PCA in flow cytometry was proposed initially in 1984 and again in 1987 (Mann et al., 1984; Kosugi et al., 1988). Kosugi et al. (1988) successfully identified four separate clusters of cells in a bivariate plot resulting from the PCA of cells, identified based on two parameters, FSC and SSC. With advancements in flow cytometry, high-dimensional data can be challenging to interpret visually. Dimensionality reduction techniques, such as t-distributed stochastic neighbor embedding (t-SNE) and uniform manifold approximation and projection (UMAP), are now widely used to visualize multiparameter data in two-dimensional space, allowing for clearer identification of cell clusters (Becht et al., 2019). These tools facilitate more nuanced interpretations of cellular heterogeneity and complex datasets. High-quality data representation enhances analysis precision and reliability, making it a cornerstone of flow cytometry research and diagnostics.

5.10 Applications

Flow cytometry has multidisciplinary applications with high sensitivity and accuracy. It is a rapid tool for various clinical applications. A comprehensive description is beyond the scope of this chapter, and it can be summarized in the following section:

Cell cycle analysis, proliferation assay, and apoptosis

Neoplastic hematologic disorders:

- Diagnosis and classification of hematologic neoplasms
- DNA ploidy analysis
- Monitoring (MRD assay)

Nonneoplastic hematologic disorders:

- PNH screening on leukocytes and RBCs
- Test for hereditary spherocytosis
- Osmotic fragility tests
- EMA binding test
- *G-6PD deficiency assay*: An extension of the MethHb reduction test
- Fetal RBCs in PB
- *Inherited platelet disorders*. Surface glycoprotein deficiency (in Bernard Soulier Syndrome and Glanzman Thrombasthenia)
- Platelet aggregation and activation studies

Immunological disorders: Inborn errors of immunity

Transplant immunology

- CD34 Enumeration
- CD3 Enumeration for Donor Lymphocyte Infusion
- HLA Cross match

Solid tumors

- Circulating tumor cells
- Tumor-infiltrating lymphocytes
- Identifying metastasis in effusion fluids and cerebrospinal fluid (CSF)
- Ploidy analysis.

Transfusion medicine

- QC of apheresis Products/Blood components

Cytokine analysis (CBA): quantification of a panel of cytokines
Clinical microbiology: Rapid classification of bacteria

Cell sorting

Immunophenotyping via flow cytometry (FCM) is often necessary to identify and characterize abnormal cell populations. FCM provides insights into the maturation patterns of normal cells, antigen expression, and abnormalities. Knowledge of normal leukocyte antigenic profiles is essential for accurate diagnosis. Normal cells exhibit antigen expression patterns based on their lineage and maturation stage, while abnormal cells may show antigen loss, overexpression, or lineage-inappropriate expression. In hematolymphoid neoplasms, FCM first determines whether abnormal cells are immature (precursors) or mature, a critical distinction in diagnosing leukemia (immature) or lymphoma (mature). Immature cells typically express CD34, bright CD99, nuclear TdT, or dim/negative CD45, while mature cells exhibit bright CD45 and are negative for CD34, CD99, and TdT. This differentiation is vital since treatments for immature and mature neoplasms differ significantly. Additionally, FCM assigns lineage to atypical cells based on specific antigens, aiding in the classification of neoplasms and guiding appropriate clinical management. For example, expression of MPO defines a myeloid lineage, bright CD19 along with bright expression of CD10, CD22, CD79A defines B lymphoid lineage, expression of cytoplasmic CD3 defines T cell lineage, etc. Similarly, for the clonality assessment of mature B cell neoplasm, expression of kappa and lambda, and for T cell clonality assessment, the TCR v Beta repertoire or TRBC1/2 expression is taken into account.

FCM is also used for confirmation of lymphomatous/leukemic involvement of cerebrospinal and other body fluids. The FNA material can also be subjected to FCM using the same set of antibody panels for the diagnosis and classification of hematolymphoid neoplasms. Paucity of cells in some body fluids, especially the CSF, makes it difficult to analyze correctly using the smears or the cell block sections. FCM plays a crucial role in this situation as it requires a small amount of sample and can analyze the expression of 8–12 antibodies in a single tube on a routine basis.

Expression of certain antigens also predicts prognosis; for example, CD56 and CD7 indicate poor outcomes in AML, and CD38 and ZAP70 are markers for poor prognosis in CLL. Targeted therapies are also guided by antigen expression, such as CD38 for daratumumab, CD20 for Rituximab, CD19 for Blinatuzumab, CD22 for ionotuzumab, CD52 for alemtuzumab etc. Panels for hematolymphoid neoplasm evaluation should include antibodies for diagnosis, prognosis, therapy, and disease monitoring.

FCM also analyzes immunophenotypes to identify a minuscule proportion of abnormal cells against a background of normal cells, the MRD. The aberrant immunophenotype known as leukemia-associated immunophenotype (LAIP) and difference from the normal population (DFN) are the two common approaches that are used to correctly identify the MRD. Using the standardized set of antibody panels to stain a sufficient number of cells on the first pull, marrow aspirate can detect the MRD to a sensitivity of 10^{-5} or better. MRD analysis is vital for hematologic neoplasms like AML, ALL (B and T cell), CLL, and MM, and it helps in the correct prediction of relapses and timely intervention in these patients.

Flow cytometric analysis of leukocytes, RBCs, and platelets can be done for the diagnosis of several nonneoplastic hematologic disorders. Screening for the presence of a GPI-deficient clone in leukocytes and RBCs helps in the diagnosis of PNH (Rahman et al., 2018, 2009). Assessment of residual red blood cell proportion after the lysis of red cells in distilled water using a standardized approach so that a similar number of RBCs are acquired over a certain period of time, and their comparison with the normal samples, helps in assessing the increased osmotic fragility of RBCs in cases with hereditary spherocytosis (Won & Suh, 2009). A better assessment would be using the Eosin 5 malemide dye to stain the RBC membrane, specifically the band 3 protein. A significant decrease in MFI of the FL1, the channel which captures the emission from EMA, denotes a loss of band 3 protein and is highly suggestive of HS (King et al., 2000). Another FCM method that extends the classic methemoglobin reduction test, assessing G6PD deficiency at the level of an individual erythrocyte, was described by Shah et al. (2012). There was strong concordance between genetic and biochemical techniques. The principle lies in the generation of ferryl HB (bright in color) or cyanmethhemoglobin (dull color) after the extended methhemoglobin reduction test in normal or G6PD-deficient individuals, respectively. Similarly, flow cytometric analysis can help in evaluating the platelet counts, diagnosis of Bernard Soulier syndrome and Glanzman thrombasthenia (loss of surface expression of CD42b and CD41/61; respectively), diagnosing platelet function defects (absence of platelet aggregation and absence of expression of markers associated with platelet activation following some agonists) (De Cuyper et al., 2013; Dave et al., 2022; Halliez et al., 2015).

Analysis of the different subsets of lymphoid cells in terms of their relative proportion and absolute count is one of the cornerstones for the screening and diagnosis of inborn errors of immunity. Other evaluations include the leukocyte adhesion defects, the dihydro-rhodamine assay (DHR), the activated T cell profile, and evaluation of sepsis. FCM has also been used regularly for the CD34+ stem cell enumeration during the HSC transplantation, as well as the HLA cross-match.

FCM also has a significant role in identifying metastases in serous cavity effusions. Commonly used antigenic markers for epithelial cancers include EpCAM, BerEP4, CD326, and MOC31. It is recommended to analyze a large volume of serous fluid effusion, prepare a cell pellet via centrifugation, and stain it with a panel of antibodies. A definitive result requires a clustered population of at least 10–20 events to rule out nonspecific binding.

FCM is also useful for pediatric round cell tumors, identifying distinct immunophenotypic profiles: Neuroblastoma (CD56hi/GD2+/CD81hi), primitive neuroectodermal tumors (CD271hi/CD99+), Wilms' tumor (CD56+, multiple populations), rhabdomyosarcoma (MYOD1+/myogenin+), germ cell tumors (CD56+/CD45−/NG2+/CD10+), and hemangiopericytomas (CD45−/CD34+). While definitive diagnosis requires histopathological and genetic analysis, FCM provides valuable differential diagnostic information, distinguishing these tumors from leukemia/lymphoblastic lymphoma, which requires entirely different treatment strategies. Thus, FCM complements traditional diagnostic methods in evaluating nonhematolymphoid tumors and metastatic conditions.

Antigen-specific responses can be measured by specific antigenic stimulation of cells and subsequent analysis of the cytokine response, proliferation, activation, memory, or antigenic recognition through MHC multimers. These MHC molecules are tagged with the desired antigen and are used to stimulate T cells, and the level of response is assessed (McKinnon, 2018). Applicable in vaccine-related studies. Intracellular cytokine analysis is another much-explored application. The cells are treated with transport inhibitors and stimulated by a suitable antigen during the incubation period, ensuring all the cytokines produced accumulate within the cells. Cells are then stained with viability markers and cell surface markers, fixed, and permeated with anticytokine antibodies. Immune response can therefore be assessed.

Flow-based detection of apoptosis is done by targeting several components of the cascade. Annexin V binds to phosphatidylserine once it translocates to the outer membrane during apoptosis. TdT dUTP nick end labeling (TUNEL) technique uses TdT to bind the ends of DNA breaks with dUTP, which is labeled with a fluorochrome for detection. Specific antibodies against the activated form of caspase 3 are available; intracellular binding helps detect the caspase signaling cascade. Antibody against APO2.7, located on the mitochondrial membrane, expressed exclusively during ongoing apoptosis, helps identify mitochondrial apoptosis (McKinnon, 2018).

Various methods are available to measure cell proliferation, for example, incorporation of thymidine analogs (BrdU) in replicating DNA, generation tracing of inheritable permanent dyes (CFSE), and analyzing proliferation-associated antigen expression (Ki67) (McKinnon, 2018).

Phases of the cell cycle can be assessed using flow cytometry. Cells are commonly fixed with 70% ethanol; however, live cells may also be stained using special dyes. Commonly used dyes are PI, 7AAD, DAPI, and Hoescht 33342. The events acquired are linearly amplified and analyzed using ploidy modeling software. Fluorescent proteins such as mCherry, mRuby, etc., along with a promoter sequence, all inside a plasmid, are used to transfect a cell. The expression of fluorescent protein is an indicator of gene expression. Various applications include tracking transplanted cells, bacterial or viral infections. Signal transduction studies allow for the identification of signaling pathways in a mixed population of cells using antibodies directed against signaling molecules.

Cells may be sorted based on single or multiple parameters. The flow cell sorters work on the principle of "electrostatic deflection of charged droplets." The cells travel through a nozzle,

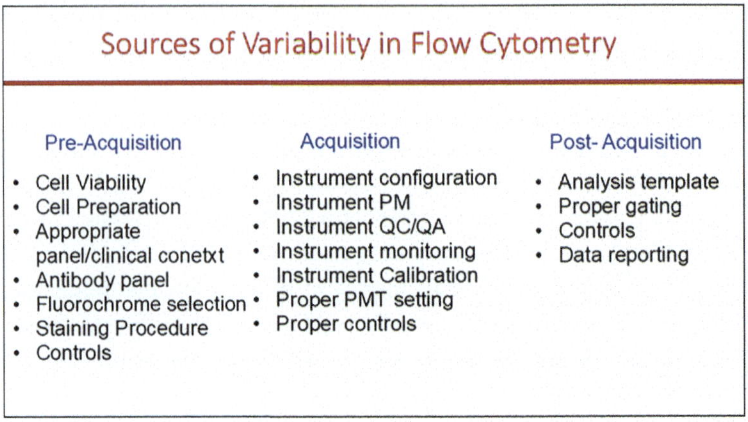

FIGURE 5.6 Sources of variation in flow cytometry. Sources of variation in flow cytometry.

forming a stream of regular droplets passing through laser beams, simultaneously getting charged by electrodes. Positive and negative charged droplets are deflected toward oppositely charged platinum plate; uncharged droplets are transferred to the waste container.

5.11 Quality assurance in flow cytometry

As in any other diagnostic process, quality assurance is of utmost importance for flow cytometry as well. It ensures that the correct sample is tested through the correct method on the correct instrument using the correct reagents to generate the correct result.

There can be several reasons for variation in flow cytometric results. This can be summarized later in the form of preanalytical/preacquisition, analytical/acquisitional, and postanalytical/postacquisitional variables (Fig. 5.6).

Robust QA practices help standardize and validate results across different laboratories, ensuring that patient outcomes are consistent and reliable. The very first step in this process is the preparation of a robust standard operating procedure (SOPs). This is vital to maintaining uniformity in sample collection, preparation, and data analysis. SOPs include guidelines on sample handling, reagent use, instrument setup, and calibration. Standardized procedures reduce variability in results, making cross-laboratory comparisons possible and more reliable (Keeney et al., 2012).

Instruments must be regularly calibrated and maintained to ensure accurate performance. This is usually achieved by running the QC beads. There are different types of QC beads with the assigned target values, based on which the equipment is calibrated and performance checked. Regular maintenance helps prevent mechanical or electronic faults, reducing downtime and the risk of inaccurate results (Perfetto, Chattopadhyay, & Lamoreaux, 2006).

The use of high-quality and stable reagents is essential to ensure the reproducibility of results. Reagents, including antibodies, must be checked for lot-to-lot consistency, stored under recommended conditions, and tested periodically to confirm they work as expected. Reagent quality impacts staining quality and, consequently, data accuracy (Stelzer et al., 1997).

Ultimately, the analysis and interpretation of data by trained personnel following rigorous and standardized protocols are important to generate fruitful results. Proficiency testing and interlaboratory comparisons can further validate data interpretation consistency (Cossarizza, Chang, & Radbruch, 2017).

This helps laboratories benchmark their performance against other facilities. EQA programs provide unidentified samples for analysis, with results compared across multiple labs.

5.12 Recent advances

There has been continuous development in all aspects of clinical flow cytometry, right from the sample collection and transportation solutions to the final analysis and result generation. Newer solutions for the preparation of a single-cell suspension from fresh tissue using mechanical grinding and enzymatic breakdown are available on board. Automation in the sample preparation, cocktail preparation, and staining has led to some interesting developments. Development of newer dyes/fluorophores and robust antibodies using recombinant DNA technology is easily available. Droplet microfluidics is a relatively new field that focuses on the formation, manipulation, and analysis of discrete droplets containing cells or DNA. One additional advantage of microfluidics systems over classical flow cytometers is the utilization of single-use fluidic and identification of minute particles like exosomes, bacteria, etc. The small footprint of the flow cytometer is due to the advancement of better solid-state lasers, and small detectors like APDs are easily available nowadays. Overall, the equipment now has a high acquisition rate, with better CVs, reduced carryovers, and better resolutions. These are now integrated with different loader options and automated processing systems. The development of a spectral cytometer with the capability of analyzing the whole spectral signature of the fluorophore, rather than capturing the narrow band of emission max, has really changed the paradigm. Now the same available reagents with overlapping spectral features can be dissected out easily for a 30–40 color experiment. Better and automated ways of spectral unmixing have made it easier to use in clinics. The development of cell sorters integrated with image analyzers has made it possible to see the cells while sorting them. These image-integrated cell sorters can sort 4–6 types of cells with good accuracy and real-time morphological evaluation.

Another sea of change has been in terms of advancement in data science and informatics that helps in analyzing the multidimensional complex data by reducing the dimension and population clustering.

Integration of artificial intelligence (AI) has gained an enormous amount of recognition and is expected to revolutionize data analysis and interpretation (Liechti et al., 2021; O'Neill et al., 2013; Rahim et al., 2018). Data visualization, which aims at data reduction, has come up very well. Automated gating and cell population identification in a supervised or unsupervised setup are being built, which pick up expert-defined gates and apply them across large datasets. This improves efficacy and reduces human bias. Integration of multiomics into the flow data with the help of AI tools is providing a more comprehensive understanding of cell function and disease nowadays.

5.13 Summary

To summarize, flow cytometry is a powerful, versatile technology widely used in biology and medicine to analyze the physical and chemical characteristics of cells or particles in a fluid as they pass through a laser beam. The fundamental principle involves suspending cells in a stream of fluid and passing them one by one through a series of lasers and detectors. The way each cell scatters light or emits fluorescence (from labeled antibodies) reveals information on size, complexity, and various molecular markers, making it valuable for immunology, oncology, and diagnostics. There is a wide variety of clinical and research applications of flow cytometry. This is one of the most important and accurate techniques for the rapid diagnosis and monitoring of various disorders. Development, especially in the form of spectral cytometry, high-dimensional data analysis tools, and integration of AI, is paving its path further.

5.14 Annexures

5.14.1 Sample preparation for surface staining using stain-lyse-wash protocol

1. *Equipment and reagents*
    - Benchtop flowcytometer
    - Swing-out table-top centrifuge
    - Vortex mixer
    - Plastic racks for glass tubes
    - 12 x 75 mm round-bottom tubes for assays and for acquisition
    - Micropipettes: adjustable with different volume ranges 0.2–2.0 μL, 0.5–10 μL, 2–20 μL, 10–100 μL, 20–200 μL, 100–1000 μL, and microtips
    - Low retention pipette tips
    - Phosphate buffer saline (PBS) pH 7.2
    - Erythrocyte lysing reagent-Commercial kits or Home-brewed reagent-ammonium chloride-based
    - 0.5% paraformaldehyde
    - Tissue paper
    - Fluorescent-labeled antibodies
    - Label sticker
    - Marker pens
2. **Procedure: Sample preparation**
    1. Label the tube with the corresponding lab number, along with the panel name for which staining is to be done.
    2. Take 100 μL patient whole blood/bone marrow (~1 million Leukocytes) in 12 x 75 mm round-bottom polystyrene tubes.

3. Add the titrated amount of the antibody cocktail or individual antibodies as per the panel requested.
4. Vortex the tube gently.
5. Incubate in dark at RT for ~20 min.
6. Add 2ml RBC lysing solution.
7. Vortex the tube.
8. Incubate in dark at RT for 10–15 min.
9. Spin down the cells at 2000 RPM for 5 min.
10. Aspirate/discard the supernatant.
11. Resuspend the pellet by gentle flicking.
12. Wash cells once/twice with PBS by adding 2mL of PBS (steps from 8 to 10), vortex gently.
13. Finally, resuspend the cells in 0.5mL of 0.5% paraformaldehyde.
14. Run on a precalibrated flow cytometer using the templates as per the panel used.

5.14.2 Sample preparation for intracellular staining using the stain-lyse-wash protocol

1. *Equipment and reagents*: Same as discussed earlier
2. *Procedure:* Sample preparation
3. Label the tube with the corresponding lab number, along with the panel name for which staining is to be done.
4. Take 100 μL patient whole blood/bone marrow (~1 million Leukocytes) in 12 x 75 mm round-bottom polystyrene tubes
5. Add a titrated amount of the antibody cocktail or individual antibodies as per the panel requested (Only antibodies meant for surface staining are to be added at this point)
6. Vortex the tube gently
7. Incubate in dark at RT for ~20 min
8. Add 2mL RBC lysing solution
9. Vortex the tube
10. Incubate in dark at RT for 10–15 min
11. Spin down the cells at 2000 RPM for 5 min
12. Aspirate/discard the supernatant
13. Resuspend the pellet by gentle flicking
14. Wash the cells once using 2 mL of PBS
15. Add 500uL FACSTM permeabilizing solution (Perm 2) working (1X) Cat. No. 340973, BD Biosciences or a similar reagent, and incubate for 8–10 min
16. Wash the cells once using 2 mL of PBS
17. Add the titrated volume of antibodies/antibody cocktail
18. Vortex gently and incubate in the dark for 15–20 min
19. Wash cells once/twice with PBS by adding 2 mL of PBS (steps from 8 to 10), vortex gently
20. Finally, resuspend the cells in 0.5 mL of 0.5% paraformaldehyde

5.14.3 Sample preparation for bulk lyse stain wash protocol for measurable residual disease analysis

1. White blood cell count is determined on the CBC analyzer, or a rough estimate is made on the smear prepared. It is to be noted that to avoid the blockage of the nozzle of the CBC analyzer by the marrow particle, passing it through a 25-gauge needle syringe may be needed.
2. An appropriate amount of sample containing at least twice the number of cells targeted during an MRD analysis is taken (this is to compensate for the cell loss during processing).
3. This volume of sample (blood/bone marrow) is taken in a 20 mL conical tube.
4. A maximum of 1 mL sample is taken per tube. If more than that is needed, multiple tubes are used.
5. Fixative-free ammonium chloride-based RBC lysing agent is added to the tube (~15–20 times the volume of blood) and incubated for 15 min at room temperature after mixing it well.
6. After the incubation, it is centrifuged at 800 g for 5 min, and the supernatant is removed without disturbing the pellet. This is done either using a pipette or through gentle decanting. Typically, 100–300 μL of cell suspension remains in the tube.
7. 2 mL of PBS/wash buffer is added to the pellet, and the sample is vortexed. An additional PBS/Wash buffer is added to bring the volume to 20 mL.
8. It is mixed well by inverting the tube 3–5 times.
9. The tube is centrifuged at 800 g for 5 min, and the supernatant is removed without disturbing the cell pellet.
10. The pellet is resuspended in 2 mL of PBS/wash buffer and mixed well with gentle vortexing.
11. The whole volume is pipetted out and transferred to the 5 mL round-bottom polystyrene tube.
12. The conical tube is washed again with 2 mL PBS to recover any leftover cells, and this is again transferred to a 5 mL polystyrene round-bottom tube.
13. The sample in the 5 mL tube is mixed by gentle vortexing.
14. The final count is taken on a CBC analyzer. This will allow us to determine the number of cells present in that tube.
15. The sample is centrifuged at 540 g for 5 min, and the supernatant is removed to leave behind ~200 μL of pellet suspension.
16. Appropriate volume of this pellet is taken to stain the appropriate number of cells as per the MRD panel (typically 2–5 million cells).
17. Add the titrated amount of the antibody cocktail or individual antibodies as per the panel requested (Only antibodies meant for surface staining are to be added at this point).
18. Vortex the tube gently.
19. Incubate in dark at RT for ~20 min.
20. Resuspend the pellet by gentle flicking.
21. Wash the cells once using 2 mL of PBS.

22. Add 500 uL FACSTM permeabilizing solution (Perm 2) working (1X) Cat. No. 340973 BD Biosciences or a similar reagent and incubate for 8–10 min.
23. Wash the cells once using 2 mL of PBS.
24. Add the titrated volume of antibodies/antibody cocktail.
25. Vortex gently and incubate in the dark for 15–20 min.
26. Wash cells once/twice with PBS by adding 2 mL of PBS (steps from 8 to 10), vortex gently.
27. Finally, resuspend the cells in 0.5 mL of 0.5% paraformaldehyde.

References

Adan, A., Alizada, G., Kiraz, Y., Baran, Y., & Nalbant, A. (2017). Flow cytometry: Basic principles and applications. *Critical Reviews in Biotechnology, 37*(2), 163–176. doi:10.3109/07388551.2015.1128876.

Becht, E., McInnes, L., Healy, J., Dutertre, C.-A., Kwok, I. W. H., Ng, L. G., Ginhoux, F., & Newell, E. W. (2019). Dimensionality reduction for visualizing single-cell data using UMAP. *Nature Biotechnology, 37*(1), 38–44. doi:10.1038/nbt.4314.

Bonner, W. A., Hulett, H. R., Sweet, R. G., & Herzenberg, L. A. (1972). Fluorescence activated cell sorting. *Review of Scientific Instruments, 43*(3), 404–409. doi:10.1063/1.1685647.

Brecher, G., Schneiderman, M., & Williams, G. Z. (1956). Evaluation of electronic red blood cell counter. *American Journal of Clinical Pathology, 26*(12), 1439–1449. doi:10.1093/ajcp/26.12.1439.

Coons, A. H., Creech, H. J., Jones, R. N., & Berliner, E. (1942). The demonstration of pneumococcal antigen in tissues by the use of fluorescent antibody. *Journal of Immunology, 45*, 159–170.

Coons, A. H., & Kaplan, M. H. (1950). Localization of antigen in tissue cells. *Journal of Experimental Medicine, 91*(1), 1–13. doi:10.1084/jem.91.1.1.

Cossarizza, A., Chang, H. D., & Radbruch, A. (2017). Standardization guidelines for flow cytometry. *European Journal of Immunology, 47*(10), 1584–1797.

Dave, R. G., Geevar, T., Chellaiya, G. K., Mammen, J. J., Vijayan, R., Samuel, A., Gowri, M., & Nair, S. C. (2022). Stability and utility of flow cytometric platelet activation tests: A modality to bridge the gap between diagnostic demand and supply. *Platelets, 33*(7), 1043–1051. doi:10.1080/09537104.2022.2042232.

De Cuyper, I. M., Meinders, M., Van De Vijver, E., De Korte, D., Porcelijn, L., De Haas, M., Eble, J. A., Seeger, K., Rutella, S., Pagliara, D., Kuijpers, T. W., Verhoeven, A. J., Van Den Berg, T. K., & Gutíerrez, L. (2013). A novel flow cytometry-based platelet aggregation assay. *Blood, 121*(10), e70. doi:10.1182/blood-2012-06-437723.

Halliez, M., Fouassier, M., Robillard, N., Ternisien, C., Sigaud, M., Trossaert, M., & Bene, M. C. (2015). Detection of phosphatidyl serine on activated platelets' surface by flow cytometry in whole blood: A simpler test for the diagnosis of Scott syndrome. *British Journal of Haematology, 171*(2), 290–292. doi:10.1111/bjh.13391France.

Herzenberg, L. A., Tung, J., Moore, W. A., Herzenberg, L. A., & Parks, D. R. (2006). Interpreting flow cytometry data: A guide for the perplexed. *Nature Immunology, 7*(7), 681–685. doi:10.1038/ni0706-681.

Purvis, N.B., Jr. and Oldaker, T. (2012). Validation and Quality Control in Clinical Flow Cytometry. In Laboratory Hematology Practice (eds. K. Kottke-Marchant and B.H. Davis). https://doi.org/10.1002/9781444398595.ch11.

King, M. J., Behrens, J., Rogers, C., Flynn, C., Greenwood, D., & Chambers, K. (2000). Rapid flow cytometric test for the diagnosis of membrane cytoskeleton-associated haemolytic anaemia. *British Journal of Haematology, 111*(3), 924–933. doi:10.1046/j.1365-2141.2000.02416.x.

Kosugi, Y., Sato, R., Genka, S., Shitara, N., & Takakura, K. (1988). An interactive multivariate analysis of FCM data. *Cytometry, 9*(4), 405–408. doi:10.1002/cyto.990090419.

Liechti, T., Weber, L. M., Ashhurst, T. M., Stanley, N., Prlic, M., Van Gassen, S., & Mair, F. (2021). An updated guide for the perplexed: Cytometry in the high-dimensional era. *Nature Immunology, 22*(10), 1190–1197. doi:10.1038/s41590-021-01006-z.

Lugli, E., Roederer, M., & Cossarizza, A. (2010). Data analysis in flow cytometry: The future just started. *Cytometry Part A, 77*(7), 705–713. doi:10.1002/cyto.a.20901.

Lugli, E., Troiano, L., & Cossarizza, A. (2009). Investigating T cells by polychromatic flow cytometry. *Springer Science and Business Media LLC, 514*, 47–63. doi:10.1007/978-1-60327-527-9_5.

Macey, M. G. (2007). Principles of flow cytometry. In *Flow cytometry: Principles and applications* (pp. 1–5). Humana Press.

Maecker, H., & Trotter, J. (2009). Selecting reagents for multicolor BD flow cytometry. *Postepy Biochemii, 55*(4), 461–467.

Maecker, H. T., Frey, T., Nomura, L. E., & Trotter, J. (2004). Selecting fluorochrome conjugates for maximum sensitivity. *Cytometry Part A, 62*(2), 169–173. doi:10.1002/cyto.a.20092.

Mann, R. C., Popp, D. M., & Hand, R. E. (1984). The use of projections for dimensionality reduction of flow cytometric data. *Cytometry, 5*(3), 304–307. doi:10.1002/cyto.990050314.

McKinnon, K. M. (2018). Flow cytometry: An overview. *Current Protocols in Immunology, 120*(1). doi:10.1002/cpim.40.

Moldavan, A. (1934). Photo-electric technique for the counting of microscopical cells. *Science, 80*(2069), 188–189. doi:10.1126/science.80.2069.188.

O'Neill, K., Aghaeepour, N., Špidlen, J., Brinkman, R., & Wodak, S. (2013). Flow cytometry bioinformatics. *PLoS Computational Biology, 9*(12), e1003365. doi:10.1371/journal.pcbi.1003365.

Ormerod, M. G. (2008). *Flow cytometry: A basic introduction (free online-version*. Michael G. Ormerod.

Perfetto, S.P., Chattopadhyay, P.K., & Lamoreaux, L. (2006). Quality assurance in flow cytometry. Nat Protoc. 1(3), 1552-30.

Picot, J., Guerin, C. L., Le Van Kim, C., & Boulanger, C. M. (2012). Flow cytometry: Retrospective, fundamentals and recent instrumentation. *Cytotechnology, 64*(2), 109–130. doi:10.1007/s10616-011-9415-0.

Rahim, A., Meskas, J., Drissler, S., Yue, A., Lorenc, A., Laing, A., Saran, N., White, J., Abeler-Dörner, L., Hayday, A., & Brinkman, R. R. (2018). High throughput automated analysis of big flow cytometry data. *Methods, 134-135*, 164–176. doi:10.1016/j.ymeth.2017.12.015.

Rahman, K., Gupta, R., Hussein, N., Yadav, G., Singh, M. K, Kumar, P., & Nityanand, S. (2009). Fluorescent aerolysin (FLAER) based parox-ysmal nocturnal hemoglobinuria (PNH) screening—A single centre experience from India. *International Journal of Laboratory Hematology, 39*(3), 135–141.

Rahman, K., Gupta, R., Harankhedkar, S., Gupta, T., Sarkar, M. K., & Nityanand, S. (2018). Utility of CD157 as a common leukocytes marker for paroxysmal nocturnal hemoglobinuria screening in a single tube five color combination. *Indian Journal of Hematology and Blood Transfusion, 34*(2), 304–309. doi:10.1007/s12288-017-0867-z.

Reggeti, F., & Bienzle, D. (2011). Flow cytometry in veterinary oncology. *Veterinary Pathology, 48*(1), 223–235. doi:10.1177/0300985810379435.

Shah, S. S., Diakite, S. A. S., Traore, K., Diakite, M., Kwiatkowski, D. P., Rockett, K. A., Wellems, T. E., & Fairhurst, R. M. (2012). A novel cytofluorometric assay for the detection and quantification of glucose-6-phosphate dehydrogenase deficiency. *Scientific Reports, 2*. doi:10.1038/srep00299.

Shapiro, H. M, & Telford, W. G. (2009). Lasers for flow cytometry. *Current Protocols in Cytometry, 49*, 1–9.

Snow, C. (2004). Flow cytometer electronics. *Cytometry Part A, 57*(2), 63–69. doi:10.1002/cyto.a.10120.

Steinkamp, J. A., Fulwyler, M. J., Coulter, J. R., Hiebert, R. D., Horney, J. L., & Mullaney, P. F. (1973). A new multiparameter separator for microscopic particles and biological cells. *Review of Scientific Instruments, 44*(9), 1301–1310. doi:10.1063/1.1686375.

Stelzer, G. T., Marti, G., Hurley, A., McCoy, P., Jr., Lovett, E. J., & Schwartz, A. (1997). U.S.-Canadian consensus recommendations on the immunophenotypic analysis of hematologic neoplasia by flow cytometry: standardization and validation of laboratory procedures. *Cytometry, 30*, 214–230.

Wilkerson, M. J. (2012). Principles and applications of flow cytometry and cell sorting in companion animal medicine. *Veterinary Clinics of North America - Small Animal Practice, 42*(1), 53–71. doi:10.1016/j.cvsm.2011.09.012.

Won, D. I., & Suh, J. S. (2009). Flow cytometric detection of erythrocyte osmotic fragility. *Cytometry Part B - Clinical Cytometry, 76*(2), 135–141. doi:10.1002/cyto.b.20448.

6

Flow cytometry in effusions: Lessons from the literature

Namrata P. Awasthi and Gauri Gaur

DEPARTMENT OF PATHOLOGY, DR. RAM MANOHAR LOHIA INSTITUTE OF MEDICAL SCIENCES (DR. RMLIMS), LUCKNOW, UTTAR PRADESH, INDIA

6.1 Introduction: Clinical significance of analyzing serous effusions

Serous cavities represent specialized potential spaces within the human body, specifically the pleural cavity surrounding the lungs, the pericardial cavity enclosing the heart, and the peritoneal cavity lining the abdominal organs. These cavities are uniquely characterized by their lining of mesothelial cells supported by vascular connective tissue, which plays a pivotal role in their physiological function. The primary function of these membranes is the continuous secretion of a small, lubricating volume of serous fluid. This fluid is essential for facilitating frictionless movement of vital organs within their respective compartments, such as the rhythmic expansion and contraction of the lungs during respiration, the constant contractile cycles of the heart, and the peristaltic movements of the intestines. The formation of serous fluid entails the passage of this fluid from capillaries in the serous membranes into connective tissue. This fluid enters the serous cavity via cell junctions and the cytoplasm of the pinocytic vesicles of the mesothelial cells (Krausz & Barker, 2003).

Under normal conditions, each serous cavity is a potential cavity, usually collapsed and containing only a small amount of fluid to allow movement of the underlying organs. An abnormal accumulation of fluid within these typically minimal serous spaces is termed a serous effusion. The very presence of an effusion signifies a deviation from the normal physiological state and invariably points to an underlying pathological process that demands thorough and immediate investigation to ascertain its cause and clinical implications. The inherent nature of these cavities as "potential spaces," designed for minimal fluid, means that any significant abnormal fluid accumulation can rapidly compromise the function of vital organs. For instance, in the largely nondistensible pericardial and pleural cavities, even moderate fluid volumes can exert considerable pressure on the heart, potentially leading to life-threatening cardiac tamponade, or on the lungs, causing severe respiratory distress. This direct link between fluid volume and organ dysfunction underscores why effusion analysis is frequently not just a diagnostic procedure but an urgent clinical intervention, often requiring rapid diagnosis and therapeutic drainage

Advancing Body Fluid Diagnostics with Flow Cytometry: A Comprehensive Guide to Effusions and CSF.
DOI: https://doi.org/10.1016/B978-0-443-27426-8.00006-5

to alleviate symptoms and prevent life-threatening complications. Infection, inflammation, and malignancy can all cause an increase in the production of this serous fluid.

The analysis of serous effusion fluids to determine the underlying nature of a disease has a rich diagnostic history grounded in both cytomorphology and biochemistry. Microscopic examination of body fluids dates back to the 19th century; investigators Lucke and Klebs (1867) are credited as the first to recognize malignant cells in ascitic fluid, which was followed by Quincke's (1882) detailed descriptions of lung and ovarian cancer cells in serous effusions (Nguyen, 2010). Since then, routine serous effusion cytology has remained a minimally invasive cornerstone for differentiating benign, reactive, and malignant conditions, heavily relying on conventional stains like Papanicolaou (Pap) and May-Grünwald Giemsa (MGG) to evaluate cellular architecture. Biochemically, the evaluation of effusions was completely revolutionized when Light et al. (1972) published their landmark diagnostic criteria. By measuring the ratios of pleural fluid to serum protein and lactate dehydrogenase (LDH), Light's criteria provided a highly sensitive, standardized method to classify fluids into transudates and exudates based on their biochemical profiles (Light et al., 1972). This distinction is critical for determining the pathophysiological mechanism of the fluid accumulation: transudates generally reflect systemic pressure imbalances (such as congestive heart failure), whereas exudates are caused by local disease processes that trigger increased capillary permeability, such as pleural infection or malignancy. Together, this historical integration of microscopic cellular visualization (cytology) and pathophysiological profiling (biochemistry) formed the traditional diagnostic foundation for effusion analysis. It is precisely the limitations of these two historical pillars—such as the morphological overlap between reactive mesothelial cells and malignant cells in cytology—that set the stage for advanced techniques like flow cytometry (FCM).

The pleural, pericardial, and peritoneal cavities are frequently involved in epithelial malignancies that cause symptomatic effusions, and these effusions are most commonly caused by metastasis from ovarian, breast, lung, stomach, and hepatobiliary malignancies (Azar et al., 2011). A false-positive result may result in unnecessary surgical treatment, while a false-negative result may prevent a necessary surgery and cause the postponement of otherwise indicated medical treatments.

6.2 Conventional cytology and ancillary techniques for effusion fluid analysis

Traditionally, cytological examination of effusion fluids is considered the gold standard for detecting malignancy. Cytomorphology involves visually inspecting stained smears or cell blocks (CBs) to identify cellular features consistent with malignancy: nuclear enlargement, prominent nucleoli, high nuclear-cytoplasmic ratio, irregular nuclear contours, and abnormal mitosis are among the typical findings (Motherby et al., 1999b; Pillai et al., 2013; Thunnissen et al., 1993).

The specificity of cytology is high, with reported rates between 91% and 100%, averaging 97%. Sensitivity, however, demonstrates a wide range—22% to 81.2%, averaging 58.2%—reflecting several intrinsic challenges (Pillai et al., 2013). Recent comprehensive meta-analyses involving over 19,000 cases have confirmed these performance characteristics, with the International System for Reporting Serous Fluid Cytopathology (ISRSFC) providing standardized diagnostic categories that demonstrate sensitivities of 57% to 77% depending on the diagnostic threshold applied (Ahuja et al., 2024).

False negative rates in cytology arise from issues such as sampling error (malignant cells may be sparse, unevenly distributed, or lost during processing), interpretational variability, and technical limitations. Morphological overlap between reactive mesothelial cells and malignant epithelial cells remains a persistent challenge. Cytology has a limited role in hemorrhagic fluids, mucinous fluids, coagulum-containing fluids, paucicellular smears, or smears with abundant inflammatory aggregates that mask the malignant cells or cells of interest (Gabriel et al., 2004). Before diagnosing a case as malignant, sufficient evidence of malignancy is necessary. Therefore, in uncertain cases, it is not advisable to rely solely on cytomorphology to report a case as positive for malignancy. An ancillary workup is recommended to substantiate a benign or malignant diagnosis (Mohanty & Dey, 2003; Shidham & Falzon, 2010).

6.3 Ancillary techniques to augment cytological diagnosis

To overcome the diagnostic challenges of cytospreading examination, ancillary techniques have become integral. These comprise immunocytochemistry (ICC), CB preparation, automated image morphometry, DNA ploidy analysis, fluorescent in-situ hybridization (FISH), polymerase chain reaction (PCR) based methods, and flow cytometric immunophenotyping (FCI). Each brings unique strengths compensating for the shortcomings of morphology alone.

6.3.1 Immunocytochemistry

ICC utilizes monoclonal or polyclonal antibodies to identify lineage-specific antigens, enabling the distinction between carcinoma, mesothelioma, and reactive mesothelial cells. ICC can be performed on alcohol fixed direct smears, liquid-based cytology smears, or more commonly and preferably, on paraffin-embedded CB sections prepared from effusion sediments. ICC on CB sections is considered optimum, as ICC on direct smears suffers from limiting factors like possible interference with immunoreactivity, increased background staining, and lack of opportunity to evaluate coordinate immunoexpression in the same cells (Shidham & Janikowski, 2022). Key epithelial markers include Ber-EP4, MOC-31, and epithelial membrane antigen (EMA), which are sensitive for adenocarcinoma, while mesothelial markers such as calretinin, WT1, mesothelin, CK5/6, and D2-40 help confirm mesothelial origin. Beyond this primary distinction, ICC can also help identify the primary site of a metastatic adenocarcinoma. For example, TTF-1 is highly specific for lung and thyroid primaries, PAX8 is characteristic of gynecological (ovarian) and renal cancers, CDX2 is a marker for colorectal primaries, and GATA3 is used for breast and urothelial carcinomas (Ordóñez, 2013; Shidham & Janikowski, 2022). This information is invaluable for directing further clinical investigation and treatment.

6.3.2 Cell block preparation

CBs are paraffin-embedded reconstructions prepared from centrifuged sediments of effusion fluids. They mimic histological sections, providing improved cellular preservation and facilitating ICC and molecular studies. CBs often yield superior morphological detail over smears and allow for repeated testing. A recent study in 2024 demonstrated that combining conventional smears with CBs achieves diagnostic accuracies of 92.27%, compared to 90.61% for conventional smears alone and 90.73% for CBs alone (Mandava et al., 2024). Limitations include potential cell loss during processing and variable yield depending on specimen volume and cellularity. Yet, the CB approach substantially expands diagnostic capabilities in effusion cytology when combined with ancillary assays.

6.3.3 Automated image morphometry

Automated image morphometry utilizes digital microscopy and software tools to objectively measure cellular parameters: nuclear area, diameter, perimeter, roundness, convex perimeter, and texture features are quantitatively analyzed. Early studies highlighted that nuclear measurements—such as area, perimeter, and roundness—can distinguish benign from malignant cells based on typical tumor-associated anaplasia. Marchevsky et al. (1987) demonstrated the utility of these quantitative indices in effusion samples, showing morphometric differences in nuclear profile. Subsequent work by Walts et al. (1993) refined the model further, suggesting certain nuclear parameters could delineate malignancy with reasonable accuracy (Marchevsky et al., 1987; Walts et al., 1993).

Nevertheless, the potential for overlap between reactive and malignant cell populations remains, particularly with roundness indices, which may not be sufficiently discriminatory. The presence of vesicular nuclei, degenerate or necrotic cells, and overlapping debris can obscure accurate measurement. Moreover, distinguishing artifacts—such as red blood cells or lymphocytes—from tumor cells by morphometric means alone requires sophisticated algorithms and sometimes manual correction—limiting the practicality in routine settings. Automated morphometry is best applied as a complementary approach and benefits from integration with ancillary ICC and molecular diagnostics.

6.3.4 DNA ploidy analysis

DNA ploidy analysis is an ancillary technique used to measure the DNA content of cells, primarily to detect aneuploidy—an abnormal amount of DNA—which is a strong hallmark of malignancy. This objective, quantitative data can be a powerful adjunct to subjective cytomorphology, especially in diagnostically challenging cases. The analysis is typically performed using one of two main technologies: image cytometry (ICM) or FCM.

ICM: ICM has proven to be a sensitive and reliable method for assessing DNA content and tumor cell proliferation. In serous fluids, background mesothelial cells are typically diploid; therefore, the detection of aneuploid cells can be indicative of malignancy (Motherby et al., 1999a). DNA aneuploidy—representing abnormal DNA content—identified through ICM (ICM) is internationally recognized as a marker of malignant cell transformation and serves as the cytometric counterpart to chromosomal aneuploidy (Wang et al., 2020). Both FCM and ICM can analyze a large number of cells, increasing the likelihood of detecting a small subpopulation of malignant cells that might be missed during routine microscopic screening. This is particularly useful in paucicellular samples or when tumor cells are obscured by inflammation or blood. ICM offers the distinct advantage of preserving cellular architecture on a slide. This allows the cytopathologist to directly correlate the DNA measurement with the morphology of the cell being analyzed, ensuring that aneuploidy is detected in the atypical cells of interest and not in benign or inflammatory cells (Böcking et al., 1984). In a large comparative series evaluating 200 effusions, authors noted the performance of DNA flow and ICM and concluded that while searching for DNA aneuploidy as a diagnostic marker for neoplastic cells in serous effusions, ICM revealed superior sensitivity as compared with monoparametric FCM (Motherby et al., 2002).

Despite its utility, DNA ploidy analysis has significant limitations that prevent it from being a standalone diagnostic test. A substantial percentage of cancers, particularly well-differentiated adenocarcinomas, are diploid, having a normal amount of DNA. Ploidy analysis will fail to detect these malignancies, leading to a false negative result. Dense clusters of cancer cells are often difficult to identify. Metastatic adenocarcinoma, the most common malignancy associated with serous effusion, typically presents with clustered adenocarcinoma cells. Consequently, a diploid DNA result does not rule out the presence of cancer. ICM can produce false positives in inflammatory effusions, likely due to degeneration or nuclear pyknosis making some cells appear aneuploid (Wang et al., 2020). DNA ploidy analysis is a valuable tool in the pathologist's arsenal for effusion fluid analysis. However, due to the significant rate of diploid malignancies, its diagnostic power is greatest when a positive (aneuploid) result is obtained. A negative (diploid) result must be interpreted with caution and in the full context of the cytomorphologic and immunocytochemical findings. Additionally, aneuploidy/hyperploidy should be treated as supportive, not definitive, for malignancy in serous fluids. Inflammatory, degenerative, paucicellular, or blood-rich samples where spurious peaks and measurement artifacts are more likely should be treated with caution. Correlation with cytomorphology, ICC, and clinical-imaging data, and repeat sampling when results are discordant would be valuable in challenging cases.

FCM: FCM is a valuable technique for enhancing the detection of malignant cells in various effusion fluids. It plays a crucial role in establishing a definitive diagnosis, particularly in cases where traditional diagnostic methods yield indeterminate results.

FCM becomes crucial in cases where the diagnosis remains unclear after cytological examination or when findings are labeled as suspicious but not definitively malignant. For instance, exfoliated mesothelial cells or reactive mesothelial (RM) cells can exhibit changes that mimic cancer. There is considerable interobserver variability when interpreting the cytomorphology of these cells and classifying them as either reactive or malignant, especially when a few malignant cells are mixed with RM and inflammatory cells. Due to their morphological overlap, distinguishing between carcinoma cells, inflammatory cells, RM cells, and malignant cells in serous cavity effusions is particularly challenging (Davidson et al., 2002; Motherby et al., 1999b; Thunnissen et al., 1993).

Cytological examinations are sometimes reported as suspicious or atypical based on the presence of rare cells that are not sufficient for a positive diagnosis. For clinical decision-making, the cytological categories such as suspicious and atypia of undetermined significance are considered negative based on the presence of occasional atypical cells which do not qualify as definite positive. Therefore, in such cases, FCM can analyze a larger number of cells and hence has the potential to detect even the rare malignant cells, allowing these cases to be designated as positive or negative with confidence (Pillai & Dorfman, 2016). The ability of FCM to analyze millions of cells and multiple antigens simultaneously makes it a desirable tool for detecting epithelial malignancies and also provides prognostic information based on the expression of particular proteins on the cell surface. A definite edge of FCM is the rapid turnaround time, providing conclusive results within a few hours.

FCM for diagnostic workup of hematolymphoid malignancies is well established and forms the basis for detecting malignant hematolymphoid cells in effusions. FCM has a role in determining clonality in lymphocytic effusions, detecting leukemic/lymphomatous infiltration, diagnosing plasma cell neoplasms, and primary effusion lymphomas. Rather than revisiting the widely reported literature on hematolymphoid malignancies in effusions, this chapter focuses on the lesser explored subject of application of FCM in effusions caused by epithelial malignancies.

6.3.5 Applications of flow cytometry in nonhaematological malignancies

Assessment of DNA ploidy and S phase fraction: One of the most widely acknowledged uses of FCM till now was DNA content measurement, which allowed for the identification of cell aneuploidy and provided information on cell proliferation by examining the various cell cycle phases (Shapiro, 2005) (Fig. 6.1). DNA content in G0-G1 phase is diploid (2n), increases to 2n-4n in S phase and is 4n in G2-M phase. In solid tumors, there is a strong correlation between aneuploidy and neoplastic cells (Unger et al., 1983). FCM provides positive results based on aneuploidy, whereas cytology provides positive results by detecting aberrant morphology.

FCM can detect a small population of aneuploid cells within a few hours of sample collection. Based upon previous studies and literature, aneuploidy is described as a disjunct extra G0/G1 peak, distinct from the normal G0/G1 peak with a DNA index of >1.1, or a G2M fraction >20% of the analyzed cells. The ploidy obtained also assesses the patient's prognosis; diploidy indicates a good prognosis, while aneuploidy indicates a poor prognosis. As reported by Puri et al. (2020), the sensitivity of DNA FCM was 96.6%, and the specificity was 100%. The sensitivity of FCM was greater than that of cytomorphological diagnosis (94.4 vs. 96.6%), while the specificity of both DNA FCM and cytology was the same (Puri et al., 2020). The discrepancy in sensitivity and specificity of FCM in various studies may be attributed to variable sample size and type of cases, definition of aneuploidy, and variances in FCM sample processing, acquisition, and analysis through different gating strategies. The false negativity can be explained by three factors: (1) the tumor's diploid nature, (2) a small aneuploid peak overshadowed by a large number of RM cells, which are diploid on DNA FCM, and (3) minor chromosomal abnormalities (1%–3%) that DNA FCM does not detect (Puri et al., 2020). Another study by Krishan et al. (2010) combined DNA

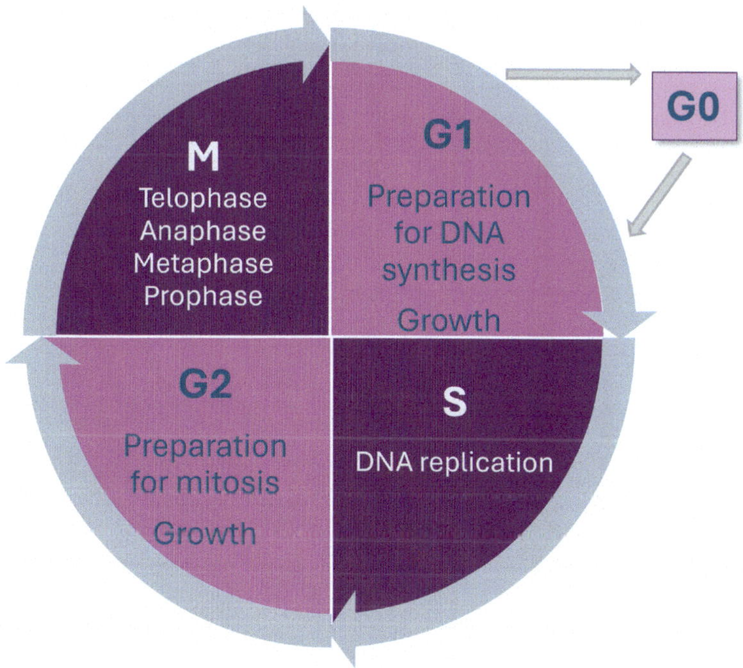

FIGURE 6.1 Cell cycle phases. G1 (Gap1phase): cell grows and prepares for DNA replication; S (Synthetic phase): DNA is replicated; G2 (Gap2 phase): cell continues to grow and prepares for division; M (Mitosis): nucleus divides, cytoplasm splits, creating two daughter cells. Some cells enter G0 phase: resting phase.

FCM with flow cytometric detection of marker expression to analyze cells in serous effusions. This study concluded that by combining data from aneuploidy with positive marker expression, the sensitivity was increased from 58.5% to 100%, whereas the specificity was reduced from 74.7% to 31.6% (Krishan et al., 2010). FCM offers the benefit of quick multiparametric analysis of marker expression and DNA aneuploidy in cells from serous effusions based on the analysis of a large number of cells without observer bias. Previous studies have shown that DNA flow analysis could help in reducing the number of false negatives (Evans et al., 1983; Lovecchio et al., 1986; Stonesifer et al., 1987). On the contrary, other studies demonstrated that DNA FCM led to a large number of false positives, limiting its usefulness for detecting malignant cells (Both et al., 2001; Hedley et al., 1984; Jones et al., 1991). False positivity can be attributed to tetraploidy in mesothelial cells, as they can have several nuclei and may demonstrate mitotic activity as well as a high S phase fraction (Lazcano et al., 2000). Technical errors in sample preparation, more flexible diagnostic criteria for aneuploid DNA, and errors in histogram interpretation can also lead to subjectivity in interpretation and may contribute to false positives (Both et al., 2001).

FCM can be used to determine cell cycle distribution kinetics, chromosomal analysis, and cellular DNA contents. S phase fraction measurement by FCM is an important prognostic predictor in some malignancies, as it indicates a highly proliferative tumor (Fig. 6.2). The process from sampling to DNA histogram is about 20 to 30 min. Unlike mitotic karyotyping, this technique

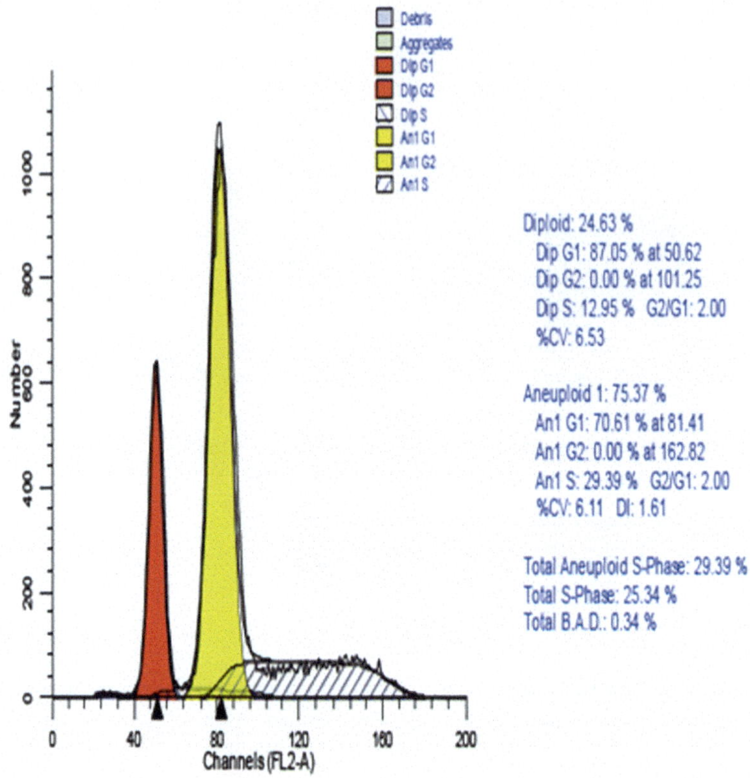

FIGURE 6.2 DNA ploidy and S phase fraction measurement by FCM in a case of diffuse large B cell lymphoma with malignant ascites. Diploid peak (red) 24.63%, aneuploid peak with a DNA index of 1.6 (yellow) 75.37%, S phase fraction of aneuploid 29.39%.

allows investigation of interphase cells in addition to the proliferative status of cells (Hittelman et al., 1977). All these features make DNA analysis by FCM a potentially attractive diagnostic tool for malignancy.

FCI: Besides being used for DNA ploidy determination of cells in serous effusions, FCM has also been used for immunophenotyping tumor cells in malignant serous effusions. Although FCI is primarily used for diagnosis and treatment planning of hematopoietic malignancies, it has also been utilized for cell detection and DNA ploidy analysis of malignant effusions (Craig & Foon, 2008; Krishan et al., 2010; Kentrou et al., 2011; Risberg et al., 2000). Besides being a sensitive, reproducible, fast, and cost-effective method, FCI has the additional advantage of simultaneous analysis of the expression of multiple antigens on the surface of millions of cells, thus providing diagnostic information on other nonmalignant cell populations, such as macrophages, lymphocytes, and monocytes. FCI also has an excellent correlation with cytology (Wong-Arteta et al., 2020). FCI is a very quick procedure that can analyze over 50,000 cells per minute. Furthermore, with the advent of advanced FCM and the availability of newer fluorochromes, multiple antibodies can now be used to study the presence of a specific population. It is possible to study the expression of multiple antigens in the same cell using selective gating (Roca et al., 2016).

Despite the high specificity and good sensitivity of FCM, there are only limited studies available in the literature (Davidson et al., 2002; Gaur et al., 2023; Krishan et al., 2010; Kentrou et al., 2011; Pillai et al., 2013; Risberg et al., 2000; Ramalingam et al., 2025; Sahu et al., 2021; Sayed et al., 2009; Wong-Arteta et al., 2020). Studies conducted previously focused on the role of DNA ploidy by FCM to detect aneuploid cell populations. However, the more recent ones highlighted the potential use of multicolored FCM along with cytological examination in the diagnosis of metastatic carcinoma in serous effusion fluids, particularly those belonging to the atypical category. Gaur et al. (2023) reported the sensitivity, specificity, positive predictive value (PPV), negative predictive value (NPV), and diagnostic accuracy of FCM using Epithelial Cell Adhesion Molecule (EpCAM) to be 92.06%, 96.9%, 98.3%, 86.48%, and 93.75%, respectively. The sensitivity, specificity, PPV, and NPV of FCM in atypical cytological diagnosis were 90.9%, 100%, 100%, and 66.7%, respectively. The authors concluded that by combining the cytological diagnoses and EpCAM results, the sensitivity, specificity, PPV, and NPV were increased to 95.3%, 100%, 100%, and 91.4%, respectively. The authors also utilized MUC 1 as an epithelial marker in a subgroup of 40 samples and reported that assessing dual expression of EpCAM and MUC 1 (without cytomorphology) resulted in 100% specificity (Gaur et al., 2023). In a similar study conducted by Sahu et al. (2021), the sensitivity, specificity, PPV, and NPV in all cases and in atypical cytological diagnosis were 87%, 100%, 100%, 74%, and 83%, 100%, 100%, and 40%, respectively. On combining the cytological diagnoses and EpCAM results, the values improved to 96%, 100%, 100%, and 90%, respectively (Sahu et al., 2021).

Wong Arteta et al. (2020) emphasized the role of FCI as a confirmatory tool in diagnosing malignant effusions, based on the expression of EpCAM in malignant cells and comparing the results with the cytological evaluation. In this study, 101 fluids (68 pleural and 33 ascites) underwent FCM analysis and cytological evaluation. 73 samples (46 pleural and 27 ascites) were considered malignant as determined by clinical, cytological, and radiological criteria. 75% (55/73) of these samples were positive by Cytology/IHC and 74% (54/73) by FCM. The authors observed that, although the sensitivity, specificity, and area under the curve (AUC) were similar, the turnaround time was shorter when using FCM with a median of 1 day, as compared to cytology alone (5 days). Moreover, these results improved by combining both techniques, giving an overall sensitivity of 86.3%, specificity of 100%, and an improved AUC of 0.93. They concluded that FCM provides information about malignant effusions faster than ICC and that performing both techniques simultaneously would improve diagnostic performance (Wong-Arteta et al., 2020). The authors utilized CD200 to identify mesothelial cells in addition to several markers to identify epithelial cells; however, they emphasized EpCAM-based identification of epithelial cells.

Pillai et al. (2013) enlisted the qualitative and quantitative flow cytometric data to distinguish between benign and malignant effusions using a simple, rapid 3-color flow cytometric panel with Ber-EP4 to detect epithelial cells in 193 effusions (both benign and malignant). The study reported that FCM has a sensitivity of 88.15% and specificity of 97.64% compared with 73.68% and 100% on cytologic examination alone for the presence of malignant effusion. The author found that in many of the atypical/suspicious cases seen on cytology, FCM was either clearly negative or positive and hence concluded that FCM would add clinical value if concurrently performed (Pillai et al., 2013). Kentrou et al. (2011) used an elaborate FCM panel comprising EMA, BerEP4, CD66, CD56, Desmin, CD71, and CD45 along with DNA ploidy assessment and compared the results obtained on cytomorphological analysis. In their study, the sensitivity, specificity, PPV, and NPV for FCI using BerEp4 were 80.6%, 90.7%, 94%, and 74%, respectively, while cytology showed a sensitivity of 93.2% and specificity of 95.6% (Kentrou et al., 2011). Krishan et al. (2010) combined DNA FCM with FCI to analyze seventy-nine body cavity fluids. The FCM panel included Ber-EP4, MUC4, progesterone (PR), and thyroid transcription factor-1 (TTF-1). Data from DNA aneuploidy, expression of at least one of the markers, and cytology were combined in a statistical analysis of the serous effusion fluids, and the sensitivity was raised from 58.5% to 100% by combining data from positive marker expression with that of aneuploidy. Due to the presence of aneuploidy and marker expression in these samples, the specificity was reduced from 74.7% to 31.6% (out of the 38 samples categorized as nonmalignant by diagnostic cytology, nine had aneuploid DNA content, and 16 of the diploid samples had a positive marker expression) (Krishan et al., 2010). Sayed et al. (2009) reported FCI (BerEP4, CK, CD3 and CD45) to have a higher sensitivity (73%), higher specificity (95.5%), a higher PPV (94%), and higher NPV (80%) as compared to cytology which had a sensitivity, specificity, PPV, and NPV of 57%, 94%, 90%, and 70% respectively (Sayed et al., 2009). Davidson et al. (2002) emphasized the role of FCI in diagnosing malignant effusions and classifying them into epithelial and mesothelial categories. The authors analyzed a total of 92 cases. CB immunohistochemistry, clinico-radiological investigations, and clinical follow-up were combined with FCM (CD45, CD14, BerEP4, AH6, B72.3, HBTN-1, EMA, and N-cadherin) to investigate the contribution of FCM in the diagnosis, prediction, and prognosis of such cases. 36 cases were detected as positive for malignancy on FCM out of the 43 malignant cases on cytomorphology and CB ICC; 12 cases were detected as positive for malignancy out of the 42 cases labeled as benign based on cytomorphology and CB immunohistochemistry. They found cytology to have a sensitivity of 89% and specificity of 90%, while FCI had a sensitivity of 93% and specificity of 93%. A significant association was seen for coexpression. Ber-EP4 and AH6 were the most sensitive; Ber-EP4 was the most specific (Davidson et al., 2002). Risberg et al. (2000) performed FCI of serous effusions and peritoneal washings in 49 cases and compared the results with ICC and morphological findings. The authors reported that FCM (CD45, BerEP4, and N-cadherin) was able to detect 16 out of 17 malignant cases, five cases as malignant out of the 25 cases labeled as benign on cytomorphology and ICC, and five cases as malignant out of the seven cases labeled as an indeterminate category. The authors concluded that FCM is a rapid and highly effective method for evaluating effusions and peritoneal washings. The detection of Ber-EP4-positive cells using FCM strongly indicates the presence of carcinoma cells in effusions and peritoneal washings (Risberg et al., 2000). In Davidson's (2016) review, the author has advocated FCI of epithelial and mesothelial cells in effusions as a quantitative and highly informative method. FCM may be an efficient tool in identifying isolated tumor cells in a reactive background using reliable epithelial markers. It may also be applied for quantitative detection of cancer-associated molecules and critical cellular processes, such as proliferation and apoptosis. These tests, although not in routine use currently, may have relevance for monitoring treatment response in the presence of metastatic disease

Table 6.1 Diagnostics of studies related to flow cytometry in malignant carcinomatous effusions.

S.no.	Author	Number of cases (n)	Cyto Sens%	Cyto Spec%	FCI Sens (%)	FCI Spec (%)	Combined (Cyto+FCI) sens (%)	Combined (Cyto+FCI) spec (%)	FCI antibody panel	Study design
1.	Gostomczyk et al. (2024)	119	70.2	98.7	91.8	98.7	–	–	CD45, CD14, and EpCAM	FCI results compared with cytomorphology
2.	Gaur et al. (2023)	96	84.6	100	92.06	96.9	95.3	100	CD45, CD14, EpCAM, and MUC 1	FCI results compared with cytology and ICC
3.	Sahu et al. (2021)	75	96.36	90	87	100	96	100	CD45,CD14, and EpCAM	FCI results compared with cytology and ICC
4.	Wong-Arteta et al. (2020)	101	75	100	74	100	86.3	100	CD9, CD45, CD71, CD99, CD56, CD 10, CD 81, NG2, EpCAM, CD200, and CD14	FCI results compared with cytology and ICC
5.	Acosta et al. (2015)	134	–	–	96.7	99.3	–	–	CD45, CD33, and CD326(BerEP4)	FCI results compared with final clinical diagnosis. The study included bone marrow, FNA, lymph node biopsies, other tissues, and Cerebrospinal fluid (CSF) in addition to effusion (total n = 238, effusion n = 134)

(*continued on next page*)

Table 6.1 Diagnostics of studies related to flow cytometry in malignant carcinomatous effusions—cont'd

S.no.	Author	Number of cases (n)	Cyto Sens%	Cyto Spec%	FCI Sens (%)	FCI Spec (%)	Combined (Cyto+FCI) sens (%)	Combined (Cyto+FCI) spec (%)	FCI antibody panel	Study design
6.	Pillai et al. (2013)	195	73.68	100	88.15	97.64	–	–	CD45, CD14, and BerEP4	FCI results compared with cytology and ICC
7.	Kentrou et al. (2011)	125	93.2	95.6	80.6	90.7	–	–	EMA, BerEP4, CD66, CD56, desmin, CD71, and CD45	FCI results compared with cytology. DNA ploidy was also assessed
8.	Krishan et al. (2010)	79	4 out of 79 were positive	–	–	–	–	–	BerEP4, PR, MUC4, and TTF-1	FCI results compared with cytology. DNA ploidy was also assessed
9.	Sayed et al., (2009)	119	57	94	73	95.5	–	–	BerEP4, CK, CD3, and CD45	FCI results compared with cytology. DNA ploidy was also assessed
10.	Davidson et al. (2002)	92	89	90	93	93	–	–	CD45, CD14, Ber-EP4, AH6, TN, B72.3, EMA, and N-cadherin	FCI results compared with cytomorphology

Cyto, cytology; *FCI*, flow cytometric immunophenotyping; *ICC*, immunocytochemistry; *Sens*, sensitivity; *Spec*, specificity; –, not reported.Modified from Gaur, G., Awasthi, N.P., Gupta, A., Agarwal, A., Sachan, R., Malhotra, K.P., Shukla, S., Singh, A.K., Singh, P. & Husain, N. (2023). Diagnostic accuracy of flow cytometry in detecting malignant epithelial cells in serous effusions. Journal of the American Society of Cytopathology, 12(6), 423–435. https://doi.org/10.1016/j.jasc.2023.09.003.

(Davidson, 2016). However, peritoneal washings might lower specificity as some of the benign ovarian and uterine epithelial cells may be shed into these washings, and epithelial markers such as EpCAM might detect these benign epithelial cells and highlight them. Limited research has been conducted to evaluate the diagnostic efficacy of FCI in detecting epithelial malignancies in serous effusions. This may be attributed to limited availability of flow cytometers and experts in FCM, expenses involved, and the fact that flow cytometers are usually considered domains of hematology or research laboratories rather than cytology laboratories. The studies have been summarized in Table 6.1.

In a recent study by Ramalingham et al. (2025), the authors utilized a nuclear dye SYTO 13 and a viability dye 7-AAD in addition to CD45, EpCAM, and CD200 (for mesothelial cells) to improve sensitivity and reduce false positives. A total of 121 samples were studied by FCM and cytology/CB. The authors discussed nine discordant results and emphasized clustered EpCAM-positive events with high scatter properties as definitive for positive diagnosis by FCM (Ramalingam et al., 2025).

Bridging the gap between image morphometry and FCM, Long et al. (2024) demonstrated a novel diagnostic framework using intelligent imaging FCM powered by optofluidic time stretch (OTS) microscopy. Unlike conventional CMOS-based sensors, OTS microscopy utilizes a femtosecond pulse laser to map cellular spatial information into the optical spectrum, enabling an ultra-high event rate of 100,000 events per second with a spatial resolution of less than 1 μm. This label-free approach eliminates the need for time-consuming and costly staining processes. To analyze the massive volume of data generated, the researchers implemented a lightweight convolutional neural network (CNN) based on the MobileViT architecture, which incorporates an attentional mechanism to distinguish between cancer cells and benign mesothelial cells. This is particularly critical as both cell types often share similar sizes, making manual differentiation difficult. The study validated the model using clinical samples, achieving a malignant cell detection accuracy of 90.53%. Furthermore, the CNN's attention maps revealed that the model identifies cancer cells by focusing on irregular nuclear shapes and uneven cytoplasm distribution, whereas benign cells exhibit more circular, centered features. By combining high-speed optical imaging with deep learning, this method provides a potential clinical solution for high-throughput, objective, and accurate cancer diagnosis from serous effusion samples, even in cases with low malignant cell concentrations (Long et al., 2024).

6.4 Conclusion

Multicolored FCI is a rapid and efficient technique for detecting metastatic carcinoma in effusion samples. Some authors have utilized limited panels, whereas others have used extensive panels for detection of malignant epithelial cells with or without DNA ploidy assessment. The focus in the past few years has been more on FCI than DNA ploidy assessment. EpCAM has emerged as a common marker for epithelial cells being used in most studies. Incorporation of viability dyes, nuclear dyes, and newer epithelial markers is leading to further refinement in techniques. Combined use of FCI and cytological evaluation, especially for specimens with atypical or suspicious morphology, appears to be a promising approach in clinical practice. Intelligent imaging flow cytometers represent the futuristic tool poised to redefine the objectivity and accuracy of effusion fluid diagnosis.

AI disclosure

During the preparation of this work, the author(s) used Gemini Pro to improve language. After using this tool/service, the author(s) reviewed and edited the content as needed and take(s) full responsibility for the content of the publication.

References

Acosta, M., Pereira, J., & Arroz, M. (2016). Screening of carcinoma metastasis by flow cytometry: A study of 238 cases. *Cytometry. Part B, Clinical cytometry.*, *90*(3), 289–294. doi:10.1002/cyto.b.21258.

Ahuja, S., Ahuja, R., Pandey, S., & Zaheer, S. (2024). Diagnostic accuracy of International System for Reporting Serous Fluid Cytopathology: A systematic review and meta-analysis in malignancy diagnosis. *Cancer Cytopathology, 132*(10), 609–620. http://onlinelibrary.wiley.com/journal/10.1002/(ISSN)1934-6638. doi:10.1002/cncy.22822.

Azar, G., Jorda, M., & Krishan, A. (2011). *Effusion cytology, A practical guide to cancer diagnosis.* DemosMedical Publishing.

Böcking, A., Hilgarth, M., Auffermann, W., Hack-Werdier, C., Fischer, T., & von Maur, E. (1984). DNA-cytometric diagnosis of prospective malignancy in borderline lesions of the uterine cervix. *Acta Cytologica, 28*(2), 154–160.

Both, C. T., Mattos, A. A. D., Neumann, J., & Reis, M. D. (2001). Flow cytometry in the diagnosis of peritoneal carcinomatosis. *American Journal of Gastroenterology, 96*(5), 1605–1609. http://www.nature.com/ajg/archive/index.html. doi:10.1016/S0002-9270(01)02348-6.

Craig, F. E., & Foon, K. A. (2008). Flow cytometric immunophenotyping for hematologic neoplasms. *Blood, 111*(8), 3941–3967. http://bloodjournal.hematologylibrary.org/cgi/reprint/111/8/3941. doi:10.1182/blood-2007-11-120535.

Davidson, B. (2016). malignant nonhematological effusion characterization by flow cytometry. *Acta Cytologica, 60*(4), 365–371. doi:10.1159/000447687.

Davidson, B., Dong, H. P., Berner, A., Christensen, J., Nielsen, S., Johansen, P., Bryne, M., Asschenfeldt, P., & Risberg, B. (2002). Detection of malignant epithelial cells in effusions using flow cytometric immunophenotyping: An analysis of 92 cases. *American Journal of Clinical Pathology, 118*(1), 85–92. doi:10.1309/M877-QABM-D9GB-FJAX.

Evans, D. A., Thornthwaite, J. T., Ng, A. B. P., & Sugarbaker, E. V. (1983). DNA flow cytometry of pleural effusions. Comparison with pathology for the diagnosis of malignancy. *Analytical and Quantitative Cytology, 5*(1), 19–27.

Gabriel, C., Achten, R., & Drijkoningen, M. (2004). Use of liquid-based cytology in serous fluids: A comparison with conventional cytopreparatory techniques. *Acta Cytologica, 48*(6), 825–835. http://content.karger.com/ProdukteDB/produkte.asp?Aktion=JournalHome&ProduktNr=254338&ContentOnly=false. doi:10.1159/000326453.

Gaur, G., Awasthi, N. P., Gupta, A., Agarwal, A., Sachan, R., Malhotra, K. P., Shukla, S., Singh, A. K., Singh, P., & Husain, N. (2023). Diagnostic accuracy of flow cytometry in detecting malignant epithelial cells in serous effusions. *Journal of the American Society of Cytopathology, 12*(6), 423–435. https://www.sciencedirect.com/science/journal/22132945. doi:10.1016/j.jasc.2023.09.003.

Gostomczyk, K., Łukaszewska, E., Borowczak, J. ę, Bator, A., Zdrenka, M., Bodnar, M., & Szylberg, Ł (2024). Flow cytometry in the detection of circulating tumor cells in neoplastic effusions. *Clinica Chimica Acta, 552*, 117651. doi:10.1016/j.cca.2023.117651.

Hedley, D. W., Philips, J., Rugg, C. A., & Taylor, I. W. (1984). Measurement of cellular DNA content as an adjunct to diagnostic cytology in malignant effusions. *European Journal of Cancer and Clinical Oncology, 20*(6), 749–752. doi:10.1016/0277-5379(84)90211-6.

Hittelman, W., Spitzer, G., Trujillo, J. M., Hart, J. S., Smallwood, L., & Drewinko, B. (1977). Correlation of DNA distribution abnormalities with cytogenetic findings in human adult leukemia and lymphoma. *Cancer Research, 37*(12), 4400–4407.

Jones, M. A., Hitchcox, S., D'Ascanio, P., Papillo, J., & Tarraza, H. M. (1991). Flow cytometric DNA analysis versus cytology in the evaluation of peritoneal fluids. *Gynecologic Oncology, 43*(3), 226–232. doi:10.1016/0090-8258(91)90025-Z.

Kentrou, N. A., Tsagarakis, N. J., Tzanetou, K., Damala, M., Papadimitriou, K. A., Skoumi, D., Stratigaki, A., Anagnostopoulos, N. I., Malamou-Lada, E., Athanassiadou, P., & Paterakis, G. (2011). An improved flow cytometric assay for detection and discrimination between malignant cells and atypical mesothelial cells, in serous cavity effusions. *Cytometry Part B - Clinical Cytometry, 80*(5), 324–334. doi:10.1002/cyto.b.20608.

Krausz, F., & Barker (2003). Reactive effusions. In W. Gray, & T. G. McKee (Eds.), *Diagnostic cytopathology* (pp. P135–P152). Elsevier Science Limited.

Krishan, A., Ganjei-Azar, P., Hamelik, R., Sharma, D., Reis, I., & Nadji, M. (2010). Flow immunocytochemistry of marker expression in cells from body cavity fluids. *Cytometry Part A, 77*(2), 132–143. http://www3.interscience.wiley.com/cgi-bin/fulltext/122679530/PDFSTART. doi:10.1002/cyto.a.20824.

Lazcano, O., Chen, L. M., Tsai, C., Li, C. Y., Katzmann, J. A., Sebo, T. J., Kimlinger, T. K., & Baker, J. (2000). Image analysis and flow cytometric DNA studies of benign and malignant body cavity fluids: Reappraisal of the role of current methods in the differential diagnosis of reactive versus malignant conditions. *Modern Pathology, 13*(7), 788–796. http://www.nature.com/modpathol/index.html. doi:10.1038/modpathol.3880137.

Light, R. W., Macgregor, M. I., Luchsinger, P. C., & Ball, W. C. (1972). Pleural effusions: The diagnostic separation of transudates and exudates. *Annals of Internal Medicine, 77*(4), 507–513. doi:10.7326/0003-4819-77-4-507.

Long, M., Weng, Y., Mei, L., Yang, D., Wei, S., Meng, G., Zhao, W., Liu, S., Wang, D., Liu, Y., Shen, H., Hou, J., Xu, Y., Tao, L., Zhou, F., Chen, H., Hu, T., & Lei, C. (2024). Diagnosis of serous effusion with intelligent imaging flow cytometry. *Advanced Sensor Research, 3*(8). https://advanced.onlinelibrary.wiley.com/journal/27511219. doi:10.1002/adsr.202300183.

Lovecchio, J. L., Budman, D. R., Susin, M., Grueneberg, D., & Fenton, A. N. (1986). Flow cytometry of peritoneal washings in gynecologic neoplasia. *Obstetrics and Gynecology, 67*(5), 675–679. doi:10.1097/00006250-198605000-00014.

Mandava, H., Venkata Renuka, I., Potti, R., Mounica, B., & Kalla, I. (2024). Deciphering serous effusions using the new International System for Reporting Serous Fluid Cytopathology. *Cureus, 16*(5). doi:10.7759/cureus.60042.

Marchevsky, A. M., Hauntman, E., Gil, J., & Watson, C. (1987). Computerized interactive morphometry as an aid in the diagnosis of pleural effusions. *Acta Cytologica, 31*(2), 131–136.

Mohanty, S. K., & Dey, P. (2003). Serous effusions: Diagnosis of malignancy beyond cytomorphology. An analytic review. *Postgraduate Medical Journal, 79*(936), 569–574. doi:10.1136/pmj.79.936.569.

Motherby, H., Friedrichs, N., Kube, M., Nadjari, B., Knops, K., Donner, A., Baschiera, B., Dalquen, P., & Böcking, A. (1999a). Immunocytochemistry and DNA-image cytometry in diagnostic effusion cytology. II. Diagnostic accuracy in equivocal smears. *Analytical Cellular Pathology, 19*(2), 59–66. doi:10.1155/1999/934146.

Motherby, H., Nadjari, B., Friegel, P., Kohaus, J., Ramp, U., & Böcking, A. (1999b). Diagnostic accuracy of effusion cytology. *Diagnostic Cytopathology, 20*(6), 350–357. doi:10.1002/(SICI)1097-0339(199906)20:6⟨350::AID-DC5⟩3.0.CO;2-7.

Motherby, H., Pomjanski, N., Kube, M., Boros, A., Heiden, T., Tribukait, B., & Böcking, A. (2002). Diagnostic DNA-Flow- vs. -image-cytometry in effusion cytology. *Analytical Cellular Pathology, 24*(1), 5–15. doi:10.1155/2002/840210.

Nguyen, G. K. (2010). *Essentials of fluid cytology* https://pathology.ubc.ca/files/2012/06/FLUIDCYTOLOGYBook09R1.pdf.

Ordóñez, N. G. (2013). Application of immunohistochemistry in the diagnosis of epithelioid mesothelioma: A review and update. *Human Pathology, 44*(1), 1–19. doi:10.1016/j.humpath.2012.05.014.

Pillai, V., Cibas, E. S., & Dorfman, D. M. (2013). A simplified flow cytometric immunophenotyping procedure for the diagnosis of effusions caused by epithelial malignancies. *American Journal of Clinical Pathology, 139*(5), 672–681. http://ajcp.ascpjournals.org/content/139/5/672.full.pdf+html. doi:10.1309/AJCP4HIFSHMO9WTK.

Pillai, V., & Dorfman, D. M. (2016). Flow cytometry of nonhematopoietic neoplasms. *Acta Cytologica, 60*(4), 336–343. doi:10.1159/000448371.

Puri, M., Sen, R., Gupta, M., Kalra, R., Bhargava, S., & Puri, K. (2020). DNA ploidy analysis and cell block immunohistochemistry in the diagnosis of malignant effusions. *Acta Cytologica, 64*(3), 256–264. doi:10.1159/000501790.

Ramalingam, T. R., Mani, S., Narla, S., Lakshmanan, A., Muthu, A., Nk, H. R., Mohanraj, S., Ramachandran, E. D., Subramanyam, A., & Vaidhyanathan, L. (2025). Simple and rapid flow cytometry assay for the detection of malignant epithelial cells in body fluids. *Cytopathology, 36*(1), 23–30. http://onlinelibrary.wiley.com/journal/10.1111/(ISSN)1365-2303. doi:10.1111/cyt.13408.

Risberg, B., Davidson, B., Dong, H. P., Nesland, J. M., & Berner, A. (2000). Flow cytometric immunophenotyping of serous effusions and peritoneal washings: Comparison with immunocytochemistry and morphological findings. *Journal of Clinical Pathology, 53*(7), 513–517. doi:10.1136/jcp.53.7.513.

Roca, E., Lacroix, R., Judicone, C., Laroumagne, S., Robert, S., Cointe, S., Muller, A., Kaspi, E., Roll, P., Brisson, A. R., Tantucci, C., Astoul, P., & Dignat-George, F. (2016). Detection of EpCAM-positive microparticles in pleural fluid: A new approach to mini-invasively identify patients with malignant pleural effusions. *Oncotarget, 7*(3), 3357–3366. http://www.impactjournals.com/oncotarget/index.php?journal=oncotarget&page=article&op=download&path%5B%5D=6581&path%5B%5D=18280. doi:10.18632/oncotarget.6581.

Sahu, S., Gupta, P., Susheilia, S., Gautam, U., & Dey, P. (2021). Application of multicolour flow cytometry in the detection of metastatic carcinoma in serous effusions: Special emphasis in atypical cytology. *Cytopathology, 32*(2), 169–179. http://onlinelibrary.wiley.com/journal/10.1111/(ISSN)1365-2303. doi:10.1111/cyt.12922.

Sayed, D. M., El-Attar, M. M., & Mohamed Hussein, A. A. R. (2009). Evaluation of flow cytometric immunophenotyping and DNA analysis for detection of malignant cells in serosal cavity fluids. *Diagnostic Cytopathology, 37*(7), 498–504. http://www3.interscience.wiley.com/cgi-bin/fulltext/122204515/PDFSTART. doi:10.1002/dc.21047Egypt.

Shapiro, H. M. (2005). *Practical flow cytometry*. Wiley. doi:10.1002/0471722731.

Shidham, V. B., & Falzon, M. (2010). *Serous effusions* (pp. 115–175). Elsevier BV. doi:10.1016/b978-0-7020-3154-0.00003-x.

Shidham, V. B., & Janikowski, B. (2022). Immunocytochemistry of effusions: Processing and commonly used immunomarkers. *CytoJournal, 19*. https://cytojournal.com/immunocytochemistry-of-effusions-processing-and-commonly-used-immunomarkers/. doi:10.25259/CMAS_02_15_2021.

Stonesifer, K. J., Xiang, J., Wilkinson, E. J., Benson, N. A., & Braylan, R. C. (1987). Flow cytometric analysis and cytopathology of body cavity fluids. *Acta Cytologica, 31*(2), 125–130.

Thunnissen, F. B. J. M., Peterse, J. L., van Pel, R., Jansen, J.-W., Laddé, B. E., Arends, J.-W., & Volovics, A. (1993). Reliability of fine needle aspiration cytology for distinguishing between carcinoma, lymphoma and sarcoma; the influence of clinical information. *Cytopathology, 4*(2), 107–114. doi:10.1111/j.1365-2303.1993.tb00522.x.

Unger, K. M., Raber, M., Bedrossian, C. W. M., Stein, D. A., & Barlogie, B. (1983). Analysis of pleural effusions using automated flow cytometry. *Cancer, 52*(5), 873–877. doi:10.1002/1097-0142(19830901)52:5⟨873::AID-CNCR2820520522⟩3.0.CO;2-V.

Walts, A. E., Morimoto, R., & Marchevsky, A. M. (1993). Computerized interactive morphometry and the diagnosis of lymphoid-rich effusions. *American Journal of Clinical Pathology, 99*(5), 570–575. doi:10.1093/ajcp/99.5.570.

Wang, S., Li, D., Wang, J., Wan, Y., Huang, C., Yang, B., Tang, X., Tian, G., & Yang, Z. (2020). A comparative study of liquid-based cytology and DNA image cytometry in the diagnosis of serous effusion. *Technology in Cancer Research and Treatment, 19*. https://journals.sagepub.com/home/TCT. doi:10.1177/1533033820942298.

Wong-Arteta, J., Rey, M., Aragón, L., Gil-Rodríguez, E., & Bujanda, L. (2020). The utility of flow cytometry in the diagnostic workup of malignant effusions due to nonhematopoietic neoplasms. *Cytometry Part B - Clinical Cytometry, 98*(6), 504–515. http://onlinelibrary.wiley.com/journal/10.1002/(ISSN)1552-4957. doi:10.1002/cyto.b.21886.

7

Flow cytometry as a tool for detecting malignant epithelial cells in effusion fluids-technique and assay standardization

Namrata P. Awasthi[a], Thulasi Raman Ramalingam[b] and Gauri Gaur[a]

[a]DEPARTMENT OF PATHOLOGY, DR. RAM MANOHAR LOHIA INSTITUTE OF MEDICAL SCIENCES (DR. RMLIMS), LUCKNOW, UTTAR PRADESH, INDIA [b]APOLLO HOSPITALS, CHENNAI, TAMIL NADU, INDIA

7.1 Introduction

Development of serous effusions involving the pleural, peritoneal, pericardial, and cerebrospinal cavities in patients with malignancies carries profound diagnostic and prognostic implications, often signifying metastatic dissemination. This has implications for staging, making therapeutic decisions, and in selected settings determining surgical eligibility (Heffner & Klein, 2008; Kehoe, 1991; Michael & Davidson, 2016; Otero et al., 2024). In solid tumors including carcinomas of the lung, breast, ovary, gastrointestinal tract, and urothelium—the presence of malignant effusion may alter the clinical course of the patient (Heffner & Klein, 2008; Heffner & Klein, 2008; Otero et al., 2024).

Despite its clinical importance, effusion cytology remains diagnostically challenging. Conventional cytomorphological examination continues to serve as the first-line as well as gold standard method for evaluation (Heffner & Klein, 2008; Michael & Davidson, 2016). However, its sensitivity is highly variable and influenced by several biological and technical factors. Interpretation depends to a great extent on morphological expertise and subjectivity (Koss, 2006; Michael & Davidson, 2016).

One of the most problematic areas in effusion cytology is the distinction between malignant epithelial cells and reactive mesothelial cells. The diagnostic gray zones underscore the need for objective ancillary techniques that can complement morphology.

Cell block preparation followed by immunohistochemistry (IHC) has improved diagnostic accuracy by enabling lineage-specific antigen detection. But even this approach has inherent

Advancing Body Fluid Diagnostics with Flow Cytometry: A Comprehensive Guide to Effusions and CSF.
DOI: https://doi.org/10.1016/B978-0-443-27426-8.00001-6

limitations, which include the requirement of a cellular specimen, optimal fixation, and processing. Inability to perform multiparametric antigen expression evaluation and a longer turnaround time are additional factors of practical importance (Shidham & Atkinson 2007). Furthermore, immunohistochemical interpretation remains semiquantitative and subjective. In cases with scant cellularity, material exhaustion may preclude additional testing.

Flow cytometry (FCM) offers a powerful, rapid, and objective alternative (Borowitz et al., 2010; Gaur et al., 2023; Maecker et al., 2012; Shapiro, 2005; Wood, 2016; Zimmerlin et al., 2011). By analyzing thousands to millions of individual cells within minutes, FCM simultaneously assesses cellular light scatter properties and antigen expression patterns at the single-cell level. Unlike morphology-based evaluation, which depends on visual assessment, FCM generates quantitative, reproducible, and statistically robust data. The technology is particularly suited for rare-event detection, making it attractive for identifying small populations of malignant epithelial cells within heterogeneous effusion samples (Borowitz et al., 2010; Zimmerlin et al., 2011).

FCM is well established in hematopathology but remains underutilized in nonhematologic malignancies. Key limitations have included the lack of reliable epithelial markers, difficulties in processing nonhematologic specimens, interference from debris and nonspecific binding, and the absence of standardized protocols. However, recent advances—particularly the development of robust epithelial markers such as EpCAM, along with improvements in fluorochrome chemistry, multicolor instrumentation, viability dyes, and refined gating strategies—have revitalized interest in the application of FCM to effusion cytopathology (Menz et al., 2024; Ramalingam et al., 2025; Went et al., 2006).

The fundamental analytical approach to effusion FCM diverges significantly depending on whether the suspected malignancy is epithelial or hematolymphoid in origin. Hematolymphoid neoplasms naturally exist in liquid suspension and readily shed as single cells into serous cavities, making them inherently amenable to immediate cytometric interrogation. The diagnostic strategy for these hematopoietic lesions relies heavily on multiparameter surface and intracellular immunophenotyping (e.g., assessing immunoglobulin light-chain restriction, aberrant antigen loss, or lineage infidelity) to definitively establish monoclonality. Conversely, epithelial malignancies are characterized by strong intercellular adhesion resulting in cell balls and clusters. Furthermore, because carcinomas lack a universal, easily measurable surface marker of absolute clonality analogous to B-cell light chains, the diagnostic algorithm fundamentally shifts. The approach relies on identifying a distinct nonhematopoietic (CD45-negative), epithelial-associated (e.g., EpCAM-positive) population. The chapter therefore focuses on methodological aspects of FCM of effusion fluids in epithelial malignancies.

This chapter is structured to serve as both a practical manual for beginners and a comprehensive framework for advanced high-sensitivity diagnostic standardization. It is organized into two primary methodologies: a **routine approach** designed to guide beginners through the fundamental, step-by-step execution of FCM on effusion fluids, and a **high-sensitivity, rare-event approach** that details the rigorous protocols required to detect minute tumor populations often missed by conventional means. The text consistently integrates the scientific and biological rationale behind each procedural choice—explaining how markers and technical refinements transform subjective assessments into a quantitative, reproducible diagnostic assay. For beginners, the routine approach provides a reliable way to confirm clinical suspicion when tumor burden is high. However, as the laboratory seeks to reduce false negatives and handle equivocal cases, transitioning to the high-sensitivity approach is essential.

7.2 Biological rationale for flow cytometric analysis of effusions

The biological foundation for applying FCM to effusion fluids is straightforward yet powerful: normal serous cavities do not contain epithelial cells (Heffner and Klein, 2008; Michael and Davidson, 2016). Physiologic effusions typically consist of mesothelial cells lining the serosal surfaces, macrophages, lymphocytes, neutrophils, and other inflammatory components. Therefore, the identification of epithelial cells within these cavities is inherently abnormal and strongly suggestive of metastatic carcinoma.

This biological principle allows the diagnostic strategy to focus not merely on cellular atypia but on lineage-specific detection. Instead of relying solely on morphologic suspicion, FCM can objectively determine whether epithelial lineage cells are present within the fluid.

Reactive mesothelial cells present the greatest diagnostic challenge. In inflammatory states, tuberculosis, autoimmune disease, or malignancy-associated irritation, mesothelial cells can undergo reactive proliferation and display striking atypia. Morphologically, these cells may mimic carcinoma. However, immunophenotypically they differ from epithelial tumor cells. Mesothelial cells typically lack expression of epithelial adhesion molecules such as EpCAM and may express markers like CD200 or other mesothelial-associated antigens, whereas carcinoma cells show surface expression of epithelial markers and lack hematopoietic antigens such as CD45.

EpCAM (CD326), a transmembrane glycoprotein involved in cell adhesion, is strongly expressed on most epithelial cells and overexpressed in many carcinomas (Menz et al., 2024; Ramalingam et al., 2025). Importantly, hematopoietic cells, macrophages, and mesothelial cells do not express EpCAM (Menz et al., 2024; Went et al., 2006). This differential expression forms the immunologic backbone of flow cytometric detection of malignant epithelial cells. Additionally, EpCAM overexpression has been associated in several tumor types with advanced stage and poorer prognosis, adding potential prognostic relevance to its detection (Chen et al., 2020; Han et al., 2017; Spizzo et al., 2011).

Beyond antigen expression, FCM leverages physical characteristics. Malignant epithelial cells often demonstrate increased forward scatter (FSC) (reflecting larger cell size) and increased side scatter (SSC) (reflecting internal complexity) relative to lymphocytes and debris. When these physical parameters are integrated with antigen expression patterns, multiparametric discrimination becomes highly reliable (Shapiro, 2005; Wood, 2016).

Effusion samples are inherently heterogeneous and may contain significant debris, apoptotic cells, erythrocytes, and inflammatory elements. A major innovation reflected in the transition from cytoplasmic markers such as cytokeratin (CK)—requiring fixation and permeabilization and prone to generating debris—to a surface-marker-based panel. Surface-only panels reduce processing steps, minimize debris formation, preserve cell integrity, and enable inclusion of viability dyes such as 7-AAD. The addition of nuclear dyes such as SYTO 13 further refines gating by excluding nonnucleated debris that may bind anti-EpCAM nonspecifically (Ramalingam et al., 2025).

Another critical biological consideration is the rarity of malignant cells in some effusions. Tumor cells may constitute far less than 1% of total events. Therefore, the assay must function as a rare-event detection platform. This requires acquisition of very high total event counts, optimized signal-to-noise ratio, validated limits of detection and quantification, and objective gating templates derived from benign effusions (Borowitz et al., 2010; Ramalingam et al., 2022; Zimmerlin et al., 2011). Without these measures, low-frequency EpCAM-positive events may be misinterpreted as background or noise.

In this context, FCM transforms effusion analysis from a predominantly subjective morphologic exercise into a quantitative, reproducible, and statistically validated diagnostic assay. When standardized appropriately, it allows:

- Objective identification of epithelial lineage cells
- Reliable discrimination from reactive mesothelial and inflammatory populations
- Rare-event detection with defined analytical sensitivity
- Rapid turnaround time compared with cell block IHC
- Quantitative assessment of tumor burden within the fluid

Thus, the biological rationale for flow cytometric analysis of effusions is rooted in lineage specificity, objective antigen quantification, and rare-event detection capability. By integrating immunophenotypic discrimination with high-throughput single-cell analysis, FCM provides a scientifically robust and clinically meaningful adjunct to conventional cytology.

7.3 Sample collection, transport and handling

Sample adequacy: As the fluid specimen is divided between cytology and FCM laboratories, a volume of 30–50 mL is preferred. At least 10 mL of effusion sample should be dedicated for FCM evaluation. However, in a highly cellular specimen, an amount as low as 2 mL may be sufficient for FCM.

Sample collection: Ideally, the effusion sample should be collected in an anticoagulated container. The preferred anticoagulants include ethylene diamine tetra acetic acid (EDTA), acid citrate dextrose, 1% ammonium oxalate, and heparin (3 units/mL). This helps in preventing coagulation and preserving cellular morphology. The commonly used anticoagulant is ammonium oxalate, in a ratio of 9 parts fluid and 1 part anticoagulant. If the sample is not collected in an anticoagulant, it may have micro- clots that are not visible to the naked eye but may result in poor sensitivity for detecting epithelial cells. Additionally, it may also block the fluidics of the flow cytometer.

Sample transport and handling: The sample should be transported immediately and ideally processed within 24 to 48 h of collection, as cellular degradation over time increases apoptosis, debris formation, and nonspecific antibody binding (Koss, 2006; Michael & Davidson, 2016). An attempt should be made to transport the sample at 2–8°C to give the best results. Extremes of temperature can damage cells; hence, transportation and storage at very high temperatures or freezing temperatures should be avoided. It has been suggested that preparing cells for FCM on ice with gentle manipulation might help in keeping the malignant cells intact and prevent the cells from losing their antigenicity (Pillai et al., 2013).

7.4 Sample preparation for flow cytometry applications

Effusion cytology often encounters diagnostic challenges due to low tumor burden, reactive mesothelial proliferation, and sampling variability. Unlike peripheral blood, effusions are highly heterogeneous specimens composed of variable proportions of leukocytes, reactive mesothelial cells, macrophages, proteinaceous debris, apoptotic cells, erythrocytes, and, when present, malignant epithelial cells. Therefore, processing methods must not only preserve antigen integrity but also reduce background interference and optimize recovery of rare tumor populations.

Single-cell suspension: The first step for a flow cytometric analysis is to ensure a single-cell suspension of the sample. The body fluids are inherently in the fluid state and hence do not need additional steps for preparing a single-cell suspension. However, in case tiny clots or microclots are visible, it is better to filter the sample through a cell strainer or nylon mesh filters, which are available in various pore sizes.

Process the whole sample: In low-volume or low-cellularity fluids, it is critical to process the entire specimen rather than an aliquot. In toto processing maximizes the probability of detecting rare malignant epithelial events and reduces sampling bias. This principle parallels rare-event detection strategies used in intra-operative and minimal residual disease–type assays, where sensitivity is directly proportional to the number of analyzable nucleated events (Borowitz et al., 2010; Zimmerlin et al., 2011).

Concentration and washing: The next step involves concentration and washing of cells to get optimum numbers before staining. Following centrifugation, adequate washing reduces proteinaceous background that may interfere with staining.

i. Centrifuge the entire sample or the maximum available volume in 15 mL conical tubes at 1000 rpm for 10 min (or 2500 rpm for 5 min depending on your lab's settings) to gather the cells at the bottom.
ii. Discard the supernatant carefully; the resulting cell pellets are gently resuspended and pooled into a single test tube to obtain a concentrated cellular sample for further analysis.
iii. Wash the sediment with phosphate-buffered saline (PBS) solution at 1000 rpm for 10 min (or as per your lab's settings). Discard the supernatant. This step can be repeated if the suspension appears dirty.
iv. Count the number of cells in the PBS-cell suspension on Neubauer's chamber manually or on an automated cell counter. In the author's experience, a cell count of atleast $10 \times 10^3/\mu L$ in a total resuspended volume of 100–200 μL gives satisfactory results. The actual numbers

may vary, and with experience the laboratory may establish its own adequacy criteria. For a high-sensitivity rare-event analysis, even higher numbers are warranted.

Red cell lysis: Hemorrhagic samples require red blood cell (RBC) lysis to eliminate erythrocytes and platelets that can increase background fluorescence and obscure rare epithelial populations.

Two principal processing strategies are commonly employed in flow cytometric analysis of body fluids: the **stain-lyse** method and the **lyse–stain** method. Although both approaches aim to achieve selective erythrocyte removal and adequate immunostaining, they differ in sequence and analytical implications.

In the **stain-lyse approach**, antibodies are added directly to the unprocessed specimen, followed by incubation and subsequent RBC lysis. This technique is frequently used in peripheral blood immunophenotyping because cellular composition is predictable and leukocyte recovery is prioritized. In effusion samples, however, this strategy presents several limitations. First, the presence of abundant erythrocytes and plasma proteins during the staining phase may increase nonspecific antibody binding. Second, antibody consumption may be inefficient, as antibodies may bind nonspecifically to debris or protein aggregates before lysis is performed. Third, delayed removal of RBCs may contribute to increased background fluorescence and autofluorescence, especially in hemorrhagic effusions. Finally, when rare-event detection is the goal, any increase in background noise can artificially elevate the limit of blank (LOB), thereby reducing assay sensitivity.

In contrast, the **lyse–stain approach** involves initial removal of erythrocytes, followed by washing and subsequent antibody staining of the enriched nucleated cell population. This sequence offers several technical and analytical advantages in the context of effusion diagnostics. Early RBC lysis reduces sample complexity and decreases protein-mediated nonspecific binding. It also improves antibody accessibility to target epitopes by eliminating steric hindrance caused by high red cell burden (Shapiro, 2005; Wood, 2016).

From a rare-event detection perspective, the lyse–stain method contributes directly to improved analytical sensitivity (Borowitz et al., 2010; Zimmerlin et al., 2011). Because the limit of detection (LOD) is influenced by background variability, minimizing nonspecific fluorescence and debris lowers the baseline noise against which EpCAM-positive events are measured. Cleaner backgrounds allow more accurate determination of the LOB and facilitate reliable identification of small, clustered malignant populations. This is particularly important in effusions where tumor burden may be extremely low and where statistical confidence depends on clear separation of true events from artifacts.

Another important consideration is viability and antigen preservation. Excessive protein and hemoglobin breakdown products in unprocessed hemorrhagic effusions may adversely affect staining quality. Early lysis reduces hemoglobin interference and permits better performance of viability dyes such as 7-AAD. Because dead cells exhibit increased autofluorescence and nonspecific antibody uptake, effective removal of interfering components indirectly enhances downstream gating accuracy.

Operationally, the lyse-stain method also supports standardized, high event acquisition strategies. After RBC lysis and washing, the final nucleated cell pellet can be concentrated, resuspended in a defined volume, and processed in toto. This enables acquisition of maximal gated events and improves reproducibility across samples with variable initial cellularity. Furthermore, concentrating the nucleated fraction ensures that rare malignant epithelial cells are not diluted in excessive background elements, thereby strengthening both analytical and functional sensitivity.

The only potential concern with lyse-stain processing is the theoretical risk of cell loss during washing steps. However, when centrifugation parameters are optimized—typically moderate-speed centrifugation sufficient to pellet both mesothelial and epithelial cells—cell recovery remains high. In practice, the benefits of reduced background and improved staining resolution outweigh minimal procedural loss, particularly in laboratories emphasizing high-sensitivity rare-event detection.

For these reasons, the lyse-stain approach is recommended for flow cytometric detection of malignant epithelial cells in effusion fluids. It offers superior control of background noise, improved signal-to-noise ratio, enhanced antibody efficiency, better compatibility with viability and nuclear dyes, and stronger statistical performance in rare-event analysis. When combined with in toto processing, high event acquisition, and template-based gating, lyse-stain methodology provides a robust and reproducible foundation for standardized effusion FCM.

In summary, while both stain-lyse and lyse-stain techniques are technically feasible, the lyse-stain method aligns more closely with the analytical requirements of effusion diagnostics, particularly when the objective is sensitive detection of low-frequency malignant epithelial populations.

7.5 Selection of markers and panel design

The selection of appropriate markers and rational panel design are critical determinants of the sensitivity, specificity, and reproducibility of FCM-based assays for detecting malignant epithelial cells in effusion fluids. Unlike hematolymphoid immunophenotyping, effusion analysis presents unique challenges due to cellular heterogeneity, low target cell frequency, background debris, and antigenic overlap between reactive mesothelial cells and carcinoma cells. Therefore, marker selection must be biologically justified, technically feasible, and compatible with the available FCM platform.

A surface-marker-based staining approach is generally preferable for routine effusion diagnostics. Cytoplasmic staining requires fixation and permeabilization, which introduces additional debris, increases turnaround time, and may alter scatter characteristics. Avoiding intracellular staining preserves native cell morphology in scatter plots and improves discrimination of tumor clusters. Furthermore, restricting the assay to surface markers enables simultaneous incorporation of viability dyes and nuclear stains, both of which enhance analytical specificity. The availability of markers for FCM is currently limited as compared to IHC in the context of non-hematological malignancies. However, as an increasing number of fluorochrome-conjugated antibodies become commercially available, the technique will continue to evolve.

Table 7.1 Immunophenotypic markers for lineage identification in serous effusions based on published literature.

Marker	Cell of interest/purpose
CD45 (Leukocyte common antigen)	All leucocytes
CD14	Monocytes/macrophages
CD64	Monocytes/macrophages and neutrophils (low level)
CD16	Neutrophils, monocytes, NK cells, and some T Cell subsets
CD200	Mesothelial cells
Desmin	Mesothelial cells
Calretinin	Mesothelial cells
WT1	Mesothelial cells
EpCAM, BerEP4, MOC31, and CD326	Epithelial cells
MUC 1 and MUC 4	Epithelial cells
B72.3	Epithelial cells
EMA	Epithelial cells
Cytokeratin	Epithelial cells, mesothelial cells
CD56	NK cells, neuroectodermal-derived tumors
CD99	Ewing sarcoma/primitive neuroectodermal tumor, T- and B-lymphoblastic leukemias/lymphomas, desmoplastic small round cell tumor, mesenchymal chondrosarcoma and rhabdomyosarcoma
CD10	Clear cell renal cell carcinoma, hepatocellular carcinoma, some gynecologic tumors, and germ cell tumors
CD71	Upregulated in proliferating cells and reactive mesothelial cells

7.5.1 Marker selection

For flow cytometric immunophenotyping on effusion specimens targeted for malignant epithelial cells, a panel directed toward identifying epithelial cells, leucocytes, mesothelial cells and/or monocyte/macrophages would be ideal. Based on published literature, a broad range of markers—including leukocytes (CD45), macrophages (CD14 and CD68), mesothelial markers (CD200, D2-40, Calretinin, WT1, and Desmin), and diverse epithelial markers (EpCAM, BerEP4, MOC31, MUC1, MUC4, epithelial membrane antigen (EMA), B72.3, and cytokeratin)—have been utilized to characterize effusion fluids (Table 7.1). While modern high-sensitivity protocols prioritize surface markers to maintain cell integrity and simplify the workflow, several markers in the literature require specialized intracellular processing. Specifically, markers that require cytoplasmic staining (necessitating fixation and permeabilization) include cytokeratin, Desmin, Calretinin, and MUC4. Additionally, markers such as WT1 and TTF-1 typically require nuclear staining protocols due to their localization within the cell nucleus. However, based on the authors' experience, the preferred high-sensitivity screening panel should consist of only surface markers.

Among the epithelial-associated antigens evaluated over time, epithelial cell adhesion molecule (EpCAM and CD326) has emerged as the most reliable and widely accepted marker for identifying epithelial cells by FCM (Gaur et al., 2023; Menz et al., 2024; Ramalingam et al., 2025; Went et al., 2006). EpCAM is a 40 kDa transmembrane glycoprotein involved in intercellular adhesion and epithelial tissue organization. During tumor progression, EpCAM modulates cadherin-mediated adhesion, conferring antiadhesive properties that facilitate tumor cell migration and metastasis (Gires et al., 2020; Munz et al., 2009).

Importantly, EpCAM is not expressed on hematopoietic cells, macrophages, or normal/reactive mesothelial cells, including neoplastic mesothelial proliferations. As normal serous effusions physiologically lack epithelial cells, detection of an EpCAM-positive population in effusion samples is, by itself, a strong indicator of malignant epithelial involvement. In addition to its diagnostic utility, EpCAM overexpression has been shown to have prognostic significance in several epithelial malignancies, including breast carcinoma and urothelial carcinoma of the bladder, where higher expression correlates with advanced disease stage and poorer overall survival.

Other epithelial-associated antigens such as carcinoembryonic antigen (CEA), EMA, and mucin-1 (MUC-1) have also been explored in effusion analysis (Gaur et al., 2023; Horm and Schroeder, 2013; Jansen et al., 2024). However, these markers may be variably expressed, can show nonspecific staining, and are less consistently discriminatory when used as single markers. Consequently, EpCAM remains the backbone of most flow cytometric panels designed for epithelial cell detection in effusions.

7.5.2 Importance of exclusion markers

Accurate identification of malignant epithelial cells requires not only positive selection but also robust exclusion of non-neoplastic cell populations.

- CD45 is indispensable for excluding hematolymphoid cells, which often constitute the majority of cells in effusion samples (Wood, 2016).
- Macrophages, which may show high forward and SSC properties and occasional nonspecific antibody binding, represent a major source of false-positive events. Prior studies have demonstrated the utility of macrophage-associated markers such as CD14 or CD68 to avoid misclassification (Sharygin et al., 2023).
- Mesothelial cells, particularly when reactive, can overlap morphologically and phenotypically with carcinoma cells. Identification of reliable mesothelial markers like CD200 and D2-40 (podoplanin) is therefore essential in panel design (He et al., 2017; Wong-Arteta et al., 2020).

7.5.3 Evolution from cytoplasmic to surface marker-based panels

Cytokeratin is the most commonly used epithelial marker for lineage assignment in immunocytochemistry and IHC and hence was a natural choice in earlier studies. In our experience, during the initial phase of assay development, a three-marker panel comprising CD45, EpCAM, and cytokeratin was evaluated. Cytokeratin was included to enhance specificity. However, intracellular cytokeratin staining in FCM necessitates fixation and permeabilization, which introduced several practical limitations:

- Increased processing time
- Higher levels of cellular debris
- Nonspecific antibody binding
- Reduced assay reproducibility

Additionally, cytokeratin expression was observed in both malignant epithelial cells and mesothelial cells, limiting its discriminatory value in this context. Based on these observations, cytokeratin was excluded from subsequent panels.

To improve specificity while simplifying workflow, CD200 was incorporated as a mesothelial-associated marker. This allowed clear separation of mesothelial cells from EpCAM-positive epithelial cells using surface staining alone. The transition to an exclusively surface-marker-based panel significantly improved assay robustness and enabled the inclusion of viability assessment Shapiro, 2005; Wood, 2016).

7.5.4 Incorporation of viability and nuclear dyes

Viability dyes such as 7-amino-actinomycin D (7-AAD), 4′,6-diamidino-2-phenylindole (DAPI), and propidium iodide (PI) may be included in the panel for distinguishing live and dead cells, with 7-AAD being most common. These dyes selectively penetrate cells with compromised membranes (nonviable), binding to DNA and emitting fluorescence, allowing researchers to assess cell viability in various experimental conditions. Cells stained by these dyes are excluded from analysis in the context of effusion fluid FCM. Viability assessment is particularly important in effusion samples, which often contain apoptotic cells and debris that may bind antibodies nonspecifically. However, it has been observed that certain nonnucleated debris events may fall within the 7-AAD–negative gate and show nonspecific EpCAM binding.

Another set of dyes termed SYTO dyes are cell-permeant nucleic acid stains that bind to RNA and DNA, producing fluorescence upon interaction. These dyes can enter both live and dead cells and are usually used in combination with PI or 7AAD. They are also widely used in FCM to distinguish live and dead cells (Tárnok, 2008; Wang et al., 2022). Addition of SYTO-13 has been shown to confirm nucleated events, thereby improving analytical accuracy and reducing false-positive interpretation in the author's experience.

Predictive biomarkers like PD-L1 (CD274) may be added to the panel if the experimental design permits. Beyond its diagnostic role, PD-L1 expression on malignant epithelial cells provides clinically relevant information regarding eligibility for immune checkpoint inhibitor therapy, particularly as a second-line treatment option (Han et al., 2020; Jiang et al., 2019; Mahnke et al., 2010). This illustrates the evolving role of FCM from a purely diagnostic tool to a platform capable of delivering predictive and theranostic information.

7.5.5 Fluorochrome selection principles

A well-constructed FCM panel is essential for generating accurate and reproducible results. Thoughtful selection of fluorochromes, careful minimization of spectral overlap, appropriate use of controls, and optimization of antibody concentrations all contribute to the reliable performance of the panel and the accuracy of the data. The general principles for panel design for effusion fluids remain the same as discussed in Chapter 5. To summarize again:

- High antigen density markers (e.g., CD45) are optimally paired with relatively dim fluorochromes, preserving brighter fluorochromes for low-density antigens.
- Low or variable expression markers (e.g., PD-L1) should be conjugated to bright fluorochromes (e.g., PE and Allophycocyanin (APC)) to enhance detection sensitivity.
- Viability and nuclear dyes should be spectrally compatible with the antibody panel and should minimally interfere with marker detection.

7.5.6 Panel design based on instrument capability

7.5.6.1 Four color instruments

A minimal yet effective screening panel may include:

1. CD45
2. EpCAM
3. CD200
4. Viability dye

This configuration allows basic identification of malignant epithelial cells while excluding hematopoietic cells, mesothelial cells, and nonviable events.

7.5.6.2 Five- to six-color instruments

1. CD45
2. EpCAM
3. CD200
4. Viability dye
5. CD14
6. SYTO 13

Additional parameters such as nuclear dyes or macrophage exclusion markers (e.g., CD14) can be incorporated, improving specificity and analytical confidence. SYTO-13 is excited by the 488 nm laser and emits in the green spectrum with an emission maximum around 509 nm and a broad emission profile extending across the FITC detection range and into adjacent channels. Because of this spectral spread, assigning critical antibody markers to neighboring detectors may increase compensation complexity and reduce resolution; therefore, leaving the adjacent channel unassigned or reserved for noncoexpressed markers is advisable during panel design.

7.5.6.3 10-color and higher-end instruments

High-parameter instruments enable additional epithelial markers with simultaneous detection of diagnostic, exclusion, and predictive markers. There is better resolution of rare cell populations and reduction in sample consumption through multiplexing. Table 7.2 shows an example of a panel for a 10-color instrument.

Table 7.2 Table showing panel design for a 10-color instrument.

Marker name	Rationale	Fluorochrome	Remarks
Viability dye	Exclude dead cells	e.g., Fixable viability dye (near-IR)	Bright and outside main detection
SYTO-13 (nuclear dye)	Confirm nucleated events	SYTO-13 (FITC/488 channel)	Requires careful compensation due to broad emission
CD45	Hematopoietic exclusion	BV510	High expression → dim/moderate fluorescence
EpCAM (CD326)	Epithelial detection	PE	High discrimination for carcinoma cells
CD200	Mesothelial cell marker	APC	Helps separate mesothelial from epithelial cells
CD14	Macrophage exclusion	PerCP-Cy5.5	Avoids misclassification of high scatter macrophages
PD-L1 (CD274)	Predictive biomarker	BV421	Bright fluorescence for lower-density markers
D2-40 (Podoplanin)	Mesothelial lineage	BV785	Separates mesothelial from carcinoma cells
MUC-1	Secondary epithelial marker	PE-Cy7	Adds specificity in cases with variable EpCAM
Dump or empty channel (optional)	Exclude unwanted cells	e.g., CD3/CD19/CD56 (Alexa700)	Bundle lymphoid and NK markers into one channel

7.6 Sample staining

For laboratories initiating an effusion FCM program, the routine inclusion of an unstained control tube is a fundamental practice to establish baseline analytical confidence. This aliquot, which is processed identically to the stained sample but without the addition of fluorochrome-labeled antibodies, serves as a critical reference for identifying autofluorescence—a phenomenon common in large epithelial cells, macrophages, and reactive mesothelial cells that can otherwise lead to false-positive interpretations.

Especially during the validation phase of a newly designed experiment, where testing at least 10 benign and 10 malignant samples is recommended, the unstained control provides the necessary baseline to differentiate true immunophenotypic signals from nonspecific intrinsic fluorescence. This step is particularly relevant in the "routine approach" for beginners as they learn to navigate the high levels of debris and complex cellular backgrounds typical of serous fluids.

The following section elaborates on the staining protocol by stain-lyse-wash and lyse-stain-wash techniques.

7.6.1 Stain-lyse-wash

Set aside one aliquot (100 μL) as the "unstained control tube". Use the second aliquot of 100 μL for immunostaining. Label this tube as "stained tube" and proceed as later for the stain-lyse-wash technique:

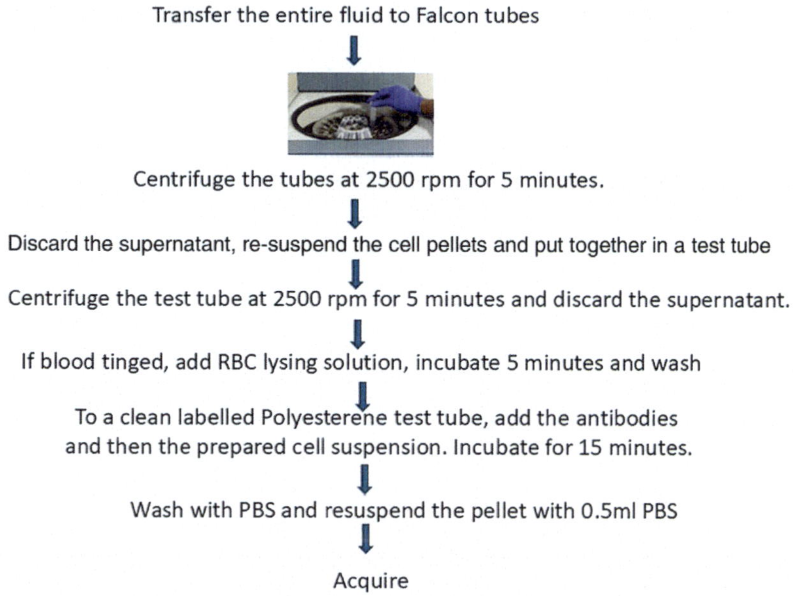

FIGURE 7.1 Sample processing and staining by lyse-stain-wash method. Flow chart for sample processing and staining by lyse-stain-wash method.

- **Staining:** Add the specified amount of fluorochrome-labeled antibody to the stained tube aliquot and keep it for 15 min at 25°C in the dark. (Note: The concentration and volume of each antibody required per test vary and should be titrated before the experiment).
- **Lysing:** RBC lysis should be performed using the commercially available RBC lysing solution or in-house ammonium chloride-based reagents. The choice of reagent depends on the experimental design (e.g., viability, downstream staining, surface marker preservation, fixation of cells, etc.). The RBC lysing agent (e.g., FACS Lyse, Becton Dickinson (BD) Biosciences) may be used in a 1:10 dilution (200 μL diluted with 1800 μL of distilled water) or as per standard lab protocol. Lysing reagent must be added to the FACS tubes and left for 15 min aftcr staining.
- **Washing:** Centrifuge the tube at 1000 rpm for 5 min. Discard thc supernatant and wash with PBS again, and finally resuspend in 500 μL of the paraformaldehyde or PBS.
- Process the unstained control tube in a similar fashion except for adding antibodies to it.

7.6.2 Lyse-stain-wash

This technique is preferred over "stain-lyse-wash" for samples which are overtly hemorrhagic or when attempting high-sensitivity rare-event analysis. The cellular component of body fluid is sedimented and washed (as explained earlier), followed by lysing. Thereafter, the antibodies are added and excess antibody washed again before acquisition. Fig. 7.1 depicts the flow chart for sample processing and staining by the lyse-stain-wash method.

7.7 Acquisition strategy

High event acquisition is a cornerstone of assay sensitivity. Effusion fluids frequently contain malignant epithelial cells at very low frequencies, sometimes well below 1% of total nucleated cells. Therefore, acquisition should not be prematurely terminated based on time or event count alone. Instead, acquisition should continue until the sample tube is fully aspirated whenever feasible. Earlier studies have reported acquisition of nearly 10,000 events. With the advancing technology, a minimum of at least 100,000 total events should be acquired for a routine assay. For a high-sensitivity assay aimed at detecting a rare cell fraction, an attempt should be made to acquire 1 million events or more. The diagnostic reliability of a negative result increases with event number, and quantitative thresholds based on absolute event counts rather than percentages provide more robust interpretation. Identification of clustered epithelial events above a validated background threshold improves both specificity and reproducibility. Establishing a LOB and LOD using benign control samples allows the laboratory to define a minimum number of EpCAM-positive clustered events required to classify a sample as suspicious or positive.

Data acquisition must be performed with careful attention to instrument stability, compensation, and artifact removal. Air bubbles and flow disturbances occurring toward the end of aspiration can generate spurious signals; therefore, a time gate should be applied to exclude terminal artifacts.

7.8 Gating and data analysis

The gating strategy should follow a logical, hierarchical structure that progressively enriches for clinically relevant events. Initial singlet gating, forward and SSC gating, excludes doublets, debris, and subcellular fragments. Subsequent gating depends upon the experimental design: routine analysis vs high-sensitivity rare-event analysis, use of viability or nuclear dyes, extent of the markers utilized, type of flow cytometer and acquisition software, etc. The following section details both strategies—ranging from the routine approach to high-sensitivity, rare-event analysis—allowing readers to select the most appropriate method for their current laboratory infrastructure or to transition from foundational techniques to advanced diagnostic frameworks as their technical expertise and clinical requirements evolve.

1. Routine approach

- **Live gating**: applied on FSC and SSC, aids in removing the debris, which usually has a very low FSC and higher SSC. However, in effusion fluids, the epithelial tumor cells (variable size and cytoplasmic content) or their fragments present in the debris area are not excluded from analysis; a broad initial gate in the FSC versus SSC dot plot is usually employed.
- **Singlet gating:** Singlet gating is an important strategy that utilizes the area and height parameters of FSC of cells. The singlet gate (on FSC-Area on the X-axis vs FSC-Height on the Y-axis) is used to exclude clusters of cells or doublets that could interfere with the analysis. A diagonal gate is drawn to select single cells, and cells/events falling off the diagonal toward the right of the main population on the X-axis are considered doublets or clusters.

Doublets have the same height as singlets but have increased pulse area. Failure to exclude doublets/clusters will result in events with high FSC and fluorescence, suggesting large cells with high antigen expression and hence false positives. Exclusion of doublets remains an essential component of DNA ploidy analysis as well as for epithelial cell detection and various other applications in recent times.

- **Sequential gating on subsets**: As the cells of interest are epithelial cells, hematopoietic cells expressing CD45 and CD14 (or other leucocyte marker) should be gated out and eliminated from the analysis. The cells that remain after gating out hematopoietic cells are either mesothelial cells or tumor cells (nonheme cells). Tumor cells expressing epithelial markers like EpCAM, CD326, or MUC1 are picked up in the next step. If a mesothelial marker like CD200 has been used, it will further improve the distinction between epithelial and mesothelial cells. Once the subset representing the cells of interest has been identified after sequential gating, the percentage of cells in the selectively gated population can be calculated.
- The proportion of EpCAM + events as numerator and "nonheme events" as denominator gives the percentage of EpCAM positivity. Based on a predecided cutoff, a decision regarding malignant effusion can be made. This method has been followed in most of the earlier studies. Gostomczyk et al. (2024) differ in their approach, calculating EpCAM-positive events as a proportion of "all cells" and concluding that it performed better as compared to EpCAM-positive events as a proportion of nonheme cells.
- Tumor cell population may range from small to medium to large in size. They can also be seen to be lying at very low FSC and SSC, suggesting cellular fragments, most likely produced during sample preparation. Different researchers adopt varying approaches to this: some consider including these cellular debris and fragments in the analysis by applying a broad initial gate, whereas others prefer to exclude them by using viability and nuclear dyes (Kentrou et al., 2011; Ramalingam et al., 2025).
- **Back-gating:** Also known as reverse gating, it involves taking a previously gated population (e.g., EpCAM +) and reapplying it to earlier scatter plots (e.g., FSC-SSC) to confirm that the selected cells align with expected characteristics. If the back-gated population clusters correctly within expected regions of FSC-SSC, the gating strategy is confirmed. If not, adjustments may be needed.
- Example of a routine percent-based analysis is shown in Fig. 7.2.

Determining the threshold for positive classification in routine approach

Deciding the threshold or cutoff beyond which an effusion fluid is considered positive for malignant epithelial cells has been approached in multiple ways by researchers:

- Based on FMO (fluorescence minus one) control in which samples are stained with all fluorophores except one, helping to determine spillover effects and set accurate gating boundaries.
- By assessing background staining in benign effusions.
- Unstained and isotype control samples to identify nonspecific fluorescence. No fluorochrome-labeled antibody is added in the unstained control as described earlier.

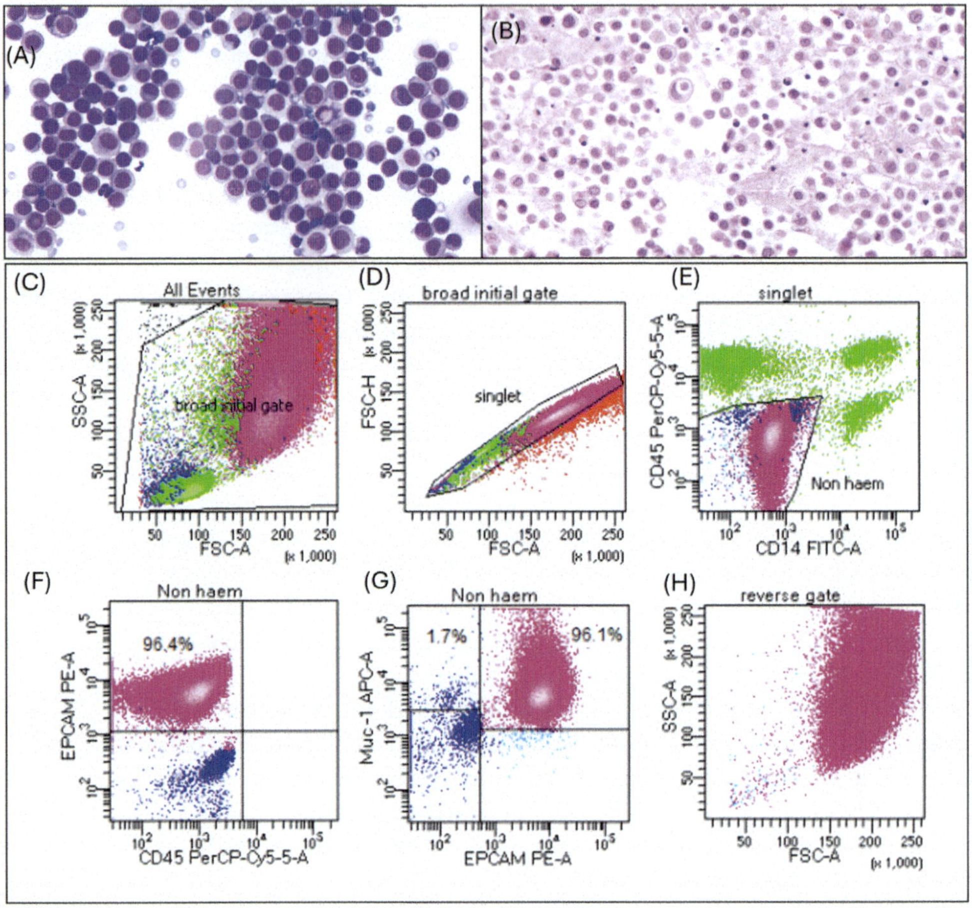

FIGURE 7.2 Cytology smears (A), cell block (B) and flow cytometry scatter plots (C–H) of a fluid involved by lobular carcinoma breast. Cytology smears (A), cell block (B) showing sheets and clusters of tumor cells in a background of sparse mesothelial cells, histiocytes and lymphocytes. Flow cytometry scatterplots with sequential gating showing broad initial gate (C), singlet gate (D), nonhaem gate (E) to exclude CD45+ and CD14+ events. Epithelial tumor cells expressing EpCAM and MUC1 shown in bright pink (F–G) have high forward and side scatter on back/reverse gate (H). *Source: Adapted and modified from Gaur, G., Awasthi, N.P., Gupta, A., Agarwal, A., Sachan, R., Malhotra, K.P., Shukla, S., Singh, A.K., Singh, P. & Husain, N. (2023). Diagnostic accuracy of flow cytometry in detecting malignant epithelial cells in serous effusions. Journal of the American Society of Cytopathology, 12(6), 423–435. https://doi.org/10.1016/j.jasc.2023.09.003.*

An isotype control is designed to match the host species, immunoglobulin subclass, and fluorophore of the primary antibody but lacks specificity for the target antigen. These controls are variably adjusted at 95%, 98%, or 99% with a 20% cutoff for a population to be considered positive for that marker. Isotype controls were commonly used in earlier studies but have become less prevalent in recent research.

- Based on receiver operating characteristic (ROC) curve analysis. In this, true positive rate (sensitivity) is plotted against the false-positive rate (1-specificity) at various threshold levels, and the point giving the best area under the curve (AUC) is considered the optimum cutoff.

2. High-sensitivity, rare-event-based approach

Initial forward and SSC gating excludes debris and subcellular fragments. Gating on nucleated events eliminates nonnucleated debris that may otherwise fall into viability-negative regions and falsely appear EpCAM-positive. Subsequent viability gating using 7-AAD excludes dead cells, which commonly exhibit increased autofluorescence and nonspecific antibody uptake. Exclusion of CD45-positive hematopoietic cells narrows the analysis to nonhematopoietic elements. Within this compartment, mesothelial cells can be identified using markers such as CD200, allowing further refinement of the epithelial gate. True malignant epithelial populations typically appear as cohesive clusters of EpCAM-positive events with characteristic intermediate to high FSC properties, rather than scattered low-intensity events. Positive classification requires both quantitative and qualitative data, with a predefined minimum number of epithelial events forming a cluster as a prerequisite. This approach utilizes principles similar to those used for measurable residual disease analyses. LOB, LOD, and limit of quantification (LOQ) may be calculated (further explained in the next section). An example of a high-sensitivity analysis is shown in Fig. 7.3.

Table 7.3 shows a summary of processing techniques, gating and analysis strategies used by various researchers for effusion fluid FCM.

(Acosta et al., 2016; Davidson et al., 2002; Gaur et al., 2023; Gostomczyk et al., 2024; Kentrou et al., 2011; Krishan et al., 2010; Pillai et al., 2013; Ramalingam et al., 2025; Risberg et al., 2000; Sahu et al., 2021; Sayed et al., 2009; Wong-Arteta et al., 2020).

7.8.1 Template generation

A critical component of assay standardization is the generation of analytical templates derived from confirmed benign effusion samples, which should be done for both routine and high-sensitivity assays. These templates define normal scatter characteristics, establish background EpCAM signal levels, and map the distribution of leukocytes and mesothelial cells. By analyzing multiple benign samples, the laboratory can determine the expected range of nonspecific Ep-CAM binding and define objective cutoffs. Once validated, the gating hierarchy and fluorescence thresholds should be locked within the template and applied uniformly across subsequent cases. This minimizes subjective gate adjustment and reduces interoperator variability. Template-driven analysis transforms FCM from a visually interpreted test into a structured quantitative assay with predefined interpretive criteria Fig. 7.4.

The advantages of template-based analysis are particularly evident when compared with conventional cytomorphology. Cytological assessment, while considered the gold standard, depends heavily on morphological interpretation and may be limited by sampling adequacy or reactive atypia. FCM, when standardized, provides objective enumeration of epithelial cells,

Table 7.3 Table summarizing sample processing technique, gating and analysis strategies in various studies.

S. no	Author	Number of cases (n)	FCI antibody panel	Technique	Gating	Average number of events acquired	Cutoff value for FCI	Study design
1	Ramalingam et al. (2025)	121	EpCAM, CD200, CD45, 7 AAD, SYTO 13	Sediment, wash, lyse-stain-wash	FSC-SSC and Viable nuclear gate	Desired minimum 100,000 events	Qualitative with Limit of detection at 500 clustered EpCAM events	FCI results compared with cytology and cell block
2	Gostomczyk et al. (2024)	119	CD45, CD14, and EpCAM	Wash, sediment, stain-lyse- wash	Borad initial gate to include debris region. Exclusion of doublets	50,000–1,00,000 events	0.5%	FCI compared with cytology. Analysis also done at various cutoffs to improve sensitivity and specificity. EpCAM +/(CD45-CD14-) & EpCAM+/'All cells' was also analyzed
3	Gaur et al. (2023)	96	CD45, CD14, EpCAM, and MUC1	Sediment, wash, count, and stain- lyse- wash	Broad initial gate to include debris region. Exclusion of doublets	50,000 events	1% for EpCAM, 2.5% for MUC 1 (only quantitative)	FCI results compared with cytology and ICC
4.	Sahu et al. (2021)	75	CD45,CD14, and EpCAM	Sediment, lyse-wash-stain-wash	Broad initial gate to include debris region. Exclusion of doublets	10,000 singlets	≥ 2% (only quantitative)	FCI results compared with cytology and ICC
5	Wong-Arteta et al. (2020)	101	CD9, CD45, CD71, CD99, CD56, CD 10, CD81, NG2, EpCAM, CD200, and CD14	Screening algorithm based on Cell count performed by automated analyzer in body fluid mode and cytology. Sediment, stain-lyse-wash	Broad initial gate to include debris region. Exclusion of doublets	Not mentioned	Not mentioned	FCI results compared with cytology and ICC

(continued on next page)

6	Acosta et al. (2016)	134	CD45, CD33, CD326(BerEP4)	Wash, sediment, stain-lyse- wash	Exclusion of debris and doublets by FSC-SSC	100,000 events	FMO control tube with CD45, CD33 in every sample to validate CD326 positive cells. At least 20 CD326 positive events forming a cluster	FCI results compared with final clinical diagnosis. The study included bone marrow, FNA, lymph node biopsies, other tissues, and CSF in addition to effusion (total n = 238, Effusion n = 134)
7	Pillai et al. (2013)	195	CD45, CD14, and BerEP4	Sediment, wash, filter (50µ mesh), count, 2 million cells in 100 µL. Stain- lyse-wash	Broad initial gate to include debris region. Exclusion of doublets	5,00,000 events	0.5% based on FMO control and background staining in benign effusions. Both quantitative and qualitative data (clustered events) considered	FCI results compared with cytology and ICC
8	Kentrou et al (2011)	125	EMA, BerEP4, CD66, CK, CD56, Desmin-33, CD71 and CD45, SYTO 16, and 7 AAD DNA ploidy	Stain-lyse-wash for surface staining. Permeabilization-fixation for Desmin 33	Initial gate based on SYTO 16, 7AAD, CD45, and scatter	60 000 events	Isotype controls to define cutoff. Positivity threshold at 20% expression	FCI results compared with cytology. DNA ploidy in CK+ cells was also assessed
9	Krishan et al (2010)	79	BerEP4, PR, MUC4 and TTF-1, DNA ploidy	Filtration, sedimentation +- layering over Ficoll-Hypaque for removing RBC, washing, sedimentation. Stain, wash for surface markers, perm fix for TTF1, MUC4	Simultaneous aneuploidy vs marker expression were studied	10,000 events	Isotype control set at 95% cutoff	FCI results compared with cytology and DNA ploidy on BerEP4+ cells were also assessed

(*continued on next page*)

Table 7.3 Table summarizing sample processing technique, gating and analysis strategies in various studies—cont'd

S. no	Author	Number of cases (n)	FCI antibody panel	Technique	Gating	Average number of events acquired	Cutoff value for FCI	Study design
10	Sayed et al. (2009)	119	BerEP4, CK, CD3, and CD45	Wash, sediment, lyse-stain- wash (CK, CD3, and CD45). Permeabilization-fixation before adding BerEP4 (cytoplasmic). Cycle test plus kit for DNA	Exclusion of debris by FSC SSC	Minimum number of cells acquired: 10,000	Isotype negative controls used to set quadrants positivity threshold at 20% expression	FCI results compared with cytology. DNA ploidy also assessed
11	Davidson et al. (2002)	92	CD45, CD14, Ber-EP4, AH6, TN, B72.3, EMA, and N-cadherin	Thawing of frozen sample, washing, staining by direct and indirect staining method, followed by filtration. No lysing	Exclusion of debris by FSC SSC	Mean number of gated cells analyzed: 6300	Isotype control set at 99% cutoff to set quadrants. Combined positivity of four antibodies-BerEP4, AH6, TN, and B72.3 in each sample was taken as positive on FCI	FCI results compared with cytology and ICC
12	Risberg et al. (2000)	49 (included peritoneal washings)	CD45, BerEP4 and N-cadherin	Thawing of frozen sample, washing, staining by direct and indirect staining method, followed by filtration. No lysing	Exclusion of debris by FSC SSC	Mean number of gated cells analyzed: 6300	Isotype control set at 99% cutoff to set quadrants	FCI results compared with cytology and ICC

FCI, flow cytometric immunophenotyping; *FMO*, fluorescence minus one control; *FNA*, fine needle aspirate; *FSC*, forward scatter; *ICC*, immunocytochemistry; *SSC*, side scatter.

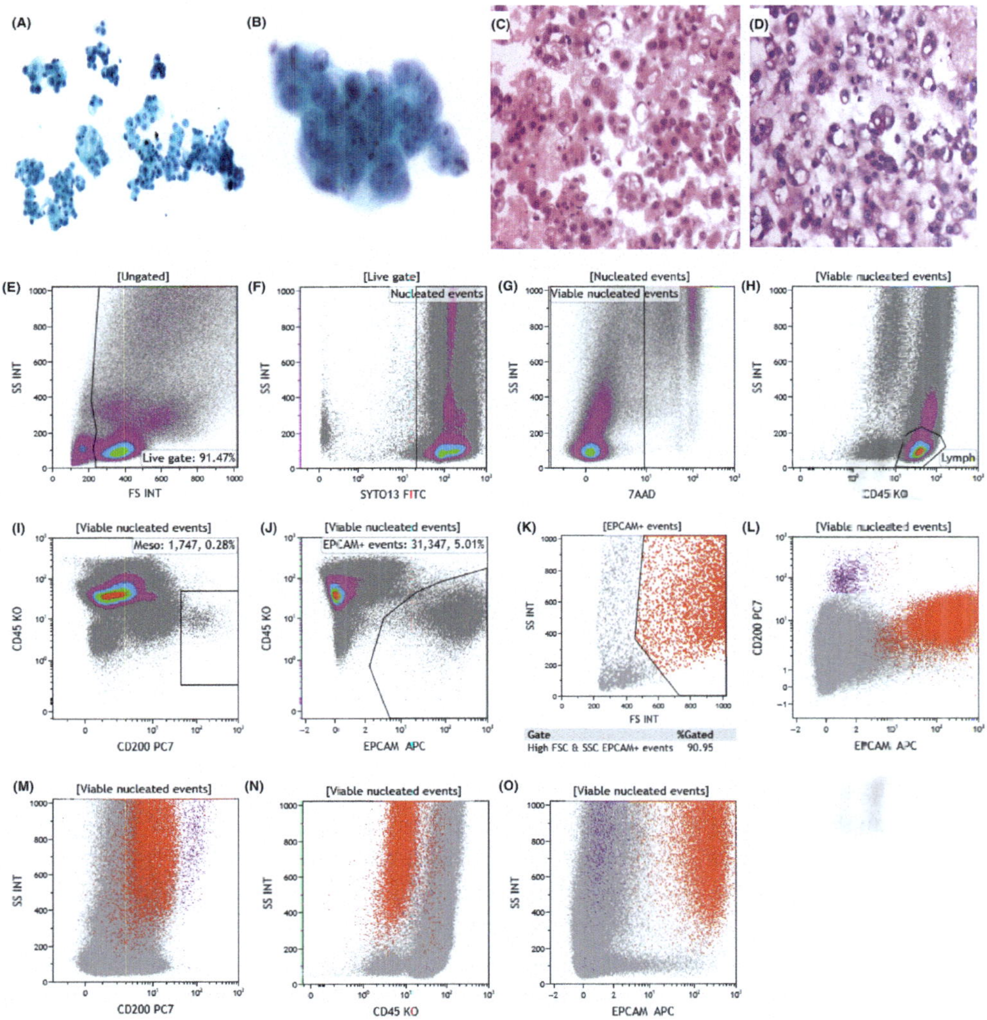

FIGURE 7.3 Liquid-based cytology (LBC) microphotographs (A–B), cell block (C–D) & flow cytometry (FCM) scatter plots (E–O) of a fluid involved by adenocarcinoma. LBC (A–B); cell block microphotographs (C–D) & representative flow cytometry (FCM) scatter plots (E–O). The sequential gating is shown to exclude the nonnucleated events and dead cells with SYTO13 (F) and 7-AAD (G), respectively. Mesothelial cells are gated with CD200 (I) and shown in purple. The EpCAM-positive events are shown in red. They show clustering in the EpCAM gate (J) and high scatter (K). *Source: Adapted and modified from Ramalingam, T.R., Mani, S., Narla, S., Lakshmanan, A., Muthu, A., Nk, H.R., Mohanraj, S., Ramachandran, E.D., Subramanyam, A. & Vaidhyanathan, L. (2025). Simple and rapid flow cytometry assay for the detection of malignant epithelial cells in body fluids. Cytopathology, 36(1), 23–30. https://doi.org/10.1111/cyt.13408.*

high-throughput analysis of large cellular populations, and reproducible quantitative thresholds. Furthermore, acquisition of high event numbers improves the negative predictive value of the assay and supports more confident exclusion of malignancy in equivocal cases. In this manner,

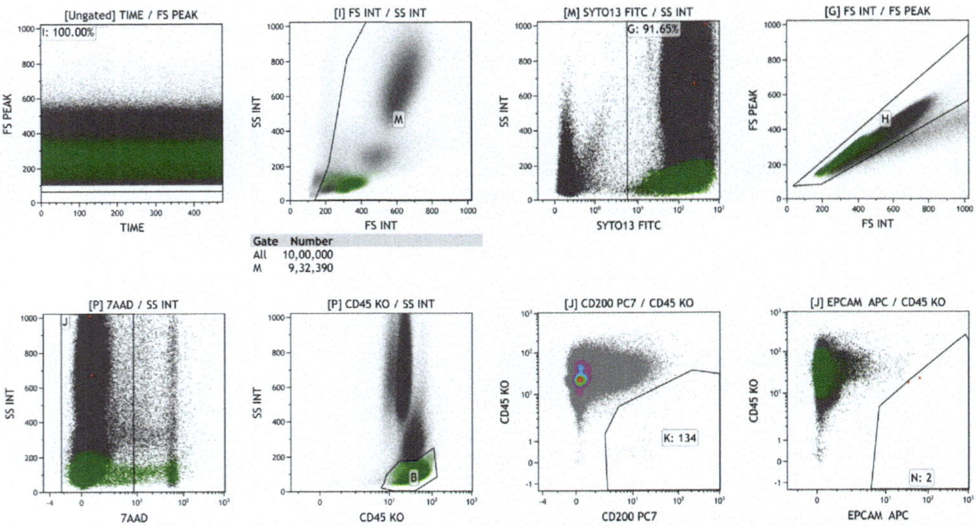

FIGURE 7.4 Analysis template for sequential gating strategy. Objective results are obtained immediately after the .fcs files and compensation settings are imported to the template. The earlier template-based analysis is that of a benign effusion sample processed for establishing LOB. *LOB*, Limit of blank.

FCM does not replace cytomorphology but rather complements it by adding sensitivity, objectivity, and reproducibility.

7.8.1.1 Establishing sensitivity in a high-sensitivity, rare-event analysis

Establishing analytical sensitivity is a critical component in the validation of flow cytometric assays designed to detect malignant epithelial cells in effusion fluids. In simple terms, sensitivity refers to the ability of an assay to distinguish true signal from background and to reliably measure very low levels of the measurand. In the context of effusion cytology, the measurand is the population of malignant epithelial cells, typically identified by EpCAM expression within viable, nucleated cells. Because malignant cells in effusions may occur at very low frequencies—often ≤1% of total nucleated cells—the assay must be validated as a rare-event detection system, similar in principle to paroxysmal nocturnal hemoglobinuria (PNH) assays and minimal residual disease methodologies. This involves establishing the LOB and LOD, which are the analytical background of the assay.

The LOB represents the highest apparent signal expected in the absence of the measurand. For effusion analysis, this is determined using confirmed benign effusion samples that lack malignant epithelial cells by cytomorphology and clinical follow-up. These benign samples are processed identically to clinical specimens, including in toto acquisition wherever feasible, and analyzed using the standardized gating template. Within the final epithelial gate—typically defined as viable, nucleated, CD45-negative, mesothelial-excluded events—the number of EpCAM-positive clustered events is recorded. The LOB is calculated as the mean background event count plus 1.645 times the standard deviation, corresponding to the 95th percentile of the blank distribution. This establishes the expected upper boundary of background noise arising from nonspecific antibody binding, autofluorescence, spectral spread, and rare reactive epithelial contaminants (Armbruster & Pry, 2008; Kim et al., 2023).

The LOD defines the lowest level of malignant epithelial cells that can be reliably distinguished from the LOB. Statistically, the LOD may be approximated as the mean background plus three standard deviations, ensuring that low-level positive samples will exceed the LOB with approximately 95% probability. Practically, determination of LOD requires analysis of multiple benign effusions—ideally at least 10 independent samples—to capture intersample variability. High event acquisition is essential during this phase. Because rare-event detection is governed by counting statistics, acquisition of large numbers of gated nucleated events, preferably in the range of several hundred thousand to one million when possible, enhances statistical confidence. The greater the number of analyzed events, the narrower the confidence intervals and the lower the detectable frequency threshold. In effusion specimens with limited cellularity, complete tube acquisition ("acquire until dry") maximizes event yield and improves assay sensitivity (Armbruster & Pry, 2008; Kim et al., 2023).

While the LOB and LOD define analytical sensitivity, clinical utility depends on functional sensitivity, expressed as the lower limit of quantification (LLOQ). The LLOQ represents the lowest concentration of malignant epithelial cells that can be detected with acceptable accuracy and precision. Establishing the LLOQ requires analysis of samples containing low but known levels of epithelial cells, ideally approaching the LOD. In practice, this may involve dilution experiments using EpCAM-positive carcinoma cell lines spiked into benign effusion matrices or serial dilutions of confirmed malignant effusion samples. At each dilution level, replicate acquisitions are performed, and the coefficient of variation (CV) of event counts is calculated. The LLOQ is defined as the lowest concentration at which the assay maintains predefined precision criteria (Davis et al., 2013). In flow cytometric biomarker assays, imprecision in the range of 20%–35% CV has been considered acceptable at the LLOQ; however, each laboratory must define tolerability limits based on the intended clinical application and document these criteria within the method validation plan.

Counting statistics also play a central role in defining minimum acceptable event numbers. Rare-event reliability improves as the absolute number of detected malignant events increases. Typically, a minimum of 50–100 clustered malignant events is considered desirable for robust quantification, though this threshold must be interpreted in light of total event acquisition and clinical context. For qualitative detection—distinguishing positive from negative—the requirement may be lower, provided that the event cluster clearly exceeds the LOB and demonstrates consistent immunophenotypic characteristics. However, when quantitative reporting is intended, such as monitoring tumor burden across serial effusions, higher event counts are necessary to ensure statistical stability.

It is important to recognize that sensitivity is influenced not only by statistical thresholds but also by biological and technical factors. Proper exclusion of debris using forward and SSC, gating on nucleated cells with SYTO-13, elimination of nonviable events using 7-AAD, and exclusion of CD45-positive hematopoietic cells collectively reduce background noise and effectively lower the LOB. Similarly, appropriate fluorochrome selection, careful compensation, and avoidance of spectral spreading into critical channels improve signal resolution and enhance true assay sensitivity. Thus, analytical sensitivity is not solely a mathematical construct but a function of optimized panel design, standardized gating templates, and rigorous preanalytical control.

When these principles are applied systematically, the assay's sensitivity can be clearly defined in terms of detectable frequency and minimum absolute malignant cell counts. By establishing the LOB, LOD, and LLOQ within the framework of rare-event analysis, laboratories can provide objective performance characteristics for effusion FCM. This structured validation approach transforms the detection of malignant epithelial cells from a subjective interpretive exercise into a quantitatively defined diagnostic method with measurable analytical performance parameters.

7.8.1.2 Compensation

Fluorescence compensation represents another critical determinant of assay accuracy. Because most fluorochromes exhibit spectral overlap, fluorescence emitted in one detector may spill into adjacent channels. For example, nuclear dyes such as SYTO-13, which are excited by the 488 nm laser and emit maximally around 509 nm, possess a broad emission spectrum that may spill into adjacent detectors. If a highly sensitive marker is placed in a neighboring channel, spreading error may reduce resolution of dim populations. Accordingly, panel configuration should account for spectral proximity, and adjacent channels may be intentionally left unassigned or reserved for noncritical markers to preserve assay sensitivity. In multiparametric panels that include EpCAM, CD45, CD200, viability dyes such as 7-AAD, and nuclear stains such as SYTO-13, compensation must be precisely calculated to prevent false-positive or false negative event assignment. Single-stained compensation controls using the same fluorochrome-antibody conjugates employed in the assay should be used for matrix calculation.

Autofluorescence controls are particularly relevant in effusion samples, where macrophages and dead cells may display intrinsic fluorescence. Improper compensation may artificially inflate background EpCAM events or reduce resolution between true epithelial cells and nonspecific signals, thereby directly affecting LOB and LOD determination.

In addition to mathematical compensation, panel design must consider fluorescence spreading error. Bright fluorochromes assigned to highly expressed markers can introduce variability into neighboring detectors, potentially masking dim antigen expression. Therefore, fluorochrome assignment should follow established principles: bright fluorochromes for low-density antigens and relatively dim fluorochromes for highly expressed markers. This approach reduces spreading error and improves rare-event discrimination.

7.8.2 Template validation

Template validation constitutes the final layer of analytical standardization. The gating hierarchy—from debris exclusion and nucleated cell gating to viability exclusion and hematopoietic removal—should be fixed within the template. Once validated, the template should be applied uniformly to subsequent clinical samples without arbitrary gate adjustment.

This "template locking" approach minimizes operator subjectivity and ensures that differences observed between cases reflect biological variation rather than analytical drift.

Periodic template revalidation is recommended, particularly when changes occur in instrument configuration, antibody lot numbers, or fluorochrome selection. Validation should include reassessment of background EpCAM signal and confirmation that LOB and LOD values remain consistent with initial performance characteristics. Documentation of these validation exercises supports regulatory compliance and strengthens assay credibility.

Importantly, instrument performance, QC consistency, compensation accuracy, and template stability are interdependent factors. A well-designed panel may fail diagnostically if daily QC is inadequate or if compensation is incorrectly applied. Similarly, precise compensation cannot compensate for unstable laser output or inconsistent voltage settings. For rare-event assays such as effusion epithelial cell detection, analytical robustness requires simultaneous control of all these variables.

When these elements are rigorously maintained, the assay demonstrates high reproducibility, stable background noise, and consistent detection thresholds. This transforms FCM from an operator-dependent technique into a standardized quantitative diagnostic method. In effusion cytology, where malignant cells may be sparse and morphologically ambiguous, such analytical precision is indispensable for achieving reliable and clinically meaningful results.

7.9 Reporting of results

A definitive positive result is characterized by a discrete, cohesive cluster of viable, nucleated, CD45-negative events demonstrating reproducible EpCAM expression with adequate fluorescence intensity and appropriate scatter characteristics. These events should localize within the expected forward and SSC region for epithelial cells, typically showing intermediate to moderately high FSC relative to debris. The cluster must be clearly separable from background noise and other cellular compartments, demonstrating a favorable signal-to-noise ratio. Importantly, the absolute event count should exceed the validated LLOQ threshold, ensuring that the detection is not only statistically distinguishable from background but also quantitatively reliable with acceptable assay precision. Caution is warranted when EpCAM-positive events fail to demonstrate appropriate scatter characteristics or appear within the debris region. Events with low FSC, irregular distribution, or proximity to the noise boundary may represent nonspecific antibody binding, apoptotic fragments, or spectral artifacts.

Situations may arise in which a discernible EpCAM-positive cluster is identified above the LOD but below the LLOQ. In such cases, the assay demonstrates analytical detection but does not meet the criteria for reliable quantification. These findings should be reported as "suspicious for malignant epithelial cells" or "EpCAM-positive events detected below the quantifiable range," with a clear statement that the event count does not meet the laboratory's established quantification threshold. This category reflects the statistical reality of rare-event assays, where detection may precede reliable enumeration. In all such instances, mandatory correlation with cytomorphological findings and clinical context is essential. Flow cytometric findings should complement, rather than replace, conventional cytology, especially in borderline cases.

Reports should clearly state the total number of events acquired, the number of gated nucleated viable events, and the absolute count or percentage of EpCAM-positive events. When feasible, reporting absolute counts in addition to percentages provides greater clarity, particularly in low-cellularity specimens. Inclusion of interpretive comments regarding assay sensitivity thresholds (e.g., LOD and LLOQ) enhances transparency and communicates analytical rigor to the clinician.

In negative cases, the report should specify that no EpCAM-positive cluster exceeding the laboratory's LOD was identified after acquisition of an adequate number of events. Mentioning the total event count reinforces the strength of the negative result, as higher acquisition numbers increase the negative predictive value. However, it is important to acknowledge that a negative flow cytometric result does not entirely exclude malignancy, particularly in cases with very low tumor burden or sampling limitations. A note of the same can be mentioned in the report.

Ultimately, standardized reporting ensures consistency, reduces interoperator variability, and enhances clinical confidence. A structured interpretive framework—requiring cluster integrity, appropriate scatter characteristics, adequate signal-to-noise ratio, and threshold-based quantification—transforms effusion FCM from a qualitative observation into a statistically supported diagnostic modality. When interpreted within the context of established sensitivity parameters and correlated with cytomorphology and clinical findings, flow cytometric reporting provides a robust and reproducible adjunct in the diagnosis of malignant effusions.

7.10 Flow cytometry of effusion fluids beyond epithelial malignancies: Hematolymphoid infiltrates

Before we come to the end of this chapter, it is essential to understand the role of FCM in hematolymphoid infiltrates in effusions. Because hematolymphoid infiltrates are shed as single cells into serous cavities and do not form three-dimensional clusters, these are inherently amenable to cytometric interrogation. FCM has a role in determining clonality in lymphocytic effusions, detecting leukemic/lymphomatous infiltration, diagnosing plasma cell neoplasms, and primary effusion lymphomas (PEL). The approach largely follows similar principles as that in blood and bone marrow specimens with respect to panel design and marker selection. While the core technology remains the same, significant differences in the cellularity, cell types, and matrix of the biological sample may impact the results. Whereas in bone marrow and blood specimens, the primary objective is to diagnose and classify the hematolymphoid tumor, in effusions it is detection of spread/ infiltration. Knowledge of the immunophenotype of the primary tumor paves the way for detecting that "specific immunophenotype" in the background of mesothelial noise. Markers to exclude mesothelial cells are usually applied to clear the noise, followed by a search for leukemic/lymphomatous infiltrate. While extensive panels are applied for various cell types (B, T, NK, and immature) in bone marrow/blood, the clinical question in effusions is usually to confirm whether "it is a lymphoma?" Demonstration of clonality or immunophenotypic aberrancy aids in this. Differentiation of reactive from malignant lymphocytosis significantly reduces the "atypical/suspicious" category by providing objective evidence of clonality. While secondary involvement by systemic nonhodgkin lymphomas (NHL) like diffuse large B-cell lymphoma (DLBCL) and Burkitt lymphoma is the most frequent scenario, the pathologist must also consider rare site-specific entities such as PEL, particularly in the setting of viral coinfections and immunosuppression. FCM can detect small clonal populations, enabling diagnosis even in cytology-negative effusions, especially in patients with early-stage, treated lymphomas and minimal residual disease. This benefit supports the use of a reflex testing strategy in lymphocyte-rich effusions.

7.11 Future horizons and emerging technological advancements in flow cytometry

The landscape of effusion diagnostics is rapidly evolving beyond traditional cytomorphology and basic immunophenotyping. As clinical demands shift toward earlier detection, greater analytical precision, and integrated therapeutic decision support, emerging technologies and applications are poised to transform how malignant cells in effusion samples are identified, characterized, and interpreted. These advancements reflect both technological innovation within FCM and broader integration with molecular, computational, and surgical disciplines.

One of the most promising directions is the application of intra-operative flow cytometry (iFC) to guide real-time clinical decisions. Originally developed to assess tumor margins and analyze cell cycle or ploidy status during surgery in solid tumors, iFC has shown value in rapidly characterizing tumor cell populations in vivo within minutes of tissue acquisition (Alexiou et al., 2015; Alexiou et al., 2023; D'Amato Figueiredo et al., 2022; Markopoulos et al., 2024; Romeo et al., 2025). Reports indicate that intra-operative cytometric analysis can distinguish malignant from benign tissue and potentially inform surgical strategy and extent of resection during procedures for intracranial, breast, hepatocellular, and head and neck cancers. Building on these principles, iFC is now being evaluated in the context of peritoneal lavage and effusion analysis. A recent study in gastric carcinoma demonstrated that intra-operative flow cytometric detection of free carcinoma cells in peritoneal lavage fluid was more sensitive than cytology, even identifying tumor cell fractions as low as 0.01% that were missed by conventional cytology (Jagric et al., 2016). Importantly, such high-sensitivity detection provides critical information about the presence of free cancer cells that may contraindicate radical cytoreductive surgery, offering an evidence-based rationale for judicious operative decision-making (Ramalingam et al., 2025). iFC can measure the burden of free cancer cells in peritoneal lavage fluid in cancer patients and may be used as an effective biomarker for outcome and effectiveness of chemotherapy (Kitayama et al., 2014). iFC was used for characterization of a bone metastasis wherein abnormal DNA ploidy and expression of cytokeratin by the tumor cells clinched the diagnosis (Markopoulos et al., 2023).

Advancements in "high-purity cell separation and sorting technologies" are expanding the utility of effusion FCM from diagnosis toward deeper biological interrogation. Microfluidics and elasto-inertial separation platforms can enrich malignant tumor cells from effusion fluids in a label-free manner, enabling subsequent downstream analyses such as drug resistance profiling, next-generation sequencing, and molecular phenotyping (Myung & Hong, 2015; Warkiani et al., 2016). These innovations facilitate the transition from purely diagnostic cytometry to functional and translational studies that probe tumor biology, tumor microenvironment interactions, and therapeutic vulnerabilities at the single-cell level. When paired with high-speed cell sorters, enriched tumor cell populations can be isolated with sufficient purity and viability for genomic and transcriptomic analysis, which opens the door to personalized medicine applications that were previously the domain of solid tissue biopsies.

FCI can provide further patient-specific data in this era of personalized therapy. In effusions, hormone status (estrogen and progesterone receptors) and Her2 status in metastatic breast carcinomas or BRAF mutant status (through mutation-specific antibodies) in metastatic lung and colon cancers can be detected. Measurement of PDL-1 (CD274) expression in malignant effusions or peritoneal lavage fluids can direct these patients toward PDL-1 inhibitor therapies. Bispecific anti-EpCAM/CD3 antibodies are in use or being researched. Catumaxomab is successful for the treatment of malignant ascites (Hodge & Badgwell, 2019), while Solitomab is being investigated in clinical studies for patients with relapsed/refractory solid tumors that are not responsive to conventional therapy (Kebenko et al., 2018). When these therapeutic antibodies are utilized, FCM could be beneficial in evaluating the degree of EpCAM expression and its link with treatment response (Patriarca et al., 2012).

FCM itself is also undergoing technological modernization. High-parameter cytometry techniques such as spectral FCM and mass cytometry (CyTOF) allow simultaneous quantification of dozens of markers per cell, dramatically expanding the resolution of immunophenotypic profiling compared with conventional multicolor assays (Bandura et al., 2009; Luo et al., 2026; Nolan, 2022). These modalities permit comprehensive phenotyping of cellular subpopulations, including rare tumor cell subsets and phenotypically aberrant mesothelial or immune populations in effusions. The integration of such high-dimensional data with sophisticated analytical frameworks—potentially incorporating machine learning and graph-based network models for single-cell classification—holds promise for distinguishing subtle malignant signatures from background noise. Broadly, these next-generation cytometric approaches are likely to produce richer datasets that can be mined not only for diagnostic classification but also for insights into tumor heterogeneity, metastatic potential, and response to therapy.

Ultimately, the future of effusion diagnostics is likely to reflect a synergistic blend of rapid, high-sensitivity detection and integrative biological characterization. In clinical practice, this will translate into more accurate and earlier detection of malignant cells, improved prognostic stratification, and more informed therapeutic decision-making, including the appropriateness of extensive surgical interventions such as cytoreductive surgery. As technology continues to advance, and as computational and molecular tools are integrated more fully with FCM, effusion analysis will transition from a primarily diagnostic endeavor to a multifaceted platform that supports precision oncology, real-time surgical guidance, and comprehensive tumor profiling.

7.12 Conclusion

The integration of FCM into the diagnostic workup of serous effusions represents a fundamental shift from subjective morphological assessment toward a quantitative, lineage-specific analysis. This chapter has provided a dual-pathway framework designed to support practitioners at every stage of their professional development. By establishing a practical "bench manual" for routine methodologies, we have equipped beginners with the foundational steps—from sample collection and anticoagulation to basic immunophenotyping—required to initiate a reliable effusion FCM program. Simultaneously, the elaboration of high-sensitivity, rare-event protocols provides a rigorous, standardized framework for advanced practitioners seeking to detect minute tumor populations often missed by conventional cytomorphology. Successful execution requires close collaboration between the cytopathology and FCM laboratories. It is essential to provide comprehensive training for flow technologists and flow cytometrists, enabling them to plan, standardize, and implement experiments tailored to their laboratory's specific requirements.

The success of either approach is predicated on a deep understanding of the scientific and biological rationale underlying each procedural choice. We have emphasized the preanalytical variables, such as the use of anticoagulants to prevent microclots, adherence to a 24-h processing window, and in toto processing of samples, which are not merely administrative suggestions but essential technical requirements for preserving cell integrity and antigenicity. Furthermore, the transition from cytoplasmic markers to surface-marker-based panels (utilizing EpCAM as the diagnostic backbone) serves to maximize signal-to-noise ratios while enabling the inclusion of critical viability and nuclear dyes like 7-AAD and SYTO 13. These refinements are crucial for distinguishing true malignant clusters from the complex backgrounds of debris and reactive mesothelial cells characteristic of body cavity fluids.

A cornerstone of this chapter is the assertion that advanced diagnostic proficiency is measured by analytical rigor. Moving beyond visual identification, laboratories must adopt statistical validation measures, including the determination of the LOB, LOD, and lower limit of quantification (LLOQ). By "locking" analytical templates and defining objective interpretive criteria for positive, suspicious, and negative results, FCM is transformed from a qualitative observation into a statistically supported, reproducible diagnostic modality.

Beyond technical standardization, clinical efficacy requires a nuanced understanding of potential diagnostic pitfalls. Understanding the causes of false negative and false-positive results is crucial for making informed decisions in complex cases. When necessary, requesting a fresh sample and repeating the test can improve accuracy.

Looking forward, the evolution of effusion diagnostics will continue to be driven by technological innovation. The emergence of iFC and high-parameter techniques such as spectral and mass cytometry will further refine our ability to characterize tumor heterogeneity and provide real-time surgical guidance. When integrated with molecular platforms, these advancements will position FCM as a multifaceted platform for precision oncology, ensuring that effusion analysis remains at the forefront of personalized patient care.

AI Disclosure

During the preparation of this work, the author(s) used Gemini Pro to refine the language. The scientific content has not been altered. After using this tool/service, the author(s) reviewed and edited the content as needed and take(s) full responsibility for the content of the publication.

References

Acosta, M., Pereira, J., & Arroz, M. (2016). Screening of carcinoma metastasis by flow cytometry: A study of 238 cases. *Cytometry. Part B, Clinical cytometry., 90*(3), 289–294. doi:10.1002/cyto.b.21258.

Alexiou, G. A., Markopoulos, G., Voulgaris, S., & Vartholomatos, G. (2023). Usefulness of intraoperative rapid flow cytometry in the surgical treatment of brain tumors. *Neuropathology, 43*(3), 273–274. http://onlinelibrary.wiley.com/journal/10.1111/(ISSN)1440-1789. doi:10.1111/neup.12883.

Alexiou, G. A., Vartholomatos, G., Stefanaki, K., Lykoudis, E. G., Patereli, A., Tseka, G., Tzoufi, M., Sfakianos, G., & Prodromou, N. (2015). The role of fast cell cycle analysis in pediatric brain tumors. *Pediatric Neurosurgery, 50*(5), 257–263. www.karger.com/journals/pne/pne_jh.htm. doi:10.1159/000439029.

Armbruster, D. A., & Pry, T. (2008). Limit of blank, limit of detection and limit of quantitation. *The Clinical Biochemist. Reviews, 29*(Suppl 1).

Bandura, D. R., Baranov, V. I., Ornatsky, O. I., Antonov, A., Kinach, R., Lou, X., Pavlov, S., Vorobiev, S., Dick, J. E., & Tanner, S. D. (2009). Mass cytometry: Technique for real time single cell multitarget immunoassay based on inductively coupled plasma time-of-flight mass spectrometry. *Analytical Chemistry, 81*(16), 6813–6822. http://pubs.acs.org/doi/pdfplus/10.1021/ac901049w. doi:10.1021/ac901049w.

Borowitz, M. J., Craig, F. E., DiGiuseppe, J. A., Illingworth, A. J., Rosse, W., Sutherland, D. R., Wittwer, C. T., & Richards, S. J. (2010). Guidelines for the diagnosis and monitoring of paroxysmal nocturnal hemoglobinuria and related disorders by flow cytometry. *Cytometry Part B - Clinical Cytometry, 78*(4), 211–230. http://onlinelibrary.wiley.com/doi/10.1002/cyto.b.20525/pdf. doi:10.1002/cyto.b.20525.

Chen, H. N., Liang, K. H., Lai, J. K., Lan, C. H., Liao, M. Y., Hung, S. H., Chuang, Y. T., Chen, K. C., Tsuei, W. W. F., & Wu, H. C. (2020). EpCAM signaling promotes tumor progression and protein stability of PD-L1 through the EGFR pathway. *Cancer Research, 80*(22), 5035–5050. ttps://cancerres.aacrjournals.org/content/80/22/5035. doi:10.1158/0008-5472.CAN-20-1264.

D'Amato Figueiredo, M. V., Alexiou, G. A., Vartholomatos, G., & Rehder, R. (2022). Advances in intraoperative flow cytometry. *International Journal of Molecular Sciences, 23*(21), 13430. doi:10.3390/ijms232113430.

Davidson, B., Dong, H. P, Berner, A., Christensen, J., Nielsen, S., Johansen, P., Bryne, M., Asschenfeldt, P., & Risberg, B. (2002). Detection of malignant epithelial cells in effusions using flow cytometric immunophenotyping: an analysis of 92 cases. *American journal of clinical pathology, 118*(1), 85–92.

Davis, B. H., Wood, B., Oldaker, T., & Barnett, D. (2013). Validation of cell-based fluorescence assays: Practice guidelines from the ICSH and ICCS - Part i - ale and aims. *Cytometry Part B - Clinical Cytometry, 84*(5), 282–285. doi:10.1002/cyto.b.21104.

Gaur, G., Awasthi, N. P., Gupta, A., Agarwal, A., Sachan, R., Malhotra, K. P., Shukla, S., Singh, A. K., Singh, P., & Husain, N. (2023). Diagnostic accuracy of flow cytometry in detecting malignant epithelial cells in serous effusions. *Journal of the American Society of Cytopathology, 12*(6), 423–435. https://www.sciencedirect.com/science/journal/22132945. doi:10.1016/j.jasc.2023.09.003.

Gires, O., Pan, M., Schinke, H., Canis, M., & Baeuerle, P. A. (2020). Expression and function of epithelial cell adhesion molecule EpCAM: where are we after 40 years? *Cancer and Metastasis Reviews, 39*(3), 969–987. doi:10.1007/s10555-020-09898-3.

Gostomczyk, K., Łukaszewska, E., Borowczak, J. ę, Bator, A., Zdrenka, M., Bodnar, M., & Szylberg, Ł (2024). Flow cytometry in the detection of circulating tumor cells in neoplastic effusions. *Clinica Chimica Acta, 552*, 117651. doi:10.1016/j.cca.2023.117651.

Han, S., Zong, S., Shi, Q., Li, H., Liu, S., Yang, W., Li, W., & Hou, F. (2017). Is Ep-CAM expression a diagnostic and prognostic biomarker for colorectal cancer? A systematic meta-analysis. *EBioMedicine, 20,* 61–69. doi:10.1016/j.ebiom.2017.05.025.

Han, Y., Liu, D., & Li, L. (2020). PD-1/PD-L1 pathway: Current researches in cancer. *American Journal of Cancer Research, 10*(3), 727–742.

He, C., Wang, B., Wan, C., Yang, T., & Shen, Y. (2017). Diagnostic value of D2-40 immunostaining for malignant mesothelioma: A meta-analysis. *Oncotarget, 8*(38), 64407–64416. doi:10.18632/oncotarget.19041.

Heffner, J. E., & Klein, J. S. (2008). Recent advances in the diagnosis and management of malignant pleural effusions. *Mayo Clinic Proceedings, 1*(2), 235–250. http://www.mayoclinicproceedings.com/pdf%2F8302%2F8302sst%2Epdf.

Hodge, C., & Badgwell, B. D. (2019). Palliation of malignant ascites. *Journal of Surgical Oncology, 120*(1), 67–73. http://onlinelibrary.wiley.com/journal/10.1002/(ISSN)1096-9098. doi:10.1002/jso.25453.

Horm, T. M., & Schroeder, J. A. (2013). MUC1 and metastatic cancer: Expression, function and therapeutic targeting. *Cell Adhesion and Migration, 7*(2), 187–198. http://www.landesbioscience.com/journals/celladhesion/article/23131/2012CAM0045R.pdf?nocache=1252700866. doi:10.4161/cam.23131.

Jagric, T., Potrc, S., Mis, K., Mars, T., Gradisnik, L., Gorenjak, M., & Goropevsek, A. (2016). Flow cytometry as the new 'gold standard' for detection of free tumour cells in abdominal lavage fluid in gastric cancer patients: A comparative study of molecular and conventional methods. *Surgical Practice, 20*(2), 73–81. http://www.blackwell-synergy.com/loi/ash. doi:10.1111/1744-1633.12173.

Jansen, K., Kornfeld, L., Lennartz, M., Rico, S. D, Kind, S., Reiswich, V., Viehweger, F., Bawahab, A. A, Fraune, C., Gorbokon, N., Luebke, A. M., Hube-Magg, C., Menz, A., Uhlig, R., Krech, T., Hinsch, A., Jacobsen, F., Burandt, E., Sauter, G., … Bernreuther, C. (2024). Carcinoembryonic antigen expression in human tumors: A tissue microarray study on 13,725 tumors. *Cancers, 16*(23), 4052. doi:10.3390/cancers16234052.

Jiang, Y., Chen, M., Nie, H., & Yuan, Y. (2019). PD-1 and PD-L1 in cancer immunotherapy: Clinical implications and future considerations. *Human Vaccines & Immunotherapeutics, 15*(5), 1111–1122. doi:10.1080/21645515.2019.1571892.

Kebenko, M., Goebeler, M. E., Wolf, M., Hasenburg, A., Seggewiss-Bernhardt, R., Ritter, B., Rautenberg, B., Atanackovic, D., Kratzer, A., Rottman, J. B., Friedrich, M., Vieser, E., Elm, S., Patzak, I., Wessiepe, D., Stienen, S., & Fiedler, W. (2018). A multicenter phase 1 study of solitomab (MT110, AMG 110), a bispecific EpCAM/CD3 T-cell engager (BiTE®) antibody construct, in patients with refractory solid tumors. *OncoImmunology, 7*(8). http://www.tandfonline.com/loi/koni20. doi:10.1080/2162402X.2018.1450710.

Kehoe, C. (1991). Malignant ascites: Etiology, diagnosis, and treatment. *Oncology Nursing Forum, 18*(3), 523–530.

Kentrou, N. A., Tsagarakis, N. J., Tzanetou, K., Damala, M., Papadimitriou, K. A., Skoumi, D., Stratigaki, A., Anagnostopoulos, N. I., Malamou-Lada, E., Athanassiadou, P., & Paterakis, G. (2011). An improved flow cytometric assay for detection and discrimination between malignant cells and atypical mesothelial cells, in serous cavity effusions. *Cytometry Part B - Clinical Cytometry, 80*(5), 324–334. doi:10.1002/cyto.b.20608.

Kim, H., Shin, S., Hwang, S. H., Kim, M., Cho, Y. U., & Jang, S. (2023). Validation of high-sensitivity flow cytometry for reliable immune cell analysis in real-world laboratory settings. *Annals of Laboratory Medicine, 43*(6), 620–624. https://www.annlabmed.org/journal/view.html?uid=3479&vmd=Full. doi:10.3343/alm.2023.43.6.620.

Kitayama, J., Emoto, S., Yamaguchi, H., Ishigami, H., Kamei, T., Yamashita, H., Seto, Y., Matsuzaki, K., & Watanabe, T. (2014). Flow cytometric quantification of intraperitoneal free tumor cells in patients with peritoneal metastasis. *Cytometry Part B - Clinical Cytometry, 86*(1), 56–62. doi:10.1002/cyto.b.21126.

Koss, L. G. (2006). *Diagnostic cytology and its histopathologic bases* (5th ed.). Lippincott Williams & Wilkins.

Krishan, A., Ganjei-Azar, P., Hamelik, R., Sharma, D., Reis, I., & Nadji, M. (2010). Flow immunocytochemistry of marker expression in cells from body cavity fluids. *Cytometry. Part A: The Journal of The International Society for Analytical Cytology., 77*(2), 132–143. doi:10.1002/cyto.a.20824.

Luo, M., Kou, T., Yin, Y., Zhou, S., Zhu, X., Zeng, X., Hu, J., & Zhu, Z.-J. (2026). Deep-coverage single-cell metabolomics enabled by ion mobility-resolved mass cytometry. *Nature Methods.* doi:10.1038/s41592-025-02970-2.

Maecker, H. T., McCoy, J. P., & Nussenblatt, R. (2012). Standardizing immunophenotyping for the human immunology project. *Nature Reviews Immunology, 12*(3), 191–200. doi:10.1038/nri3158.

Mahnke, Y., Chattopadhyay, P., & Roederer, M. (2010). Publication of optimized multicolor immunofluorescence panels. *Cytometry Part A, 77*(9), 814–818. http://onlinelibrary.wiley.com/doi/10.1002/cyto.a.20916/pdf. doi:10.1002/cyto.a.20916.

Markopoulos, G., Lampri, E., Tragani, I., Kourkoumelis, N., Vartholomatos, G., & Seretis, K. (2024). Intraoperative flow cytometry for the rapid diagnosis and validation of surgical clearance of non-melanoma skin cancer: A prospective clinical feasibility study. *Cancers, 16*(4). http://www.mdpi.com/journal/cancers/. doi:10.3390/cancers16040682.

Markopoulos, G. S., Pakos, E. E., Gavrielatos, V., Kosmas, D., Gkiatas, I., Alexiou, G. A., Batistatou, A., Lampri, E., & Vartholomatos, G. (2023). Intraoperative flow cytometry upon and beyond the cell cycle: A case study of the characterization of a bone metastasis. *Journal of Molecular Pathology, 4*(4), 225–233. https://www.mdpi.com/journal/jmp. doi:10.3390/jmp4040020.

Menz, A., Lony, N., Lennartz, M., Rico, S. D, Schlichter, R., Kind, S., Reiswich, V., Viehweger, F., Dum, D., Luebke, A. M., Kluth, M., Gorbokon, N., Hube-Magg, C., Bernreuther, C., Simon, R., Clauditz, T. S., Sauter, G., Hinsch, A., Jacobsen, F., … Weidemann, S. (2024). Epithelial cell adhesion molecule (EpCAM) expression in human tumors: A comparison with pan-cytokeratin and TROP2 in 14,832 tumors. *Diagnostics, 14*(10), 1044. doi:10.3390/diagnostics14101044.

Michael, C. W., & Davidson, B. (2016). Pre-analytical issues in effusion cytology. *Pleura and Peritoneum, 1*(1), 45–56. http://www.degruyter.com/view/j/pp. doi:10.1515/pap-2016-0001.

Munz, M., Baeuerle, P. A., & Gires, O. (2009). The emerging role of EpCAM in cancer and stem cell signaling. *Cancer Research, 69*(14), 5627–5629. https://doi.org/10.1158/0008-5472.CAN-09-0654. doi:10.1158/0008-5472.CAN-09-0654.

Myung, J. H., & Hong, S. (2015). Microfluidic devices to enrich and isolate circulating tumor cells. *Lab on a Chip, 15*(24), 4500–4511. http://pubs.rsc.org/en/journals/journal/lc. doi:10.1039/c5lc00947b.

Nolan, J. P. (2022). The evolution of spectral flow cytometry. *Cytometry Part A, 101*(10), 812–817. http://onlinelibrary.wiley.com/journal/10.1002/(ISSN)1552-4957. doi:10.1002/cyto.a.24566.

Otero, A. P., Rafecas-Codern, A., Porcel, J. M., Serra-Mitjà, P., Ferreiro, L., Botana-Rial, M., Ramos-Hernández, C., Manuel Brenes, J., Canales, L., Camacho, V., Romero-Romero, B., Trujillo, J. C, Martinez, E., Cases, E., Barba, A., Majem, M., Güell, E., & Pajares, V. (2024). Malignant pleural effusion: A multidisciplinary approach. *Open Respiratory Archives, 6*(4), 100349. doi:10.1016/j.opresp.2024.100349.

Patriarca, C., Macchi, R. M., Marschner, A. K., & Mellstedt, H. (2012). Epithelial cell adhesion molecule expression (CD326) in cancer: A short review. *Cancer Treatment Reviews, 38*(1), 68–75. doi:10.1016/j.ctrv.2011.04.002.

Pillai, V., Cibas, E. S., & Dorfman, D. M. (2013). A simplified flow cytometric immunophenotyping procedure for the diagnosis of effusions caused by epithelial malignancies. *American Journal of Clinical Pathology, 139*(5), 672–681. http://ajcp.ascpjournals.org/content/139/5/672.full.pdf+html. doi:10.1309/AJCP4HIFSHMO9WTK.

Ramalingam, T. R, Marimuthu, B., Rasheed, H. N., Lakshmanan, A., Narla, S. L, Vaidhyanathan, L., & Pai, A. (2025). Intraoperative flow cytometry in detecting free carcinoma cells in peritoneal lavage fluid of gastric carcinoma cases. *Cytometry Part B: Clinical Cytometry.* doi:10.1002/cyto.b.22238.

Ramalingam, T. R, Muthu, A., Lakshmanan, A., Narla, S., Subramanyan, A., Simon, S., Govindaraj, J., Vaidhyanathan, L., Easow, J., & Raja, T. (2022). Role of high acquisition flow cytometry in the detection of marrow involvement in patients with extramedullary B cell non-Hodgkins lymphoma: a comparison with marrow aspirate cytology, trephine biopsy, and PET. *Leukemia & Lymphoma, 63*(11), 2589–2596. doi:10.1080/10428194.2022.2092858.

Ramalingam, T. R., Mani, S., Narla, S., Lakshmanan, A., Muthu, A., Nk, H. R., Mohanraj, S., Ramachandran, E. D., Subramanyam, A., & Vaidhyanathan, L. (2025). Simple and rapid flow cytometry assay for the detection of malignant epithelial cells in body fluids. *Cytopathology, 36*(1), 23–30. http://onlinelibrary.wiley.com/journal/10.1111/(ISSN)1365-2303. doi:10.1111/cyt.13408.

Risberg, B., Davidson, B., Dong, H. P, Nesland, J. M, & Berner, A. (2000). Flow cytometric immunophenotyping of serous effusions and peritoneal washings: comparison with immunocytochemistry and morphological findings. *Journal of Clinical Pathology, 53*(7), 513–517.

Romeo, E., Markopoulos, G. S., Vartholomatos, G., Voulgaris, S., & Alexiou, G. A. (2025). The role of intraoperative flow cytometry in surgical oncology: A systematic review. *Cancers, 17*(24), 3898. doi:10.3390/cancers17243898.

Sahu, S., Gupta, P., Susheilia, S., Gautam, U., & Dey, P. (2021). Application of multicolour flow cytometry in the detection of metastatic carcinoma in serous effusions: Special emphasis in atypical cytology. *Cytopathology: Official Journal of the British Society for Clinical Cytology, 32*(2), 169–179. doi:10.1111/cyt.12922.

Sayed, D. M, el-Attar, M. M, & Hussein, A. AR. M (2009). Evaluation of flow cytometric immunophenotyping and DNA analysis for detection of malignant cells in serosal cavity fluids. *Diagnostic Cytopathology, 37*(7), 498–504. doi:10.1002/dc.21047.

Shapiro, H. M. (2005). *Practical flow cytometry*. Wiley. doi:10.1002/0471722731.

Sharygin, D., Koniaris, L. G., Wells, C., Zimmers, T. A., & Hamidi, T. (2023). Role of CD14 in human disease. *Immunology, 169*(3), 260–270. http://onlinelibrary.wiley.com/journal/10.1111/(ISSN)1365-2567. doi:10.1111/imm.13634.

Shidham, V. B, & Atkinson, B. F. (2007). *Cytopathologic diagnosis of serous fluids e-book*. Elsevier Health Sciences.

Spizzo, G., Fong, D., Wurm, M., Ensinger, C., Obrist, P., Hofer, C., Mazzoleni, G., Gastl, G., & Went, P. (2011). EpCAM expression in primary tumour tissues and metastases: An immunohistochemical analysis. *Journal of Clinical Pathology, 64*(5), 415–420. doi:10.1136/jcp.2011.090274.

Tárnok, A. (2008). SYTO dyes and histoproteins—Myriad of applications. *Cytometry Part A, 73A*(6), 477–479. doi:10.1002/cyto.a.20588.

Wang, M., Ateia, M., Hatano, Y., Miyanaga, K., & Yoshimura, C. (2022). Novel fluorescence-based method for rapid quantification of live bacteria in river water and treated wastewater. *Environmental Science: Advances, 1*(1), 30–36. doi:10.1039/D1VA00017A.

Warkiani, M. E., Khoo, B. L., Wu, L., Tay, A. K. P., Bhagat, A. A. S., Han, J., & Lim, C. T. (2016). Ultra-fast, label-free isolation of circulating tumor cells from blood using spiral microfluidics. *Nature Protocols, 11*(1), 134–148. http://www.natureprotocols.com/. doi:10.1038/NPROT.2016.003.

Went, P., Vasei, M., Bubendorf, L., Terracciano, L., Tornillo, L., Riede, U., Kononen, J., Simon, R., Sauter, G., & Baeuerle, P. A (2006). Frequent high-level expression of the immunotherapeutic target Ep-CAM in colon, stomach, prostate and lung cancers. *British Journal of Cancer, 94*(1), 128–135. doi:10.1038/sj.bjc.6602924.

Wong-Arteta, J., Rey, M., Aragón, L., Gil-Rodríguez, E., & Bujanda, L. (2020). The utility of flow cytometry in the diagnostic workup of malignant effusions due to nonhematopoietic neoplasms. *Cytometry. Part B, Clinical cytometry., 98*(6), 504–515. doi:10.1002/cyto.b.21886.

Wood, B. L. (2016). Principles of minimal residual disease detection for hematopoietic neoplasms by flow cytometry. *Cytometry Part B - Clinical Cytometry, 90*(1), 47–53. http://onlinelibrary.wiley.com/journal/10.1002/(ISSN)1552-4957. doi:10.1002/cyto.b.21239

Zimmerlin, L., Donnenberg, V. S., & Donnenberg, A. D. (2011). Rare event detection and analysis in flow cytometry: Bone marrow mesenchymal stem cells, breast cancer stem/progenitor cells in malignant effusions, and pericytes in disaggregated adipose tissue. *Methods in Molecular Biology, 699*, 251–273. http://www.springer.com/series/7651. doi:10.1007/978-1-61737-950-5_12.

8

Cerebrospinal fluid analysis for metastatic hematological malignancies by flow cytometry

Khaliqur Rahman

DEPARTMENT OF HEMATOLOGY, SANJAY GANDHI POST GRADUATE INSTITUTE OF MEDICAL SCIENCES, LUCKNOW, UTTAR PRADESH, INDIA

8.1 Introduction

Leptomeningeal involvement by metastatic cancer cells was first reported by Eberth in 1870, and the term "carcinomatous meningitis" (CM) was introduced by Beerman in 1912 (Aboulafia et al., 1996;). It represents one of the most serious complications in hematology and oncology. This is not an uncommon finding, and its incidence varies from 5% to 15% in hematological malignancies and 5%–8% in solid tumors (Bhambhvani et al., 2020). The presence of malignant cells within the cerebrospinal fluid (CSF) indicates leptomeningeal involvement and is associated with poor prognosis (Kwon et al., 2014). Accurate and timely diagnosis of CSF involvement is therefore essential for appropriate staging, therapeutic decision-making, and monitoring of treatment response.

Traditionally, leptomeningeal involvement is suspected based on the clinical features and is confirmed by radiological imaging, such as magnetic resonance imaging (MRI). The involvement is confirmed by directly visualizing the tumor cells on CSF cytology, which has been regarded as the diagnostic gold standard. However, cytology suffers from low sensitivity, high interobserver variability, and dependence on sample adequacy and cellular preservation. In the evaluation of leukemia, the total leukocyte count (TLC) of CSF is also an important modality to grade the extent of central nervous system (CNS) involvement and is conventionally labeled as CNS1, CNS2, and CNS3, based on the TLC per microliter. The looks of lymphoid cells may be deceptive microscopically. Sometimes, small lymphoid cells may have a swollen appearance and may resemble blasts, while in other instances, they may have small pyknotic blasts resembling the normal mature lymphoid cells. Immunophenotyping, which assigns the correct lineage and maturational stage of lymphoid cells, may help in the correct identification of these abnormal populations.

Advancing Body Fluid Diagnostics with Flow Cytometry: A Comprehensive Guide to Effusions and CSF.
DOI: https://doi.org/10.1016/B978-0-443-27426-8.00011-9

CSF FCM has emerged as a highly sensitive, objective, and reproducible technique capable of detecting rare malignant cells in paucicellular specimens such as CSF. It relies on the scatter properties as defined by the size and granularity of the cell, along with the immunophenotypic character. The expression of antigens is detected with the fluorescent-tagged antibodies, which define the lineage, stage of maturation, and aberrant expression of the cells analyzed. It provides an objective way of identifying abnormal cell populations based on their phenotype.

Over the last two decades, better availability of FCM instruments, improvements in multicolor FCM, antibody availability, and standardized processing protocols have established FCM as a critical tool in CSF diagnostics. This chapter provides a comprehensive overview of the role of FCM in the evaluation of CSF for metastatic malignancies, with emphasis on disease spectrum, technical considerations, panel design, and interpretation strategies.

8.2 Flow cytometry for cerebrospinal fluid: A short review

Over the last two decades, multicolor flow cytometric analysis has been widely used for the diagnosis and classification of hematological neoplasms. A wide variety of samples, like peripheral blood, bone marrow, effusion fluids, CSF, FNA material, etc., have been subjected to immunophenotypic analysis by FCM, which can sensitively and rapidly identify abnormal cells. For the paucicellular samples like CSF, it has been proposed as a more sensitive method for detecting CNS involvement, better than the conventional cytomorphological approach. A panel of antibody-fluorescein combination is used to stain the cells present in the CSF. Normally, the CSF has only a few lymphocytes and monocytes. A hemorrhagic tap may lead to the presence of some granulocytes as well. Abnormal cells are identified based on the antigenic expression patterns. This can either be the expression of some epithelial or neuroendocrine markers, positivity of which defines the presence of metastatic tumor cells in CSF. Or, these can be some immaturity markers or aberrant markers that define the presence of abnormal blasts or atypical lymphoid cells. FCM can rapidly and precisely quantify the expression of multiple markers on the cells. The swelling or shrinkage of the cells, which are seen as a common artifact on cytological smears, giving a deceptive look to these cells under the microscope, can also be overcome by FCM. It objectively identifies the lineage of the cells, their maturational stage, and antigenic aberrancies, making it a more sensitive technique in the identification of abnormal cells (Singh et al., 2025). The presence of even a few abnormal cells can be identified, making this technique highly specific (DiGiuseppe & Wood, 2019; Singh et al., 2025). In addition, the FCM evaluation for the presence of malignant cells has a high concordance (>90%) between different laboratories (de Haas et al., 2021).

The limitation of the FCM evaluation includes low cellularity of the sample, rapid degeneration of abnormal cells suspended in the CSF, a requirement for technical expertise, higher cost, and reduced sensitivity in solid tumor or large-cell lymphoma metastases. Sample transportation may be an important issue, especially for the stand-alone labs, which frequently receive transported samples. However, there is an availability of some transport media and specialized tubes where the cells can be fixed before FCM, e.g., via the collection of CSF directly into specialized transfix tubes containing an optimized formaldehyde solution with EDTA for the preservation of CSF cells (De Jongste et al., 2014). Storing the samples in a transfix tube preserves the cells, which can be optimally processed and stained for up to 72–96 h. However, there may be a chance of reduced signals from the commonly used markers owing to the presence of formaldehyde, which may cross-link the antigenic epitopes (Levinsen et al., 2016). The addition of RPMI or a serum-containing medium to CSF also helps in preserving the cells for a short duration of time (De Graaf et al., 2011). Maintenance of ambient temperature during transportation is also important. Extreme temperature leads to the rapid degeneration of cells. Despite all these limitations of the sample storage of transportation, discarding the precious CSF sample needs to be weighed. It is advised to process the sample, but to interpret the result with caution, with the background information about the sample conditions.

8.3 Cytomorphology versus flow cytometry-based analysis

The detection of neoplastic cells in CSF samples by cytomorphology represents the method traditionally used for this purpose and is still considered to be the method of choice. An increase in TLC with suspicious morphology suggests CNS involvement. The availability of a flow cytometer with a trained technologist, required for processing the CSF sample optimally, is still limited compared to cytopathology.

However, considering the paucicellular nature of these samples and the technical difficulty in distinguishing between normal and neoplastic cells on morphology, it is desirable to incorporate complementary methods in the laboratory routine. A significantly higher number of inconclusive results are observed by cytomorphology, in which the definitive characterization of suspected cells was not possible. Doubts in the morphological analysis arise due to the presence of morphological changes secondary to the chemotherapy treatment or secondary to their ex vivo suspension in CSF. There may be degeneration, shrinkage, or swelling of the cells. A study has shown that it was possible to define conclusive results by adding FCM in more than 95% of the samples, which were initially inconclusive by morphology. FCM allowed for the characterization of the aberrant antigen expression of the neoplastic cell (Rocha et al., 2023). In one of our studies on the prevalence of CSF involvement by a hematolymphoid neoplasm, we found a positivity rate of 16.6% (67 of 404 cases). Among the FCM-positive samples, only 29 (43.2%) were labeled as positive on cytomorphological assessment as well. Most of the FCI-positive and cytomorphology-negative cases (21 of 38, 55.3%) were paucicellular (Singh et al., 2025). This becomes important as even in CNS2 (where the leukocyte count is <5%), identification of bona fide blasts will be considered a poor prognostic factor, thus determining the intensification of chemotherapeutic treatment by different clinical protocols. Even in samples with low cellularity (<5 cells/μL), the presence of blasts in the CSF (anytime at diagnosis or during treatment monitoring) is associated with a higher incidence of CNS recurrence, thus justifying the adoption of FCM as an objective method to identify the blasts. Several earlier studies have reported discordant results, consistently demonstrating a significantly higher rate of positivity with FCM. This observation is expected, as FCM enables superior recovery and analysis of cells compared to conventional CSF cytocentrifugation used for slide preparation in cytomorphology. Hegde et al. analyzed 51 samples from patients with B-cell lymphomas and identified occult or subclinical CNS involvement in 11 cases using FCM, of which only one case was also detected by cytomorphology (Hegde et al., 2005). Similarly, Bromberg et al. (2007) evaluated 219 samples from patients with hematologic malignancies at diagnosis and reported 44 FCM-positive cases compared to 19 detected by cytomorphology; notably, cytomorphological positivity correlated with higher CSF cell counts. In another study, Martínez-Laperche et al. (2013) assessed 108 pediatric patients with acute lymphoblastic leukemia (ALL) at diagnosis and identified subclinical CNS involvement in 30 cases (27.8%) using FCM, whereas cytomorphology detected only three cases (2.8%). Collectively, these findings highlight the superior sensitivity of FCM in detecting low-level or occult CNS disease, particularly in samples with low cellularity, thereby underscoring its clinical utility in early and accurate disease assessment.

FCM also helps in ruling out the CNS involvement in morphologically suspicious cases. These false-positive results of the morphological analysis can be justified by morphological changes imparted to the normal lymphoid cells, which appear as blastic. In these cases, by characterizing the normal antigenic expression of the analyzed cells, the FC removes the possibility of CNS infiltration raised by the morphological analysis.

Cytomorphological evaluation of CSF in cases of lymphoma may also be tricky. The native T-cells, when increased in number, may be misinterpreted as CNS involvement by the B-cell Non Hodgkins Lymphoma. But the FCM may correctly identify these as reactive T-cells. One disadvantage of FCM is missing out the rare large B-cells, which often go off scale on the FSC-SSC plot and are not included in subsequent analysis. In these situations, the cytomorphology may be more sensitive in picking up even a few atypical large cells with the characteristic morphology of Burkitt or DLBCL.

8.4 Diseases involving the cerebrospinal fluid

A wide range of malignant disorders can involve the CSF, either as primary CNS disease or secondary leptomeningeal spread. Hematological malignancies: ALL is the most common malignancy involving the CSF, particularly in pediatric patients. Hence, as a part of the treatment protocol, the CSF tap is done and sent for cytomorphology and FCM evaluation. Acute myeloid leukemia (AML) may also involve CSF, especially in monocytic subtypes. Chronic lymphoproliferative disorders such as chronic lymphocytic leukemia and mantle cell lymphoma may infiltrate the CSF. These neoplasms often show subtle morphological features, making immunophenotypic confirmation essential. Mature lymphoid neoplasm like B-cell and T-cell nonHodgkin lymphomas, may also involve the CSF. Some diseases, like Burkitt Lymphoma, may have a propensity toward CNS spread. Primary CNS lymphoma, typically diffuse large B-cell lymphoma, frequently requires CSF FCM for diagnosis and staging. Secondary leptomeningeal involvement is common in aggressive lymphomas. Although rare, multiple myeloma and plasma cell leukemia can also involve the CSF.

CNS involvement with CSF spread is a serious complication of cancer that occurs in 1%–5% of patients with solid tumors. Primary sites of these tumors can be breast carcinoma, lung carcinoma, gastrointestinal tract, malignant melanoma, etc. Pediatric small round cell tumors like Ewings/PNET or neuroblastoma may also show CNS spread. This heralds a poor prognosis in these patients with reduced overall survival (De Graaf et al., 2011). Hence, early diagnosis of the leptomeningeal disease can be crucial because the institution of aggressive treatment, such as radiotherapy/intrathecal chemotherapy, can increase the patient's chance of survival and relief from symptoms. This is particularly true in the case of small cell carcinoma of the lung, which is associated with treatable occult cerebro-meningeal metastasis.

8.5 Calling the cerebrospinal fluid positive

To call a CSF abnormal, the mere presence of an abnormal cell is sufficient, no matter whatever the cell count. Most of the time, when a CSF is found to be involved by a hematological neoplasm, the cell count is extremely low. On the other hand, a few infectious diseases, such as tuberculosis or some viral meningitis, may lead to an increase in lymphocyte and monocyte counts. However, these do not show the presence of any clonal or phenotypically abnormal lymphoid cells.

8.6 Technical aspects of FCM based analysis of CSF

8.6.1 Cerebrospinal fluid collection and processing

CSF is a fragile, low cellularity specimen requiring meticulous handling. A delay in processing, careless transport, inability to maintain temperature, harsh processing, etc., will lead to loss of important information from this precious sample. Specimens should be processed as soon as possible, ideally within 2 h of collection. Delays result in rapid cell degeneration, leading to loss of leukocyte count, abnormal scatter properties, and loss of antigen expression. After 2 h, the leukocyte loss ranges between 15% and 62%, while after 24 h, only 39%–60% of leukocytes remain detectable (Chow & Schmidley, 1984). Use of stabilizing agents like serum-containing preservative media and transfix tubes, can be used. The serum-containing preservative tubes are not commercially available. Moreover, these have a shorter shelf life. The transfix tube has been found to preserve the cellular subset morphology as well as antigenic expression till 72 h for the CSF samples evaluated for hematolymphoid neoplasm (Johansson et al., 2014). It contains an aliphatic aldehyde and heavy metal salts and has been found to preserve a variety of samples (Del Principe et al., 2021). It has been recommended that to obtain optimal results, any delay in sample transportation or processing needs to be avoided. It is to be processed within 1–2 h (De Jongste et al., 2014; Del Principe et al., 2021). If transportation is needed, a fixative like transfix should be used, but keeping in view the fact that the intensity of expression of some markers may be compromised.

For optimal processing, a minimum volume of 1–2 mL of CSF is recommended for FCM. Samples should be transported at room temperature and gently mixed to prevent cell settling. The sample needs to be centrifuged to prepare the cell pellet, which can be stained for the leukocyte immunophenotyping. Low-speed centrifugation is used to concentrate cells, followed by careful removal of supernatant and gentle resuspension of the cell pellet in 0.2%–0.5% BSA.

Table 8.1 The normal cell count in the CSF of healthy individuals. Based on de Graaf et al. (2008).

Lymphocyte subset in CSF	Median (cells/μL)	5th–95th percentile (cells/μL)
T-cells	0.46	0.2–2.02
CD4+ T-cells	0.34	0.12–1.36
CD8+ T-cells	0.13	0.06–1
B-cells	0.005	0–0.034
NK cells	0.011	0.002–0.058
NKT cells	0.005	0–0.037

Adapted from de Graaf et al. (2008).
CSF, Cerebrospinal fluid.

8.6.2 Leukocyte count of cerebrospinal fluid

CSF is an inherently paucicellular fluid. Usually, the cell count is too low, and a count of 5 cells/μL without any abnormal cells is regarded as normal. This still remains controversial as it is not feasible and practical to evaluate the CSF cell count of a healthy individual. The conventional use of microscopic chambers to perform the manual cell count has been found to have a lot of variations in the cell count. Some high-end automated analyzers have the capability to count the leukocytes accurately in these paucicellular samples, but even for the ranges below 10–20 cells/μL, the accuracy is questionable. The bead cell count based on FCM is a feasible strategy. But still, the count obtained by the automated cell counter is the current gold standard method as of now. The leukocytes present in the CSF are usually the mature T lymphocytes and a few monocytes. The B-cells are extremely rare. In infectious conditions, like viral meningitis or tuberculosis, the lymphocyte counts are increased, and a few plasma cells may be seen. Neutrophils may be seen in bacterial meningitis. The table 8.1,provides the normal lymphoid cell count in CSF, as adapted from de Graaf et al. (2008).

8.6.3 Red blood cell contamination in cerebrospinal fluid

A traumatic puncture, even the slightest one, will lead to the presence of neutrophils and B-cells, etc. The presence of red cells reflects either CNS bleeding or a traumatic puncture. The ways to identify blood contamination can either be an enumeration of erythrocytes in the CSF or a simultaneous examination of the peripheral blood. Finding more than 1000 erythrocytes per leukocyte means blood contamination. On the other hand, a blood-contaminated CSF sample could only be interpreted as CNS involvement when these cells are not detected in a blood sample of the same drawing date (Kraan et al., 2008).

Kraan et al. (2008) described the method to find out the expected number of peripheral blood white blood cells (WBC) added to the CSF, using the red blood cell (RBC) count. This is done based on the relative number of leukocytes and erythrocytes in blood (Kraan et al., 2008). The formula is as follows:

$$WBC\ added = WBC\ blood / RBC\ blood \times RBC\ CSF$$

In a retrospective study, Szánthó et al. (2019) demonstrated that in two groups of CSF samples, collected from ALL pediatric patients, 45% and 24% of the samples were PB contaminated, respectively. The chances of contamination and iatrogenic presence of abnormal leukocytes in CSF is directly proportional to the leukocyte count. Hence, some of the experts suggest that the CSF should be tapped only if the peripheral blood TLC is in the normal range.

8.6.4 Identification of abnormal cells in cerebrospinal fluid

As mentioned previously, the CSF is inherently a paucicellular fluid. Identification of abnormal events in CSF is, however, not very difficult. Any event with an abnormal phenotype can be detected using a well-designed panel, which is usually based on the background knowledge of the diagnostic phenotype of hematolymphoid neoplasms. In our experience, we centrifuge ~1 mL of CSF sample at a speed of 400 g for 5 min to get the cell pellet, which is stained with a different antibody panel depending on the primary hematolymphoid disease. We acquire the whole tube and have noticed that as few as 10–20 abnormal events with a convincing phenotype can be picked easily. Since there is no issue of differentiating normal precursors from the abnormal ones, as in marrow, the presence of any genuine precursors defines the involvement of CSF by leukemia.

8.6.5 Antibody panel selection

Antibody panel design is critical for maximizing diagnostic yield in CSF FCM. Panels should be tailored to clinical suspicion while maintaining a core screening backbone. Since the CSF is paucicellular, it is better to use a single eight or more color tube, so that a sufficient number of events can be acquired and analyzed. Several studies have used different panels, which usually vary based on the primary disease involvement being evaluated. Popov et al. (2019) used a 4–6 color antibody combination, including CD19/CD10/CD34/CD45 for the majority of BCP-ALL and CD7/CD3/CD5/CD2/CD99/CD45 for T ALL. In B ALL, a similar combination was used by Cancela et al. (2017). In a study involving 2871 CSF samples from 357 adult ALL, the antibody panel included the following: CD19, CD10, CD34, CD38, CD20, CD58, CD33, and CD45 for B-lineage commitment and CD3, CD99, CD5, CD2, CD7, CD8, CD56, and CD45 for T-lineage commitment (Gong et al., 2018).

For the NHL infiltration, a minimum of six color combinations using CD45, CD19 or CD20, CD10, CD5, CD3 will be sufficient to suggest the presence of an increased and/or abnormal B-cell population in the CSF. The addition of kappa and lambda will allow for the identification of light chain restriction. The expression of kappa and lambda may not be optimal, and the addition of markers like CD305 (its loss being abnormal in B-cells) will help in the identification of an abnormal CD5/CD10-negative B-cell population.

Other diseases like CNS involvement in multiple myeloma or a myeloid neoplasm (AML or chronic myeloid leukemia) may be tracked using plasma cell-specific and myeloid-specific markers, respectively. In cases of plasma cell neoplasm, the processing for cytoplasmic kappa and lambda is usually avoided, as this may further lead to cell loss. The abnormality is reliably detected by the abnormal expression of surface antigens only. In our practice, we use different antibody fluorophore combinations for different diseases. These panels are shown in Table 8.2.

For the situations where the primary diagnosis is not clear, a mixed tube may be used for the evaluation. Similar is the approach of Euroflow, which suggests putting up a small sample tube (SST) designed for CSF study. This includes CD20, CD4, CD45, CD56, CD19, CD3, CD14, CD38, kappa/CD8, and lambda/CD56 (van Dongen et al., 2012). The approach is to define all the events present in CSF, which can be done by the use of these antibodies. It is to be noted that CD34 is not present in this tube, and an indication about the immaturity of these cells may be obtained by a dim CD45 expression.

8.6.6 Interpretation of Flow cytometric findings

Data acquisition should aim to collect all available events at low flow rates. Debris exclusion using forward and side scatter parameters is followed by CD45-based gating to identify the leukocyte population. Lineage-specific gates are then applied to identify abnormal populations and see their expression pattern. The abnormal phenotype may be in the form of gain or loss of expression, aberrant expression, expression of immaturity markers, light chain restriction, asynchronous maturation, and presence of a population distinct from normal CSF cells. In cases of some viral infections, encephalitis, multiple sclerosis, etc., there is an increase in leucocytes, which mainly are the CD4+ T lymphoid cells, monocytes, and sometimes a few plasma cells. However, the immunophenotype does not show any obvious abnormality in these cells.

Immaturity may be identified by a downregulation of CD45, the expression of CD34, and CD99. Similar to the leukemia lymphoma immunophenotyping, the lineage can be assigned as per the WHO criteria, where CD7, CD19, and CD117 play a vital role in assigning the T, B, and myeloid lineage, respectively. It is to be noted that the cytoplasmic or nuclear staining is technically very challenging in the CSF. Thus, MPO, cytoCD3, TdT, and cytoplasmic kappa and lambda are usually not administered. A representative case of T ALL, Myeloid sarcoma, and Burkitt Lymphoma involvement is shown in Figs. 8.1–8.3.

Interpretation must always be integrated with clinical, radiological, and cytological findings. A genuine presence of cells with an aberrant phenotype, even below the limit of quantification, should be reported as suspicious. As mentioned previously, the evidence of blood contamination should also be mentioned. Pitfalls include low event counts, therapy-related antigen modulation, and overinterpretation of small populations. Experienced interpretation is essential to avoid false-positive or false-negative results.

8.6.7 Reporting of cerebrospinal fluid sample

Just like any other precious sample, the report should reiterate the indication for CSF testing and should be organized in a way to give a clear-cut conclusion. It should mention the quality and quantity of the sample in terms of volume received, appearance, visible blood contamination,

Table 8.2 Antibody panel for CSF. Antibody panel used in our lab for CSF evaluation.

Tube name	BV 605	BV 510	BV 421/ BV450	FITC	PE	PerCP-Cy5.5/ PC5.5	PC7	APC	APC R700	APC H7
CSF tube for BALL	x	CD45	CD20	X	CD22	CD34	CD19	CD10	CD3	CD38
Clone	x	HI30	2H7	X	S-HCL-1	8G12	SJ25C1	HI10a	UCHT1	HB7
Company	x	BD	BD	x	BD	BD	BD	BD	BD	BD
CSF tube for TALL	CD7	CD8	x	CD16/CD56	CD34	CD5	CD4	CD38	CD45	CD3
Clone	M-T701	SK1	X	3G8/B159	8G12	L17F12	SFCI12T4D11	HIT2	HI30	SK7
Company	BD	BD	X	BD/BD	BD	BD	BC	BD	BD	BD
CSF tube for AML	CD15	CD45	HLA DR	CD16	CD33	CD34	CD117	CD10	CD13	CD11b
Clone	W6D3	HI30	G46-6	3G8	P67-6	8G12	104D2	HI10a	WM15	ICRF44
Company	BD	BD	BD	BD	BD	BD	BD	BD	BD	BD
CSF tube for B-NHL	x	CD8/CD10	CD20	Lambda	CD200	CD5	CD19/CD4	CD56/Kappa	CD45	CD3
Clone	x	SK1/HI10a	2H7	(1-155-2)	MRC 0X-104	L17F12	SJ25C1/ SFCI12T4D11	B159/G20-193	HI30	SK7
Company	x	BD/BD	BD	BD	BD	BC	BD/BC	BD/BD	BD	BD
CSF tube for PCN	CD117	CD27	CD138	x	CD56	CD38	CD19	X	CD45	CD81
Clone	104D2	L128	MI15	x	A07788	HIT2	SJ25C1	X	HI30	JS-81
Company	BD	BD	BD	x	BC	BD	BD	X	BD	BD

AML, acute myeloid leukemia; *BALL*, B-cell precursor acute lymphoblastic leukemia; *BC*, Beckman Coulter; *BD*, Becton Dickinson biosciences; *CSF*, cerebrospinal fluid; *NHL*, NonHodgkin lymphoma; *PCN*, plasma cell neoplasm.

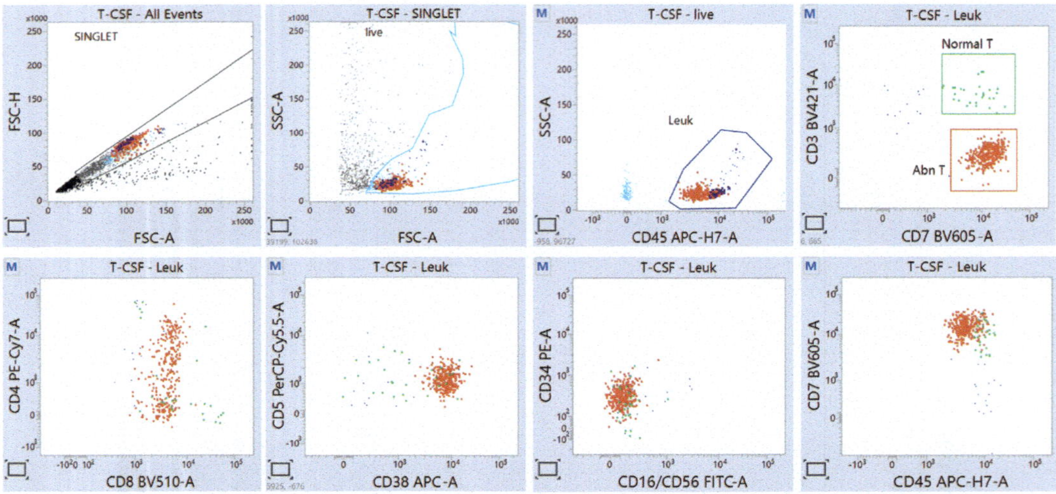

FIGURE 8.1 CSF involvement in a case of T ALL. FCM had identified an abnormal population of cells (*red color*) which are positive for CD7, CD4 (dim a7 heterogeneous), and CD38 (homogeneous bright) while being negative for surface CD3, CD5, and CD8. The green population is of normal T-cells, which show normal expression of CD3/CD5/CD7 and show normal distribution of CD4 and CD8.

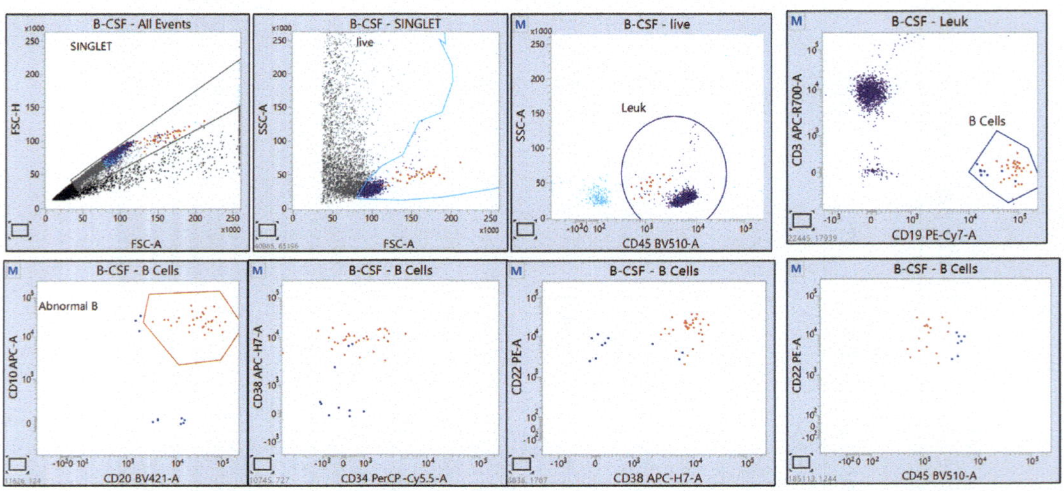

FIGURE 8.2 CSF involvement in a case of Burkitt Lymphoma. A scant population of abnormal cells was identified (*red color*), which is positive for CD19/CD10/CD20/CD38 (*bright*)/CD45(*moderate*), while being negative for CD34.

time of receipt of the sample, and any other important naked eye findings. It should mention the amount of sample processed, whether the whole of the submitted sample or the volume processed. Total number of viable, live gated leukocytes, along with the quantification of the abnormal population, along with their descriptive phenotype, should be mentioned. The clear-cut terms such as “CSF is INVOLVED by a B ALL in this known case” or “CSF is NOT INVOLVED”

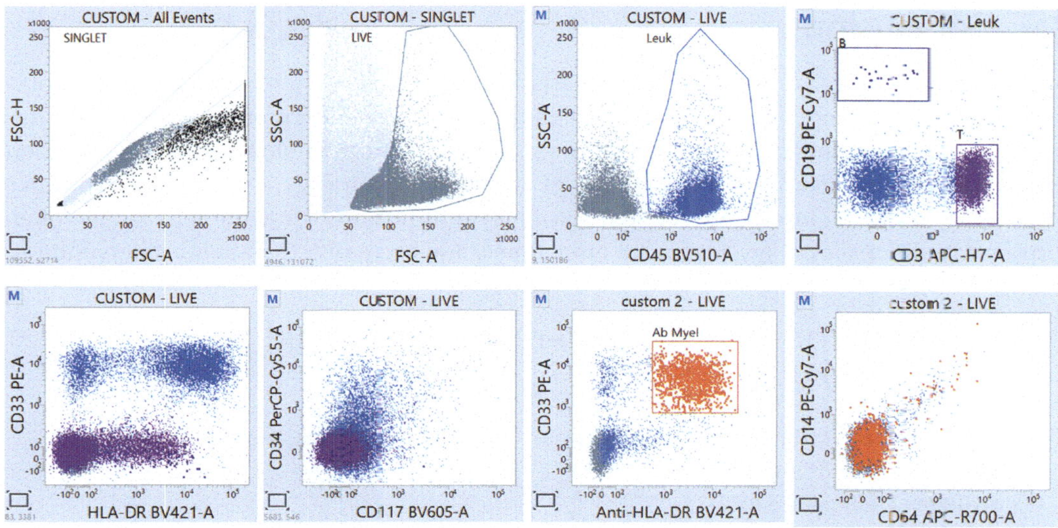

FIGURE 8.3 CSF involvement in a case of Chronic Myeloid Leukemia - myeloid blast crisis. FCM shows the presence of a large population of abnormal myeloid precursors in the moderate CD45 region, which was positive for CD33/HLADR, while being negative for CD34/CD117/CD14/CD64.

should be mentioned in the conclusion. However, the interpretation of results should, when possible, take the full clinicopathological picture into consideration.

8.6.8 Future prospects

Keeping in view the challenges in the diagnosis of CNS infiltration, the harmonization of the Multicolor flow cytometry(MFC) approach across FCM laboratories may improve the overall diagnostic yield. This could also help the evaluation and comparison of data from different centers, which might currently have different approaches. FCM also helps to evaluate the quality of response to intrathecal therapy. In other words, it could guide the number of therapeutic lumbar puncture, which may be customized based on residual disease in CSF and thus may achieve a balance between the risk of CNS relapse and avoid excess neurotoxicity.

A high rate of discordance between the cytological smears based on morphological assessment and the objective immunophenotyping may lead us to think about the approach for the evaluation of this precious and paucicellular sample. FCM has been found to overshadow the conventional cytological smears in terms of diagnostic yield for acute leukemia. Hence, it may be prudent to directly subject the CSF sample to FCM-based analysis, preventing the loss of this precious material. On the other hand, for large-cell lymphoma involvement, a cytomorphological evaluation will be better.

8.7 Conclusion

FCM has revolutionized CSF analysis for metastatic malignancies. When performed with appropriate sample handling, optimized antibody panels, and expert interpretation, it provides unparalleled sensitivity and clinical utility. Continued advances in technology and standardization will further enhance its role in neuro-oncology and hematopathology. Despite limitations like cost and technical requirements, its high sensitivity—particularly for acute leukemias—often surpasses traditional cytology, leading to more timely and aggressive therapeutic interventions.

8.8 Protocol for processing cerebrospinal fluid for flow cytometry

- Inspect the CSF for the volume, color, presence of any hemorrhage, coagulum, etc.
- Transfer to a 12 × 75 mm polysterene tube
- Centrifuge at 400 g for 5 min
- Gently aspirate the supernatant and discard
- Gently vortex the pellet so that it is suspended in 100 μL volume
- Add the antibody cocktail and incubate in the dark for 20 min
- Wash with 500 μL of Phosphate-buffered saline having 0.5% bovine serum albumin (PBS_BSA)
- Centrifuge at 550 g for 5 min
- Discard the supernatant and suspend the pellet in 0.5% PFA
- Acquire and analyze the FCM equipment

References

Aboulafia, D. M., Taylor, L. P., Crane, R. D., Yon, J. L., & Rudolph, R. H. (1996). Carcinomatous meningitis complicating cervical cancer: A Clinicopathologic Study and Literature Review. *Gynecologic Oncology, 60*(2), 313–318. doi:10.1006/gyno.1996.0045.

Beerman, W.F (1912). Meningial carcinomatosis. JAMA, 58, 1437-9.

Bhambhvani, H. P., Rodrigues, A. J., Umeh-Garcia, M. C., & Gephart, M. H. (2020). Leptomeningeal carcinomatosis: Molecular landscape, current management, and emerging therapies. *Neurosurgery Clinics of North America, 31*(4), 613–625. doi:10.1016/j.nec.2020.06.010.

Bromberg, J. E. C., Breems, D. A., Kraan, J., Bikker, G., Van Der Holt, B., Smitt, P. S., Van Den Bent, M. J., Van't Veer, M., & Gratama, J. W. (2007). CSF flow cytometry greatly improves diagnostic accuracy in CNS hematologic malignancies. *Neurology, 68*(20), 1674–1679. doi:10.1212/01.wnl.0000261909.28915.83.

Cancela, C. S. P., Murao, M., Assumpção, J. G., Souza, M. E. d. L., de Macedo, A. V., Viana, M. B., & De Oliveira, B. M. (2017). Immunophenotyping of the cerebrospinal fluid as a prognostic factor at diagnosis of acute lymphoblastic leukemia in children and adolescents. *Pediatric Hematology and Oncology, 34*(2), 53–65. doi:10.1080/08880018.2017.1313920.

Chow, G., & Schmidley, J. W. (1984). Lysis of erythrocytes and leukocytes in traumatic lumbar punctures. *Archives of Neurology, 41*(10), 1084–1085. doi:10.1001/archneur.1984.04050210082020.

De Graaf, M. T., Van Den Broek, P. D. M., Kraan, J., Luitwieler, R. L., Van Den Bent, M. J., Boonstra, J. G., Schmitz, P. I. M., Gratama, J. W., & Sillevis Smitt, P. A. E. (2011). Addition of serum-containing medium to cerebrospinal fluid prevents cellular loss over time. *Journal of Neurology, 258*(8), 1507–1512. doi:10.1007/s00415-011-5970-8.

De Jongste, A. H., Kraan, J., Van Den Broek, P. D., Brooimans, R. A., Bromberg, J. E., Van Montfort, K. A., Smitt, P. A. S., & Gratama, J. W. (2014). Use of TransFix™ cerebrospinal fluid storage tubes prevents cellular loss and enhances flow cytometric detection of malignant hematological cells after 18 hours of storage. *Cytometry Part B - Clinical Cytometry, 86*(4), 272–279. doi:10.1002/cyto.b.21097.

Del Principe, M. I., Gatti, A., Johansson, U., Buccisano, F., & Brando, B. (2021). ESCCA/ISCCA protocol for the analysis of cerebrospinal fluid by multiparametric flow-cytometry in hematological malignancies. *Cytometry Part B - Clinical Cytometry, 100*(3), 269–281. doi:10.1002/cyto.b.21981.

DiGiuseppe, J. A., & Wood, B. L. (2019). Applications of flow cytometric immunophenotyping in the diagnosis and posttreatment monitoring of B and T lymphoblastic leukemia/lymphoma. *Cytometry Part B - Clinical Cytometry, 96*(4), 256–265. doi:10.1002/cyto.b.21833.

van Dongen, J. J. M., Lhermitte, L., Böttcher, S., Almeida, J., van der Velden, V. HJ., Flores-Montero, J., Rawstron, A., Asnafi, V., Lécrevisse, Q., Lucio, P., Mejstrikova, E., Szczepański, T., Kalina, T., de Tute, R., Brüggemann, M., Sedek, L., Cullen, M., Langerak, A. W, Mendonça, A., … Orfao, A. (2012). EuroFlow antibody panels for standardized n-dimensional flow cytometric immunophenotyping of normal, reactive and malignant leukocytes. *Leukemia, 26*(9), 1908–1975. doi:10.1038/leu.2012.120.

Gong, X., Lin, D., Wang, H., Wang, Y., Liu, B., Wei, H., Zhou, C., Liu, K., Wei, S., Gong, B., Zhang, G., Liu, Y., Zhao, X., Li, Y., Gu, R., Qiu, S., Mi, Y., & Wang, J. (2018). Flow cytometric analysis of cerebrospinal fluid in adult patients with acute lymphoblastic leukemia during follow-up. *European Journal of Haematology, 100*(3), 279–285. doi:10.1111/ejh.13011.

de Graaf, M., de Beukelaar, J., Bergsma, J., Kraan, J., van den Bent, M., Klimek, M., van Norden, Y. Kusuma, A., Smitt, P. S., & Gratama, J. W. (2008). B and T cell imbalances in CSF of patients with Hu-antibody associated PNS. *Journal of Neuroimmunology, 195*(1-2), 164–170. doi:10.1016/j.jneuroim.2008.01.007.

de Haas, V., Pieters, R., van der Sluijs-Gelling, A. J., Zwaan, C. M., de Groot-Kruseman, H. A., Sonneveld, E., Stigter, R. L., & van der Velden, V. H. J. (2021). Flowcytometric evaluation of cerebrospinal fluid in childhood ALL identifies CNS involvement better then conventional cytomorphology. *Leukemia, 35*(6), 1773–1776. doi:10.1038/s41375-020-01029-9.

Hegde, U., Filie, A., Little, R. F., Janik, J. E., Grant, N., Steinberg, S. M., Dunleavy, K., Jaffe, E. S., Abati, A., Stetler-Stevenson, M., & Wilson, W. H. (2005). High incidence of occult leptomeningeal disease detected by flow cytometry in newly diagnosed aggressive B-cell lymphomas at risk for central nervous system involvement: The role of flow cytometry versus cytology. *Blood, 105*(2), 496–502. doi:10.1182/blood-2004-05-1982.

Johansson, U., Bloxham, D., Couzens, S., Jesson, J., Morilla, R., Erber, W., & Macey, M. (2014). Guidelines on the use of multicolour flow cytometry in the diagnosis of haematological neoplasms. *British Journal of Haematology, 165*(4), 455–488. doi:10.1111/bjh.12789.

Kraan, J., Gratama, J. W., Haioun, C., Orfao, A., Plonquet, A., Porwit, A., Quijano, S., Stetler-Stevenson, M., Subira, D., & Wilson, W. (2008). Flow cytometric immunophenotyping of cerebrospinal fluid. *Current Protocols in Cytometry*, (45), 6.25.16. doi:10.1002/0471142956.cy0625s45.

Kwon, J., Chie, E. K., Kim, K., Kim, H. J., Wu, H. G., Kim, I. H., Oh, D. Y., Lee, S. H., Kim, D. W., Im, S. A., Kim, T. Y., Heo, D. S., Bang, Y. J., & Ha, S. W. (2014). Impact of multimodality approach for patients with leptomeningeal metastases from solid tumors. *Journal of Korean Medical Science, 29*(8), 1094–1101. doi:10.3346/jkms.2014.29.8.1094.

Levinsen, M., Marquart, H. V., Groth-Pedersen, L., Abrahamsson, J., Albertsen, B. K., Andersen, M. K., Frandsen, T. L., Harila-Saari, A., Pronk, C., Ulvmoen, A., Vaitkevičienė, G., Lähteenmäki, P. M., Niinimäki, R., Taskinen, M., Jeppesen, M., & Schmiegelow, K. (2016). Leukemic blasts are present at low levels in spinal fluid in one-third of childhood acute lymphoblastic leukemia cases. *Pediatric Blood & Cancer, 63*(11), 1935–1942. doi:10.1002/pbc.26128.

Martínez-Laperche, C., Gómez-García, A. M., Lassaletta, A., Moscardó, C., Vivanco, J. L., Molina, J., Fuster, J. L., Couselo, J. M., De Toledo, J. S., Bureo, E., Madero, L., & Ramírez, M. (2013). Detection of occult cerebrospinal fluid involvement during maintenance therapy identifies a group of children with acute lymphoblastic leukemia at high risk for relapse. *American Journal of Hematology, 88*(5), 359–364. doi:10.1002/ajh.23407.

Popov, A., Henze, G., Verzhbitskaya, T., Roumiantseva, J., Lagoyko, S., Khlebnikova, O., Streneva, O., Bidanov, O., Tsaur, G., Inaba, H., Karachunskiy, A., & Fechina, L. (2019). Absolute count of leukemic blasts in cerebrospinal fluid as detected by flow cytometry is a relevant prognostic factor in children with acute lymphoblastic leukemia. Journal of Cancer Research and Clinical Oncology, 145(5), 1331–1339.

Rocha, J. M. C., Murao, M., Cancela, C. S. P., de Oliveira, L. P., Perim, L. P., Santos, J. P., & de Oliveira, B. M. (2023). Comparative analysis between cytomorphology and flow cytometry methods in central nervous system infiltration assessment in oncohematological patients. *Hematology, Transfusion and Cell Therapy, 45*(2), 188–195. doi:10.1016/j.htct.2021.09.016.

Singh, A., Rahman, K., Dutta, P., Singh, M. K., Chandra, D., Gupta, R., Sarkar, M. K., Aggarwal, V., Kumar, S., & Kashyap, R. (2025). Flow cytometric immunophenotyping of cerebrospinal fluid for the detection of central nervous system involvement in haematological neoplasms: An experience and the lesson learnt. *Indian Journal of Hematology and Blood Transfusion*. doi:10.1007/s12288-025-02201-w.

Szánthó, E., Kárai, B., Ivády, G., Baráth, S., Száraz-Széles, M., Kappelmayer, J., & Hevessy, Z. (2019). Evaluation of sample quality as preanalytical error in flow cytometry analysis in childhood acute lymphoblastic leukemia. *Electronic Journal of the International Federation of Clinical Chemistry and Laboratory Medicine, 30*(4), 385–395. https://www.ifcc.org/media/478199/ejifcc2019vol30no4pp385-395.pdf.

9

DNA ploidy analysis of body fluids and cerebrospinal fluid

Venkatesh Dhanasekaran[a], Namrata P. Awasthi[b], Praveen Sharma[c] and Nabhajit Mallik[c]

[a]DEPARTMENT OF PATHOLOGY, ALL INDIA INSTITUTE OF MEDICAL SCIENCES, MADURAI, TAMIL NADU, INDIA [b]DEPARTMENT OF PATHOLOGY, DR. RAM MANOHAR LOHIA INSTITUTE OF MEDICAL SCIENCES (DR. RMLIMS), LUCKNOW, UTTAR PRADESH, INDIA [c]DEPARTMENT OF HEMATOLOGY, POSTGRADUATE INSTITUTE OF MEDICAL EDUCATION AND RESEARCH (PGIMER), CHANDIGARH, INDIA

9.1 Introduction: Fundamentals of DNA ploidy, cell cycle, and liquid cytopathology

9.1.1 Fundamentals of DNA ploidy

The normal resting human somatic nondividing cell has two sets of chromosomes, one received from the father and the other from the mother. The standard physiological two sets of chromosomes are termed as "euploid" or "diploid" state of the chromosomes in the nucleus of the human somatic cells, except gametes (egg and sperm, which have a single set of chromosomes and are termed as "haploid") (Bakloushinskaya, 2022). Maintaining this normal euploid state of the chromosomes is a controlled process involving many genes, cellular factors, and regulatory processes supporting each cell's genomic stability and integrity (Bansbach & Cortez, 2011). DNA ploidy analysis refers to measuring the amount of deoxyribonucleic acid (DNA) content of each cell's nucleus.

To better understand the ploidy analysis, it is essential to understand the different stages of the cell cycle and the quantity & changes in the DNA and other cellular components. The cell cycle proceeds regularly from G0, G1, S, G2, and M phases. The G0 phase is the resting/quiescent phase of the cell cycle. Though the cells in this phase are not in cell division (resting phase), they do their physiological function and can enter the cell cycle if required or stimulated. The perfect example is a hepatocyte, usually present in the G0 phase, which can enter cell division when needed (proliferation of hepatocytes replacing the part of the liver parenchyma following partial hepatectomy). On the contrary, neurons in our body stay in the G0 phase and do not enter the cell cycle. The cells in the G0 phase have two sets of chromosomes, referred to as the diploid state (2n).

Advancing Body Fluid Diagnostics with Flow Cytometry: A Comprehensive Guide to Effusions and CSF.
DOI: https://doi.org/10.1016/B978-0-443-27426-8.00015-6

Stimulation by the mitogenic factors and nutritional signals leads the cells from the G0 phase to enter the G1 phase. In resting cells, retinoblastoma protein (RBP) keeps early 2 transcription factors (E2F) trapped within it and regulates chromatin remodeling, thus helping the cells stay in the resting phase. Under mitogenic signals, the Cyclin-dependent kinases, especially Cyclin D/CDK4 & Cyclin E/CDK2, cause phosphorylation of the RBP, leading to the progression into the G1 phase. In this phase, the cell synthesizes enzymes, proteins, energy, and mRNA, which are required for DNA replication in the S-phase. This phase also acts as part of the G1-S checkpoint, where the cell accesses its nutrition, raw materials for DNA synthesis, and any errors in the DNA present, before entering into the S-phase of the cell cycle. Hence, the G1 phase is a preparatory phase of the cell cycle. If the conditions and raw materials are not suitable, the cells return to the G0 phase (Hume et al., 2020).

Under favorable conditions with adequate nutrition, cellular survival signals, and good quality DNA without any error, the cells enter into "S" phase (Synthesis phase). In this phase, the cells cause reduplication of the DNA to produce two identical sister chromatids attached at the centromere. This increase in DNA results in "4n" DNA content when compared to the DNA "2n" in the G0 and G1 phase (but the number of chromosomes does not change). For the cell to undergo mitotic cell division, it will undergo another preparatory phase, like G1, as the "G2" phase, where it prepares its proteins and enzymes for cell division. The important protein includes the protein that forms the mitotic spindle, which helps in the separation of the DNA. In addition, the cells check for any errors in the newly produced DNA. In the G2 phase, they also have "4n" content of the DNA material; the division of DNA only happens in the next "M" phase (Hume et al., 2020).

During the M phase, after ensuring the DNA is checked for any errors once again, the cells will undergo mitosis in four distinct steps, followed by cytokinesis, which are prophase, metaphase, anaphase, and telophase. At prophase, the sister chromatids condense into the shape of the chromosome. Meanwhile, in the cytoplasm, mitotic spindles form from the centrioles and attach to the sister chromatids with the accompanying disappearance of the nuclear membrane. In metaphase, the chromosomes line up at the center of the cell with mitotic spindle fibers attached at the centromere. Till this stage, the "4n" tetraploid state is maintained. Anaphase is the stage where exact cell division occurs, with the help of spindle fibers, the sister chromatids are pulled apart, separating them to opposite sides of the cell, so that identical chromosomes are formed. At telophase, the separated chromosomes start to unwind, returning to their native chromatin mode with accompanying development of individual nuclear membranes, making the "2n" state. Finally, cytokinesis results in the division of the cell's cytoplasm, resulting in two newly formed identical cells (Hume et al., 2020).

From the above discussion of the normal cell cycle, we can understand that it is very tightly regulated and the cell maintains its normal "2n" state in G0 and G1 phase, followed by the physiological "4n" state from the S and G2 phase, followed by prophase and metaphase of the M phase. At the anaphase of mitosis, the cell's "2n" state is restored, and it is maintained up to the telophase and cytokinesis. The normal cell cycle and its varying DNA content have been illustrated in Fig. 9.1.

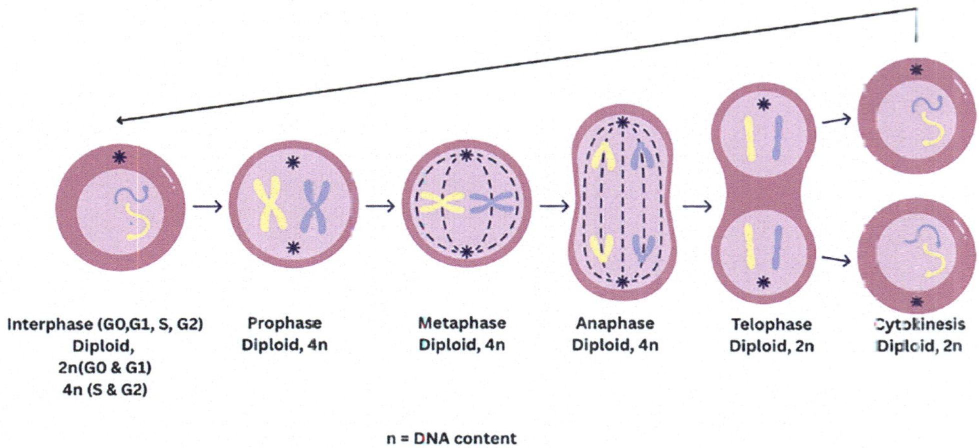

FIGURE 9.1 Normal cell cycle demonstrating varying DNA content throughout the cell cycle.

Before understanding the concept of aneuploidy, let us first understand the DNA ploidy analysis of the normal cell cycle. The normal cell cycle division at different phases and its varying amounts of the DNA content can be better visually understood by plotting in a histogram by staining the cells with DNA-binding dyes and plotting their fluorescence intensity of the DNA on the x-axis and the number of cells on the y-axis. The intensity of fluorescence is a direct measure of the DNA content of the cells. By plotting the histogram of the normal cell cycle, we will observe two peaks and a bridge/plateau-like area between the peaks. The first peak is tall, followed by a bridge/plateau-like, and a smaller second peak. The first peak represents the G0 along with the G1 phase of the cell cycle. The reason behind the formation of a tall peak is that, in G0, which is a resting phase, and G1, which is a preparatory phase, all the cells have a similar amount of DNA, and they react with fluorescence dye with the same intensity, resulting in a single tall peak. A bridge/plateau-like structure follows this in S-phase; this phenomenon is because cells at this phase are synthesizing DNA at varying proportions (e.g., a cell can produce 30%, another could produce 50%, and another could produce 75%). This bridge/plateau-like structure is followed by a second peak, representing G2 and M phases, in which cells contain double the amount of DNA (Pozarowski & Darzynkiewicz, 2004). The reason for the smaller peak than G0/G1's larger peak is due to a lesser population of cells undergoing cell division at any moment, and the larger population of cells in the resting phase in the G0 phase. Hence, a normal cell division will have a tall first peak (representing G0 and G1 phase), which is followed by a bridge/plateau-like (representing S-phase) and a small peak (representing G2 phase) (Pozarowski & Darzynkiewicz, 2004).

Aneuploidy refers to an abnormality in the DNA content of the cell's nucleus. It can be either an increase or a decrease in the amount of DNA within the cell's nucleus. It arises due to the abnormality in the chromosomal aggregation during mitosis due to pathology in the mitotic spindle formation, cell cycle regulators, and centrosomes. Aneuploidy is a marker of genomic instability (Giam & Rancati, 2015). The aneuploidy results in abnormal, uncontrolled cellular proliferation, increased protein synthesis, and inhibition of cells from undergoing normal apoptotic cell death.

Table 9.1 Common sites of malignant effusions in the body fluids and the CSF.

Type of fluid	Primary tumor site
Cerebrospinal fluid	Leptomeningeal metastasis from breast and lung carcinoma, along with melanoma, and hematolymphoid malignancies
Pleural fluid	Lungs in males, Breast in females, Gastrointestinal cancer, and hematolymphoid malignancies
Peritoneal fluid	Ovarian, colorectal, and stomach carcinomas and hematolymphoid malignancies
Pericardial fluid	Lungs, breast, and gastrointestinal tract

CSF, cerebrospinal fluid.

Aneuploid cells exhibit an increased proliferation index, which leads to an increased number of mutations within the tumor cells, which further causes heterogeneity of the tumor population, resulting in resistance to conventional forms of treatment. In addition, aneuploidy is associated with poorly differentiated histomorphology, increased rates of metastasis, and poor prognosis for the patient (Replogle et al., 2020).

A diploid population of one tall peak and one smaller, with a bridge-like area present in the DNA ploidy histogram, is that of a normal cell cycle. On the contrary, the aneuploid cells show more peaks apart from the physiological two peaks. This additional peak indicates the presence of abnormal amounts of DNA within the cell's nucleus, which can be either less (hypoploidy) or more (hyperploidy) (Huang et al., 2008). To quantify this, the DNA index (DI) has been used. Normal physiological cells have a DI of 1; anything less than or equal to physiological values indicates aneuploidy. This concept of DI is explained in detail in the upcoming Chapter

9.1.2 Body fluids as diagnostic windows

Pleural, pericardial, peritoneal, and cerebrospinal fluid (CSF) are the body fluids in which malignancies can be detected. Apart from the CSF, the pleural, pericardial, and peritoneal cavities are lined by mesothelial cells and are referred to as serous fluids. CSF and serous cavities contain minimal fluids, which serve their physiological function. The physiological range of fluids is in the pleural cavity (5 to 20 mL), peritoneal cavity (5 to 50 mL), pericardial cavity (20 to 60 mL), and CSF (approximately 150 mL), respectively. Tumor cells can metastasize into the serous cavities and CSF, causing increased amounts of fluids beyond their physiological limits, resulting in malignant effusions. The modes of tumor spread into serous cavities include direct extension, lymphatic and hematogenous spread, and increased vascular permeability resulting in tumor spread by tumor-induced inflammatory process (Karaman & Detmar, 2014). The tumor involvement of the body fluids indicates an advanced disease process and is often categorized as stage IV disease and metastatic disease in some solid malignancies. The tumor cells, which infiltrate the body cavities as single cells, can proliferate in the cavities, forming tumor cell clusters. The pathological examination of these tumor cells in the body fluids serves for diagnostic, prognostic, and staging purposes (Shidham & Layfield, 2021). The common causes of malignant effusions in the body fluids and CSF are presented in Table 9.1.

The body fluid samples serve as accessible windows of the extensive metastatic malignancy, through which the pathologist examines them through a less-invasive procedure (tapping of the fluids under sterile conditions) to demonstrate information equivalent to and superior to invasive tissue sampling.

In many patients, cytomorphological identification of malignancy in the fluid sample is often the first sign of malignancy. The clues to the epithelial origin of the metastatic diseases are three-dimensional clustering, cell-in-cell or cannibalism of the tumor cells, cohesion, nuclear irregularity, and prominence of the nucleoli (Bansal et al., 2011). The diagnosis of malignancy does not end with morphological evaluation alone; for the complete diagnosis, a constellation of clinical findings, age, gender, imaging findings, and performance of immunohistochemistry in the cell block to establish epithelial origin and specific site of origin are essential. Sometimes mesothelioma enters into the differential diagnosis of epithelial malignancies; at those times, immunohistochemistry with a panel of markers (BerEp4, MOC-31, calretinin, and WT-1) is required (Hjerpe et al., 2023).

In addition to the diagnosis, the fluid samples also help as molecular windows and aid in molecular diagnosis, staging purposes, and personalized therapy. For instance, when the lung adenocarcinoma is diagnosed on the pleural fluid, the cell block prepared from the sample can be utilized for performing PDL-1 immunochemistry (thus helping the physicians determine whether the patient is eligible for immunotherapy), and the fluid sample can be utilized for performing mutational studies, including EGFR, ALK, ROS, and KRAS. Also, peritoneal washing, which is performed intra-operatively during gynecological cancer, helps in the staging of the disease process (Mahajan et al., 2022; Yao et al., 2019).

9.2 Flow cytometry: High-throughput analysis in suspension

In the context of effusion cytology, differentiating between reactive mesothelial cells and malignant cells remains a significant diagnostic challenge due to overlapping morphological features. While traditional cytology relies on the subjective interpretation of nuclear-to-cytoplasmic ratios and chromatin patterns, flow cytometry (FCM) offers an objective, high-throughput alternative by providing quantitative DNA content analysis on a cell-by-cell basis (Mishra et al., 2017). By utilizing stoichiometric fluorochromes that bind proportionally to the amount of DNA present, FCM can precisely identify aneuploid populations, a hallmark of malignancy, that may be morphologically occult or obscured by inflammatory debris and proteinaceous fluid. This transition from qualitative visual assessment to a discrete numerical DI allows for the detection of "near diploid" clones with a level of statistical sensitivity that exceeds manual microscopy, especially when the malignant fraction is low. Furthermore, the ability to simultaneously analyze thousands of individual events ensures that the resulting DNA histogram reflects the true biological heterogeneity of the effusion, providing a more robust diagnostic and prognostic framework than morphology alone.

Review of publications related to DNA ploidy in effusions has already been covered in Chapter 6. The following sections detail the methodology, interpretation, and clinical utility of DNA ploidy analysis in general and then specifically for body fluid specimens.

9.2.1 Principles and techniques

Principles of flow cytometric measurements and cell processing techniques have been discussed in Chapters 5 and 7. While specimen handling and transport follow similar paths, specimen processing, staining, and acquisition are unique for DNA ploidy analysis, which will be emphasized in further sections.

9.2.1.1 Specimen handling: The critical first step

Body fluids and CSF are characterized by their tendency for degeneration, cellular lysis, and loss of antigenicity if not handled with care in the laboratory (Kraan et al., 2008). Hence, preanalytical steps are essential for the correct and accurate analysis of body fluids and CSF for the flow cytometric examination. In this Chapter, we summarize the evidence-based steps for evaluating body fluids and CSF to ensure proper preservation of cell viability and antigenicity for flow cytometric analysis.

Sample collection—The core step of the quality: After obtaining patient consent for the tapping of body fluids and lumbar puncture, the procedure should be performed under sterile conditions in an aseptic environment to prevent sample contamination. The samples should be transferred into a transparent plastic, leak-proof container of the appropriate size with a tight lid to prevent sample loss. Immediately, it is essential to label the specimen container with a bar code sticker that includes the patient's name, age, type of sample, unique hospital identification number, and time of collection to prevent any potential patient sample mismatches. To prevent clotting, the sample may be collected in a heparinized container or with ammonium oxalate.

Transport of the sample to sustain cellular viability: The collected sample must be transferred from the wards to the laboratory for processing at room temperature as soon as possible to prevent cell degeneration and avoid cellular clumping, especially in CSF samples, which are routinely paucicellular. In case if the samples are collected during weekends or late at night, when the samples cannot be transferred to laboratory due to logistic reasons, it is important to preserve the samples in RPMI-1640 medium (Roswell Park Memorial Institute medium) along with 10% fetal bovine serum to prevent degeneration of the cells, and the samples need to be kept in a refrigerator at a temperature of 4 to 8°C. The samples should never be frozen.

Despite taking adequate precautions, it is common to receive compromised samples due to freezing and keeping the samples for longer hours at the bedside. To reduce these mishappenings, it is essential to impart knowledge to clinicians by conducting regular, continuous medical education programs to enhance understanding and raise awareness about sample collection and transport to the laboratory. Additionally, it is crucial to maintain a logbook in the laboratory to record the specimen collection time from the patient and the specimen reception at the laboratory, thereby minimizing and accurately tracking the sample transit time. These practices will help in evaluating the root cause analysis of the preanalytical sample errors.

9.2.1.2 Sample processing

Cell concentration: Body fluids are characterized by the presence of cells along with noncellular elements, which include fibrin, mucin, and aggregates of protein material. These noncellular elements clog the flow cytometry channels, creating additional background noise and false positive events during flow cytometric evaluation. Hence, it is essential to separate them from the cells for accurate flow cytometric evaluation (Fuller & Thrall, 2016). The ideal method for the separation of these cellular and noncellular elements is by gentle, slow centrifugation for a shorter duration of time (since centrifugation at higher speeds and longer duration results in lysis and fragmentation of cells). The goal is to obtain a single cell suspension with a concentration of approximately 1 million cells /tube. The centrifugation at 400 to 600 g for approximately 5 to 10 min serves the purpose of gentle centrifugation for a shorter time and causes sedimentation of pellets at the base (Cossarizza et al., 2017). The supernatant, containing noncellular elements, can be decanted.

If the sample is heavily contaminated with blood, a short incubation with an ammonium chloride-based lysis buffer is necessary. In addition, filtration using nylon mesh and a cell strainer with 35 to 60 μm pore sizes also helps in excluding larger cellular debris. Using an automated counter or hemocytometer to adjust the final desired concentration of cells is preferred. Over-concentration leads to coincidence (doublets), while under-concentration extends acquisition time.

9.2.1.3 Fixation and permeabilization

The success of the DNA ploidy analysis by flow cytometry depends on how well we prepare the cells' nuclei for the stoichiometric staining with DNA-specific fluorochromes. Meticulous standardization of fixation and permeabilization will result in narrow G0/G1 peaks, an accurate & reliable DI, and reproducible S-phase estimation.

Rationale for fixation in DNA ploidy analysis: Fixation makes the cells stable mechanically, makes them permeable to dyes and resistant to autolysis, and prevents bacterial degradation (Fahlberg et al., 2024). In the context of DNA ploidy analysis, fixation has four important purposes:

1. Preservation of the structure of cells and their nucleus during storage, transport, and acquisition
2. Inactivates infectious organisms, thus ensuring safer handling of specimens from HIV patients
3. Stabilization of the DNA-protein complexes, so that they can withstand RNase digestion and the detergent treatment
4. Locks antigens and nucleic acids, thus allowing combined analysis of DNA content along with the surface and intracytoplasmic markers if required.

Fixation does impart some changes to the cellular content. Alcohol causes dehydration and protein precipitation, whereas aldehydes cause protein cross-linking. These alterations can result in alteration of chromatin, abnormal scattering property; the process of fixation should always be controlled (Forment & Jackson, 2015).

Permeabilization: Permeabilization is the process of creating a controlled opening in the plasma membrane and nuclear membrane of the cell, so that hydrophilic DNA dyes and the RNase enzyme can enter the cell (Viryasova et al., 2019). Permeabilization and fixation can be done together or separately. Permeabilization can be performed without fixation using nonionic detergents, pore-forming agents, or lysolecithin, which results in temporary membrane disruption. The nonionic detergents include Triton X-100 and Nonidet P-40 (NP-40). Pore-forming agents, such as nisin and gramicidin, allow only small dyes like propidium to cross, but do not allow larger molecules. Alternatively, cells can be exposed to fixatives with minimal to mild permeabilization, especially with shorter, low-concentration aldehyde exposures.

The ultimate goal is to prepare a state-of-the-cell with:

1. Effective removal or solubilization of cytoplasm, which minimizes light scatter artifacts and fluorescence associated with the RNA,
2. To open the nuclear chromatin and nuclear envelope, which is adequate to permit uniform and quick entry of DNA dye and RNase,
3. Preserved nuclear morphology without any clumping or lysis of the nuclei.

This is achieved in many protocols by using a mild fixative with a detergent (e.g., paraformaldehyde followed by Triton X). Many commercially available fixatives or fixation-permeabilization buffers are available.

Major fixatives used in DNA ploidy:

Alcohol fixatives: Ethanol, methanol, and acetone are coagulant fixatives that precipitate proteins and increase the cytoplasmic refractive index to approximately 1.54. This, in turn, increased the light scatter signals. Their important advantages for DNA ploidy analysis are:

1. Efficient permeabilization of cells and their nuclei, which results in direct and accurate staining with impermeant DNA dyes such as propidium iodide and mithramycin.
2. Good preservation of DNA content with relatively simple protocols and stable storage at 4°C.

Ethanol fixation of the cells is achieved by slowly adding cold absolute ethanol to a cold cell suspension under gentle vortexing, yielding preparations that can be stored for several days with good preservation of the DNA peak. For cytology samples such as urine and bladder washings, methanol–acetic acid mixtures (e.g., 20% acetic acid in methanol) provide a preparation suitable for both cytological evaluation and DNA content measurement.

The disadvantage of using alcohol fixation at higher concentrations is that it gives a bimodal DNA distribution in nucleated red blood cells and leucocytes due to differences in chromatin condensation. Methanol denatures some antigens; it is not suitable for evaluating surface and cytoplasmic immunophenotyping when combined with DNA ploidy analysis (Holtfreter & Cohen, 1990).

Aldehyde fixatives: Formaldehyde and paraformaldehyde are the commonly used fixatives for DNA ploidy analysis using flow cytometry worldwide. For DNA ploidy analysis, Aldehyde fixation offers several advantages:

1. Improved structural stability of nuclei during RNase treatment and detergent permeabilization.
2. Excellent preservation of surface antigen expression at low concentrations (around 0.25% to 0.5% paraformaldehyde), hence enabling combined surface immunophenotyping along with the DNA staining.

However, excessive aldehyde cross-linking reduces the accessibility of DNA dyes and can broaden DNA peaks by making chromatin condensation differences more influential on staining. More complex aldehyde mixtures containing periodate, lysine, and paraformaldehyde (buffered formaldehyde–acetone) and protocols combining paraformaldehyde with methanol have been systematically compared, and paraformaldehyde plus methanol has performed well in preserving both antigen staining and DNA histogram quality (Pollice et al., 1992).

Alternative formalin-free cross-linking agents: Dimethylsuberimidate and carbodiimide are common formalin-free cross-linking agents. These agents reduce and minimize fixation-induced fluorescence, but they are more expensive and less stable (Davies & Stark, 1970). Transfix, originally used as transport media for CD4, also maintains antigen expression and DNA integrity for several days under variable climate conditions and is increasingly used for delayed DNA ploidy measurements (Canonico et al., 2010).

Permeabilization strategies in DNA ploidy protocols: Permeabilization is usually performed using nonionic detergents in a hypotonic or isotonic buffer solution. Low-concentration Triton X-100 (around 0.1%–1%) is commonly used to break plasma membranes and solubilize cytoplasmic components while maintaining nuclei intact for several hours to perform DNA ploidy analysis (Darzynkiewicz et al., 2017). Routinely, the cells are fixed with low-concentration paraformaldehyde at 4°C, then permeabilized with 0.1% Triton X-100 in phosphate-buffered saline. This allows better cellular preservation of cell surface antigens before antibody staining and subsequent access of RNase and DNA dye to the nucleus for ploidy analysis. Alternative permeabilizing agents used for intracellular antigen staining also feature in ploidy-oriented workflows when dual measurements are required. Tween 20, saponin, digitonin, lysolecithin, and n-octyl-β-D-glucopyranoside have all been used, usually in combination with paraformaldehyde fixation at low temperature (Schmid et al., 1991). Each has a characteristic "pore size" and kinetic profile, so optimization is essential when one wishes to retain small soluble components while exposing large targets such as chromatin or nuclear antigens. A perfect DNA staining for the DNA ploidy analysis should produce a single sharp diploid peak with a coefficient of variation below 3% and also separating G0/G1 and G2/M peaks accurately, along with sturdy identification of S-phase. The combination of the fixative permeabilizer and the DNA dye interacts strongly.

Artifacts and quality control: Every method of fixation and permeabilization introduces artifacts and distorts the DNA ploidy analysis. The common problem with alcohol and aldehyde fixatives is an increase in the refractive index of the cytoplasm, which leads to falsely increased forward and side scatter, along with alterations in fluorescent intensities. It also induces variation in cell and nuclear size, leading to discrepancies among subpopulations that, in turn, complicate gating and compensation. Formaldehyde and paraformaldehyde cause aberrant antigen expression and masking, leading to false interpretations in ploidy analysis combined with immunophenotyping. A potent and accurate ploidy analysis, therefore, requires standardization of the procedure and the use of internal controls. Normal diploid cells from the patient can be used as internal controls, and external controls, such as chicken or trout erythrocytes, can be used to calculate the DI and monitor day-to-day variability. In addition, it is important to do routine assessment of peak position, symmetry, coefficient of variation, and debris content of the histogram whenever fixative and permeabilization conditions are changed.

9.2.1.4 Enzymatic treatment and staining

Stoichiometric DNA dyes: Stoichiometric DNA dyes bind with sequences of the DNA in equal proportions and emit fluorescence, which is directly proportional to the amount of DNA present within the nucleus (Darzynkiewicz et al., 2010). The standard and popular stoichiometric dyes include propidium iodide (PI) and 4'6-diamidino-2-phenyl iodide (DAPI). Both of these dyes directly attach to the adenine-thymidine (A-T) rich sequences. It is also important to note that both these dyes bind to RNA present in the nucleus and cytoplasm, resulting in falsely increased DNA estimation. Hence, it is important to treat the cells with DNase-free RNase in low concentrations (100 ug/mL for 30 min at 37°C). PI exhibits red fluorescence at 488 nm, and DAPI exhibits blue fluorescence on UV excitation. Other less commonly used DNA dyes are Hoechst 33,342, which stains without permeation, and 7-Aminoactinomycin D (7-AAD), which has naturally higher affinity for DNA with lesser affinity toward RNA.

1) PI and ethidium bromide are examples of intercalating dyes that bind with both DNA and double-stranded RNA. Because fixed cells have a property to retain substantial RNA, RNase treatment after permeabilization is routine to prevent artificial broadening of DNA peaks by cytoplasmic or nucleolar RNA. Detergent-supplemented hypotonic solutions of PI and sodium citrate, often with 0.1% Triton X-100, are commonly used in conjunction with ethanol fixation (Shen et al., 2017).
2) Mithramycin and chromomycin A3 are antibiotics with G-C binding preference. They stain ethanol or glutaraldehyde-fixed specimens in the presence of magnesium ions and normal salt concentrations. Their selectivity can improve ploidy precision (Crissman & Tobey, 1990).
3) Hoechst dyes and DAPI are A-T preferring, UV-excited fluorochromes that often yield the lowest coefficients of variation. For fixed samples, low dye concentrations are required to prevent nonspecific binding and spectral shifts. These dyes also show that chromatin condensation state influences dye access. DAPI is relatively insensitive to chromatin condensation, which explains its excellent performance in high-precision DNA ploidy work (Breusegem et al., 2002).
4) 7-Aminoactinomycin D (7-AAD) is a G-C preferring dye, red-emitting, but sensitive to chromatin structure, tending to give broader peaks. In ploidy analysis, it is used to stain the unfixed samples.
5) Newer cyanine dyes such as TO-PRO-3 and the SYTOX series are advanced, high-performance nucleic acid dyes that are impermeant and bind strongly to nucleic acids. TO-PRO-3 leverages the red diode laser (typically read in the allophycocyanin (APC) channel), unburdening the highly utilized blue laser line. The SYTOX family of dyes provides exceptional versatility across multiple laser lines and exhibits a >500-fold fluorescence enhancement upon nucleic acid binding. This massive quantum yield provides a superior signal-to-noise ratio and permits "no-wash" assay designs.

Fixable viability stains: Although strictly speaking, these are protein-binding rather than nucleic acid binding, amine-reactive fixable viability dyes are an indispensable component of DNA ploidy analysis. Unlike classical cell-impermeant intercalators (such as PI or 7-AAD), which

Table 9.2 DNA-binding dyes with excitation, emission, and their laser requirements.

Dye	Excitation (nm)	Emission (nm)	Laser requirement
Propidium Iodide (PI)	**535**	**617**	488 nm (blue, Argon)
DAPI/FxCycle violet	**350 to 358**	**460 to 465**	405 nm (violet) or 350 nm (UV)
7-AAD	**546**	**647**	488 nm or 561 nm (yellow/green)
Hoechst 33,342	**350 to 355**	**461**	UV (350 nm) or 405 nm (violet)
SYBR green I	**494**	**521**	488 nm or 505 nm (blue)

rely on equilibrium binding and wash out when cell membranes are altered, amine-reactive fluorochromes form irreversible covalent bonds with primary amines present on cellular proteins. In viable cells with intact plasma membranes, this binding is sterically restricted to a relatively low number of cell surface proteins, resulting in dim fluorescence. Conversely, in nonviable cells with compromised membranes, the dye freely permeates the cell and covalently binds to the highly abundant intracellular proteins, yielding a fluorescence intensity several magnitudes higher. Because this covalent linkage is permanent, these dyes endure the aggressive alcohol fixation and permeabilization protocols strictly required to drive primary DNA dyes into the nucleus. This unique property permits investigators to permanently "tag" dead cells before fixation, allowing for their reliable exclusion during analysis and preventing the nonspecific uptake of DNA dyes that frequently generate artifactual aneuploid peaks. Some examples of fixable viability stains are LIVE/DEAD fixable stains (Thermo Fisher/Invitrogen), Zombie Dyes (BioLegend), BD Horizon fixable viability stains (Becton, Dickinson and Company Biosciences, BD Biosciences), Ghost Dyes (Tonbo/Cytek) (Table 9.2).

Choice of DNA-binding dye for DNA ploidy analysis: The choice of DNA-binding dye for DNA ploidy analysis depends on various factors, which include the excitation property of the instrument, dye permeability, DNA selectivity of the dye, base preference of the dye, environment sensitivity, biological question (live versus fixed cell being analyzed), and sample preparation.

Light source of the flow cytometer: The dye selection begins with the availability of the flow cytometer's light source. Argon lasers are commonly used, and PI or 7-AAD is typically used. Hoechst 33,342 requires a UV excitation laser at 405 nm, whereas mithramycin requires a 457 nm excitation laser. It is important to match the emission spectra of the detectors to minimize overlap in multiparameter analysis (e.g., PI emits red, avoiding green channel bleed). TO-PRO3 is optimally excited by the red laser (633 nm or 640 nm) and emits in the far-red spectrum (around 661 nm). In a standard flow cytometer, it is read in the APC channel, freeing up the entire blue and violet laser pathways.

Base preferences: Base preferences also need to be considered, A-T binding for DAPI and Hoechst, G-C binding for Mithramycin.

Vital versus fixed cells: Hoechst dyes are ideal for staining Vital/Live intact cells. PI, DAPI, and 7-AAD are impermeable dyes and require fixation, permeabilization, and RNase treatment; hence, they are best suited for fixed/permeabilized cells.

Finally, the dye choice is based on the dye's DNA-binding stoichiometry, its DNA specificity, the flow cytometer's excitation-emission spectra, and whether live or fixed cells are being analyzed. PI remains the most widely used dye for routine usage and research. DAPI and Hoechst are excellent dyes for the Ultraviolet system or for live-cell analysis (Darzynkiewicz et al., 2010b).

Role of RNase: Routinely, PI is popularly used as a DNA-binding dye in the ploidy analysis. In addition to DNA-binding, it nonspecifically binds to RNA. RNA content varies among different cell types; resting cells have less RNA, whereas actively proliferating cells have more. Hence, the DNA histogram produced without RNase will show a broader peak for G0/G1 and blurring of distinct cell populations in S and G2/M phases, resulting in an inaccurate estimate of DNA ploidy. However, treatment with RNase will result in the lysis of RNA within the cells, which, in turn, satisfies the stoichiometric property of the DNA-binding dyes. Without RNase, the very foundation of quantitative DNA flow cytometry would be compromised.

As examples, sample staining protocols for two types of experiments have been given below. Protocol A is for PI staining of fresh cells, and protocol B is for modern clinical flow cytometry labs performing multiparametric ploidy using DAPI and EpCAM.

Protocol A: PI staining for fresh cells:

1. *Preparation*: Prepare a single cell suspension of the fluid sample as described earlier.
2. *Initial wash*: Centrifuge the cells at 300 × *g* for 5 min and safely discard the supernatant.
3. *Resuspension:* Resuspend the cell pellet in exactly 1 mL of cold phosphate-buffered saline (PBS). Gently pipette to ensure complete dispersion of the cells.
4. *Fixation*: While gently vortexing the cell suspension, add 3 mL of ice-cold absolute ethanol **dropwise**. *(Note: This specific volumetric ratio achieves an optimal final concentration of 75% ethanol).*
5. *Incubation*: Incubate the fixed cells at-20°C for at least 30 min. *(Note: For optimal peak resolution, a 2-h or overnight incubation at-20° C is highly recommended; cells are stable in this state for several weeks).*
6. *Alcohol removal*: Centrifuge the fixed cells at 400 to 500 × *g* for 5 min. Carefully decant the ethanol, being cautious not to dislodge the pellet.
7. *Rehydration*: Resuspend the pellet in 1 mL of cold PBS, centrifuge again at 400 to 500 × *g* for 5 min, and discard the supernatant. This critical step rehydrates the cells and removes residual alcohol that could interfere with downstream staining.
8. *Staining*: Resuspend the cell pellet in 0.5 mL of a freshly prepared PI/RNase staining cocktail (yielding a final concentration of approximately 50 μg/mL PI and 100 μg/mL RNase A in PBS).
9. *Equilibration*: Incubate the suspension at room temperature for 30 min in the dark to allow for complete RNA digestion and stoichiometric dye equilibration.
10. *Acquisition*: Transfer the sample to flow cytometry tubes and acquire the data immediately on a low flow rate setting.

Protocol B (DAPI Staining): Multiparametric ploidy using DAPI

Unlike PI, which indiscriminately intercalates into any nucleic acid, DAPI binds specifically to the AT-rich regions of the DNA minor groove. This intrinsic specificity eliminates the need for

RNA digestion (RNase). DAPI is excited by violet laser (405 nm) and has a relatively narrow blue emission profile. This leaves the entire blue and red laser pathways open for complex surface phenotyping (e.g., EpCAM, CD45, and CD200) without the spectral spillover caused by PI.

1. *Cell preparation*: Centrifuge the fresh fluid sample at 300 × *g* for 5 min. Wash the pellet once in protein-free PBS to remove heavy background serous proteins.
2. *Staining cocktail*: Resuspend the cell pellet in 100 μL of protein-free PBS. Add the pretitrated volumes of:
   - Amine-reactive fixable viability dye (e.g., LIVE/DEAD Fixable Near-IR or Zombie NIR).
   - Fluorescent anti-EpCAM antibody (e.g., EpCAM-FITC or EpCAM-APC).
3. *Incubation*: Incubate for 20 to 30 min in the dark at room temperature.
4. *Wash*: Add 2 mL of standard flow cytometry staining buffer (PBS with 1%–2% BSA) to quench any unbound viability dye and wash away unbound EpCAM antibodies. Centrifuge at 300 × *g* for 5 min and discard the supernatant.
5. *Commercial fix/perm*: Thoroughly vortex the cell pellet to ensure a single cell suspension. Add 250 to 500 μL of a commercial, aldehyde-based Fixation/Permeabilization solution (e.g., BD Cytofix/Cytoperm).
6. *Incubation*: Incubate at 4°C (or on ice) for 20 min in the dark.
7. *Permeabilization wash*: Add 2 mL of the manufacturer's specific permeabilization wash buffer (which contains saponin or a similar mild detergent to keep the cell membrane open). Centrifuge at 400 × *g* for 5 min and discard the supernatant.
8. *DAPI staining*: Resuspend the fully permeabilized pellet directly into 0.5 mL of a DAPI staining solution (typically diluted to 1 to 3 μg/mL in PBS or perm buffer).
9. *Equilibration*: Incubate for 15 to 30 min at room temperature in the dark to allow the DAPI to reach stoichiometric equilibrium with the DNA.
10. *Acquisition*: Acquire the sample on the flow cytometer on a low fluidic flow rate, triggering on the DAPI fluorescence channel (Violet laser) to exclude debris.

9.2.1.5 Flow cytometry instrument setup and acquisition

Instrument setup and optimization for analyzing samples with low cell counts:

Body fluids and especially CSF are characterized by a lower cell count when compared with peripheral blood, which usually contains an abundant number of cells to work with in respect to flow cytometry. Hence, along with meticulous collection, transport, and processing, the flow cytometric instrumentation setup, calibration, and analysis must be performed meticulously.

The low flow rate: During flow cytometry analysis, the stained nucleus passes through the laser interrogation point for a brief time (a few microseconds). During this extremely short time, the laser beam illuminates the nucleus, which is bound to the fluorochrome, and the fluorochrome emits fluorescence. The detector captures the pulse amplitude (height) and its area, which are directly proportional to the amount of DNA within the cell's nuclei. The number of photons collected measures precision; the more photons, the smaller the statistical variation (Poisson noise) and the tighter the distribution of values from diploid cells.

This is where the sample flow rate becomes paramount. The flow cytometer's sample fluidics are built to hydrodynamically focus the sample stream within a sheath fluid that flows faster. This produces a narrow "core" stream where cells pass a single file through the laser beam. The diameter of this core stream is directly controlled by the sample flow rate relative to the sheath pressure. A higher sample flow rate increases the stream's core diameter and pushes the cells toward the edges of the illuminating beam. This will falsely evaluate two diploid nuclei with the same amount of DNA, producing different fluorescence pulse heights, broadening the resulting G_0/G_1 peak, and increasing the coefficient of variation. Second, a wider core increases the likelihood of multiple cells passing through the beam simultaneously, which can be misinterpreted as a single cell with a higher DNA content (Kim & Sederstrom, 2015).

Conversely, a low sample flow rate limits the core stream to a very narrow diameter, ideally 5–10 μm. This ensures that every nucleus travels to the center of the core, which is the most uniformly illuminated region of the laser beam. Each diploid cell is, therefore, exposed to the same amounts of illumination, leading to a much more reproducible fluorescence signal. This reduction in optical variance is the single most important factor in achieving low CVs (typically less than 3%). This is essential to resolve subtle aneuploid populations admixed with normal cells. Hence, by sacrificing analysis speed (lowering the flow rate), we achieve the necessary precision. In practice, a low flow rate is achieved by using rather than pressure-driven systems, which fluctuate with resistance; this maintains Reynolds numbers below 2300, preventing turbulence in the interface of the sheath core (Suthanthiraraj et al., 2024).

Linear scale: DNA ploidy analysis measures the amount of DNA in a cell's nuclei. A diploid cell (in G_0/G_1) has a specific quantity of DNA, while a tetraploid cell (in G_2/M) has exactly double that amount. The relationship between the measured signals from these two groups is usually arithmetic. If the mean channel of the G_0/G_1 peak is 200 on a linear scale, the G_2/M peak should be at channel 400. This property allows for a quick and clear check of the instrument's linearity, which is a key measure of its accuracy. If the G_2/M peak is not at double the channel of the G_0/G_1 peak, the system is nonlinear, and any further calculations of the DI or S-phase fraction will be incorrect (Darzynkiewicz et al., 2010). A logarithmic scale, however, compresses the data. It helps visualize populations with very different fluorescence intensities on one plot, like a weakly positive antigen and a brightly positive one. On a log scale, a twofold difference in DNA content (e.g., from 200 to 400) does not show as a twofold difference in channel position; the relationship is exponential, not linear. Due to this, the G_2/M peak would appear much closer to the G_0/G_1 peak, masking the clear separation that is essential for proper analysis. Additionally, the visual width of a peak on a log scale depends on its position, which makes it hard to quickly assess the CVs of different populations. The main goal of ploidy analysis, resolving and counting populations with a known arithmetic relationship, is fundamentally at odds with a logarithmic display (Darzynkiewicz et al., 2010).

Triggering on the DNA channel: The trigger or threshold is the signal that tells the cytometer's electronics, "A cell is present, start processing the data." In immunophenotyping for blood cells, forward scatter acts as an appropriate trigger. However, in the case of fluids, many debris, apoptotic cells, and nonnucleated cells are present in the sample; hence, triggering forward scatter will produce a histogram with unnecessary, irrelevant events, a sloping debris hump at the low end that can obscure a true hypodiploid population. To overcome this, the fluorescence signal from the DNA-bound dye itself is set as a trigger for DNA ploidy analysis of the fluid samples. By doing this, nonnucleated debris and other unwanted cells are excluded from the analysis, resulting in a clean, accurate histogram (McKinnon, 2018).

Compensation matrix configuration: Fluorochromes emit light across a spectrum. They are designed to peak at specific wavelengths (e.g., a green-emitting fluorochrome), but they can also emit some amount of yellow and red light. The detection of light intended for one channel by an adjacent detector is called spectral overlap. When cells are tagged with multiple fluorochromes, we expect complex spectral overlap among the different colors. It is essential to mathematically distinguish these signals from each other through a process called compensation. The concept has been dealt with in Chapter 5 and is being explained again with respect to DNA dyes.

While DNA ploidy analysis often uses only one or two dyes, compensation must still be considered whenever measuring more than one parameter. For example, during ploidy analysis, it is crucial to exclude nonviable cells and debris, which nonspecifically bind dyes and produce artifactual peaks. If performing live-cell ploidy analysis, one might use a primary DNA dye (like Hoechst 33,342) alongside a viability dye like 7-AAD. If the protocol requires fixation, an amine-reactive viability dye must be used before adding the ploidy dye to prevent dye competition. Furthermore, compensation is strictly required when a DNA dye is combined with fluorochrome-conjugated antibodies to detect specific cellular populations in a fluid sample. Another classic example requiring compensation is the simultaneous measurement of DNA and RNA using Hoechst 33,342 and Pyronin Y.

The compensation matrix relies on single-color control samples as a fundamental principle. For a two-dye experiment, the required controls include an unstained sample, a sample stained solely with the DNA dye, and a sample stained solely with the viability dye or antibody.

Single-color controls: The following steps are used to determine spillover. First, the baseline autofluorescence of the cells is established by running the unstained sample, providing a background signal for every detector. Next, the single-stained DNA dye control is run. The signal is measured in detector A (the main DNA channel) and detector B (the secondary channel, e.g., viability). The signal registered in detector B represents the spectral spillover of the DNA fluorochrome into that adjacent channel. Finally, the single-stained viability control is run, and its signal is measured in detector A to determine the spillover of the second dye back into the primary DNA channel.

Spillover coefficient calculation: Modern flow cytometers calculate the spillover coefficient automatically. It is determined by analyzing the median fluorescence intensity (MFI) of the specifically stained population, accounting for the background autofluorescence. For example, the coefficient for the spillover of the DNA dye into the viability detector is calculated mathematically as:

$$\text{Spillover Coefficient for DNA dye} = \text{MFI}_{\text{Viability detector}} - \text{MFI}_{\text{unstained}}/\text{MFI}_{\text{DNA detector}} - \text{MFI}_{\text{unstained}}$$

This ratio gives the exact fraction of the DNA dye's signal that must be subtracted from the viability detector to yield an absolute, true reading.

Matrix application: The software utilizes these coefficients to construct a compensation matrix. For a two-dye system, the calculation is performed as:

Compensated DNA Signal = (Raw DNA Signal)-(Coefficient of viability dye spillover into DNA channel) × (Compensated Viability Signal)

Compensated Viability Signal = (Raw Viability Signal)-(Coefficient of DNA dye spillover into viability channel) × (Compensated DNA Signal)

Modern instruments utilize high-resolution digitization. The signals from all detectors are converted to numbers with high precision (e.g., 16- to 20-bit resolution). The computer software then applies the compensation matrix once the data is collected. This is referred to as "offline compensation" or "software compensation" (Roederer, 2002).

The critical role of doublet discrimination: To understand the doublets and their discrimination, it is essential to understand the normal working principle of flow cytometry briefly. It is well known that flow cytometry comprises fluidics, optics, and electronic systems. The fluidics system helps the cells to pass through a single file into the laser interrogation point. At the intersection point, the laser beam (optics system) interacts with the cells with attached antibodies coated with fluorophores and reflects the emitted fluorescent light into two directions, which are forward scatter and side scatter (forward scatter reflects the size of the cell being analyzed, and side scatter reflects the granularity of the cell being analyzed). The electronic system converts the scattered light into electric signals/pulses. Analysis of a single cell will result in each pulse. Each pulse, when plotted in the histogram using time and intensity in the x and y axes, will routinely have height, width, and area, which are its parameters. The width will represent the time of cell transit, the height will represent the maximum intensity of the fluorescent signal emitted, and the area will represent the overall signal. When we analyze the single cell's signal, these three parameters are proportional to each other (i.e., when one parameter increases by some factor, each of the other parameters increases by a constant multiple of that parameter).

Single cells, even if they are small and/ or large, will pass through the fluids in a consistent manner and form a narrow diagonal cluster because the pulse width is directly proportional to the pulse area. A small cell has a small area and a narrow width, while a large cell has a large area and a wider width in the forward scatter area (FSC-A) versus forward scatter height (FSC-H) plot. Sometimes the cells aggregate or cluster with one another, resulting in doublet formation (Cossarizza et al., 2017). The doublet is inaccurately measured as a single cell by the flow cytometer, which leads to an inaccurate estimation of the cell size and granularity. This can lead to serious false positive results, and there is a chance of missing a population of cells. When two cells pass through together, they result in a longer "dumbbell" shaped curve in the plot. This results in a pulse with a significantly increased width for a given area. On the FSC-A versus forward scatter width (FSC-W) plot, doublets will appear as a distinct population when compared with single cells. This is also of paramount importance in cases involving fluids and CSF, which have low cellularity. The doublets typically appear as a trailing population above or below the diagonal, characterized by a higher area but with the same height as the single cells (Cossarizza et al., 2021).

Let us understand this more precisely with the following example. In a reactive effusion, we have a mixture of reactive mesothelial cells and many T lymphocytes. Usually, T cells are positive for CD45+, CD3+, and negative for CD19. At the same time, mesothelial cells do not express hematolymphoid markers. When the flow cytometer analyses these mesothelial cells and T lymphocytes, we will get a distinct two-cell population with (T cells being CD45+, CD3+, and CD10−) and another population of mesothelial cells with (negativity for CD45, CD3, and CD19). If the flow cytometer analyses the doublet (an aggregate of mesothelial cells and T lymphocytes as a single cell), this will result in a new false population of one event of large cells with CD45 positivity, raising an erroneous suspicion of a large cell lymphoma.

In DNA ploidy analysis, the goal is to precisely quantify the total DNA content of a single cell to determine if it is diploid (2n) or aneuploid. If two diploid cells (2n + 2n) stick together, the flow cytometer detects a single event with a 4n DNA signal. The system cannot inherently distinguish between one cell in the G2/M phase (true 4n) and two diploid cells in the G1 phase stuck together (false 4n), leading to an overestimation of the "S-phase fraction" or the "G2/M" population, which are critical prognostic markers in effusion cytology. Hence, it is mandatory to exclude doublets in any experiment dealing with DNA ploidy analysis.

Sample acquisition: Following rigorous instrument configuration and the establishment of gating protocols, the physical acquisition of the sample demands meticulous attention to fluidic dynamics and statistical power. For high-resolution DNA ploidy analysis, it is an absolute imperative to maintain a low sample flow rate as discussed previously. While higher flow rates expedite the run, they invariably widen the hydrodynamic core stream, which directly broadens the coefficient of variation (CV) of the G0/G1 peaks, resulting in severely compromised resolution, making it difficult to identify subtle, near diploid aneuploid clones. The total event yield must be statistically robust. Acquisition of a minimum of 20,000 to 50,000 target nucleated events (postdebris and postdoublet exclusion) is generally recommended to ensure the reliable detection of rare malignant subpopulations and to provide sufficient data density for accurate S-phase fraction modeling. Throughout the acquisition process, operators must continuously monitor time-parameter plots (e.g., Time versus DNA fluorescence) to verify stream stability and immediately identify any micro-clogs or pressure fluctuations, ensuring that only data collected under perfect hydrodynamic focusing is utilized for final diagnostic interpretation.

9.2.1.6 Quality assurance and run validation

Quality control in DNA ploidy analysis is a continuous, ongoing process that begins with sample reception at the laboratory and continues till the final reporting. It can be better understood in three phases: preanalytical (including specimen handling and preparation), analytical (instrument performance and data acquisition), and postanalytical (data analysis and interpretation). By maintaining quality control at each step, we produce a histogram that reflects accurate tumor biology, not just a technical artifact. The aspects related to specimen collection, handling, and sample preparation (preanalytical, preacquisition) have been discussed in the previous section. We will now move ahead with analytical (acquisition) and postanalytical (postacquisition) quality assurance.

Analytical quality control: Evaluation of instrument alignment and fluidic stability must be confirmed daily using fluorescent calibration beads, which are highly uniform and brightly fluorescent plastic beads (Refer to Chapter 5).

In the specific context of DNA ploidy analysis, the ultimate benchmark of analytical quality is the CV of the diploid G0/G1 peak from a stable, homogeneous source. CV is the measure of the precision of the measurement. A low CV (e.g., less than 3%) of the G0/G1 peak indicates a well-aligned instrument with stable fluidics and uniform illumination.

For a more biologically relevant check, stained nuclei from a source of normal, nondividing cells, such as chicken or trout erythrocytes, which are available commercially, are used as **external standards**. These cells have a known, stable DNA content and produce a single, sharp peak. A well-performing instrument should produce a DNA histogram from these control cells with a CV of 2% or less. The DI for these standards should also be calculated against a normal human diploid control to verify the instrument's linearity.

Linearity: Linearity of the instrument indicates that a proportional change in the input must result in a similar proportionate change in the output. In DNA ploidy analysis, this means that if the G_0/G_1 peak of a diploid population is at channel 200 on a linear scale, the G_2/M peak (containing twice the amount of the DNA) should be at channel 400 (within a channel or two). Linearity can be checked by analyzing a mixture of the chicken or trout erythrocytes and human control cells, as their DNA content ratios are known. If the instrument is not linear, the accuracy of the S-Phase Fraction (SPF) and the ability to resolve near diploid aneuploid populations will be severely compromised.

Resolution and sensitivity: Resolution of the instrument means its ability to distinguish two different cellular populations with a slight difference in the amount of the DNA (Cusick et al., 1990). This is very important for detecting near diploid aneuploid tumors, where the abnormal stem line may have only a 2%–4% difference in DNA content from the normal diploid cells. A low CV is a prerequisite for a high-resolution of the instrument. Sensitivity, in this context, to the resolution, is the ability of the instrument to detect small populations of aneuploid cells against a background of normal diploid cells and debris. While primarily a function of sample preparation, instrument sensitivity is also affected by the stability of the light source and the efficiency of the detection system (Cusick et al., 1990).

Beyond instrument-level metrics, the sample data itself must pass internal biological quality gates. Before interpreting the ploidy status, the operator must retrospectively verify the gating hierarchy to ensure that debris, nonviable cells, and doublets were successfully excluded from the final analysis plot. Furthermore, the operator must confirm that the predefined acquisition targets were met, yielding a statistically robust population of nucleated events (typically >20,000). Only after both instrument performance and sample integrity have been definitively validated can the analysis proceed to the extraction of the DI and cell cycle kinetics.

Postanalytical (postacquisition) quality control:

Visual inspection: Before any mathematical model is applied, it is important to perform a visual examination of the histogram. The key features to be assessed are,

Peak symmetry: The G_0/G_1 peak should be symmetric. A skewed peak with a tail to the right suggests poor instrument alignment or the presence of cell aggregates, and the histogram should be repeated

Background debris: A large number of events in the low-channel region indicates a significant presence of cellular debris. While mathematical models can account for this, an excessive amount (e.g., >20% of total events) will make any subsequent SPF calculation inaccurate.

Distinct populations: Are the G_0/G_1 and G_2/M peaks clearly defined? Is there a visible S-phase continuum between them? Can an abnormal peak be clearly identified as a distinct population?

Assessment of internal standards: The best standard for determining the DI of a tumor is the patient's own normal cells present within the same sample. These are often stromal cells, lymphocytes, and nonneoplastic epithelial cells. The presence of a distinct diploid G_0/G_1 peak from these "reference" cells is essential for calculating an accurate DI for any aneuploid tumor population (DI = mean channel of aneuploid G_0/G_1 peak / mean channel of diploid G_0/G_1 peak). If no internal standard is visible (e.g., in a sample with a high tumor burden), the external standards become important for determining ploidy.

9.2.1.7 Gating and data analysis

To ensure the final sample is pure, the software uses a step-by-step process called Boolean gating. By using AND, OR, and NOT rules across different cell features, it systematically includes the cells of interest and throws out any artifacts, dead cells, or clumped cells.

After preparing a proper single cell suspension of cells or nuclei, gating typically begins with a time-parameter gate (e.g., Time versus forward scatter) to exclude any electronic noise or fluidic surges that occurred during acquisition, ensuring only data collected under stable hydrodynamic pressure is analyzed. Next, a broad morphological gate utilizing forward scatter (FSC) and side scatter (SSC) is applied to exclude acellular debris and pyknotic fragments. The lower-left region contains smaller debris, electronic noise, and severely fragmented cells; all of these must be excluded from the analysis. A polygon gate is made around the main cellular events, typically possessing higher FSC and moderate SSC compared to the debris (Cusick et al., 1990). If isolated nuclei are prepared, FSC values will be lower than those of whole cells but still distinctly separated from fine debris, enabling a similar FSC/SSC exclusion gate. It is critical that gating remains generous at this stage; if we gate too strictly on normal-looking cells, we risk missing complex, aneuploid malignant cells that inherently possess high scatter properties.

This morphological step is immediately followed by the viability gate. Depending on the preparative methodology, cells demonstrating high fluorescence for the chosen viability dye (such as an amine-reactive stain or a cell-impermeant dye like 7-AAD) are permanently excluded from downstream analysis.

Following viability, gating is performed on the primary DNA fluorescence channel to isolate intact nucleated cells and differentiate them from residual subnuclear fragments (Cossarizza et al., 2017). The lower DNA fluorescence threshold is carefully set just below the main diploid G0/G1 peak to exclude these lowest-intensity events. However, because severe truncation at the low end can mathematically bias S-phase estimation algorithms, the threshold placement must be cautiously guided by visual inspection of both the histogram and bivariate plots.

Finally, doublet discrimination is essential, as aggregated nuclei can falsely mimic tetraploid or near-triploid malignant populations. A standard, high-resolution strategy is to acquire the pulse area alongside the pulse width (or Height) for the DNA signal. On a plot of DNA, area versus DNA Height, true singlets form a tight diagonal cluster. Conversely, on the preferred plot of DNA area versus DNA width, singlets form a distinct, flat horizontal band. Doublets deviate from these singlet clusters by exhibiting a disproportionately increased area and width, causing them to skew away from the primary singlet populations. A strict gate enclosing the singlet cluster is then applied, and all subsequent mathematical purity analysis is performed exclusively within this final singlet gate (Wersto et al., 2001).

The lymphocytes and normal stromal cells serve as internal controls; they form a narrow diploid G0/G1 peak with a lower CV, which serves as a reference for comparing aneuploid tumor cells. A tighter gate can be placed around the normal cells, with a distinct cluster in FSC/SSC, leading to tumor cells being separated outside this gate due to their larger size and distinct granularity. In carcinomatous effusions, epithelial elements usually display higher FSC and distinct SSC compared with lymphocytes, enabling their accurate gating. While gating by this method, we can compare DNA histograms of "epithelial-gate" events and "nonepithelial" events to verify that the former contains the aneuploid peaks. At the same time, the latter provides diploid internal control. This comparative gating approach is particularly important when minor aneuploid populations are present in the sample. Sometimes immunophenotyping using epithelial (EpCAM) and hematolymphoid (CD45) markers is performed. At this time, a fluorescence versus DNA bivariate plot is constructed, and a gate is drawn around marker-positive events, which are then carried forward for DNA ploidy analysis. A sequential gating strategy for DNA ploidy analysis in Cerebrospinal fluid sample of a patient with B-cell acute lymphoblastic leukaemia (B-ALL) infiltration has been described in Fig. 9.2.

Software for analysis: Analysis software plays a significant role in transforming raw DNA ploidy data into clinically useful parameters like DI and SPF. These programs automate peak detection, model fitting, and debris correction. This ensures reproducibility in pathology workflows where tumors are often mixed with diverse cell populations. Here are the software options currently available.

ModFit LT: ModFit LT, from Verity Software House, is a commonly used program for DNA ploidy analysis. Its AutoDebris, AutoAggregates, and AutoLinearity features dynamically model and subtract artifacts. Users can import flow cytometry standard (FCS) files, apply batch processing for higher throughput, and generate reports with overlaid fits. It excels in multiparameter extensions by integrating with WinList for gating on scatter or immunofluorescence. WinList enhances ModFit by offering better visualization and region tools for pregating singlets and supports Cytometry by Time-of-flight (CyTOF) and spectral data (ModFit LTTM, n.d.).

Multicycle advanced version (*Multicycle AV)*: Multicycle AV, developed by Phoenix Flow Systems and integrated into FCS Express by De Novo Software, provides effective debris subtraction based on algorithms by Rabinovitch and Bagwell. It is free with FCS Express licenses. The software fits multiple ploidy levels, exports cycle phases, and manages aggregates using pulse processing models. FCS Express adds flexibility to workflows, such as automated reporting and checks for standards-based linearity, making it suitable for labs moving away from standalone tools (Katherine-Rittenbach, n.d.) https://denovosoftware.com/.

Sysmex's CyFlow: Sysmex's CyFlow ploidy analyzer works with CyView software and Ploidy reagent kits to deliver results in under 15 min. It uses push-button DNA histograms and automated peak fitting. The system supports UV LED or green laser excitation for DAPI/PI stains and includes quality control overlays for CV and index validation. Partec's Ploidy Analyzer, now

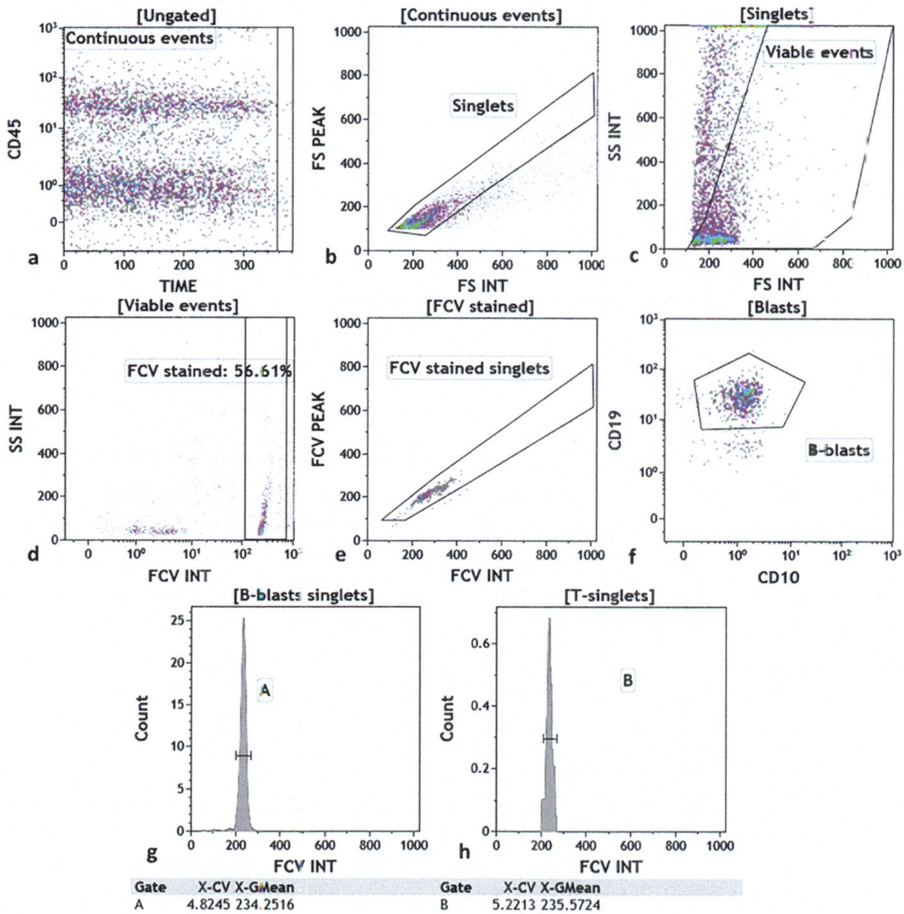

FIGURE 9.2 Sequential gating strategy for DNA ploidy analysis in Cerebrospinal fluid sample of a patient with B-cell acute lymphoblastic leukaemia (B-ALL) infiltration. Multi-parameter flow cytometry was performed on cerebrospinal fluid using a hierarchical gating strategy to quantify leukemic blast DNA content relative to a diploid internal reference population (T-cells). (a) Instrument stability gate [Ungated]. (b) Scatter singlet gate [Continuous events] – forward scatter area vs. height excludes doublets. (c) Viability gate [Singlets] – forward vs. side scatter excludes debris and dead cells. (d) FCV staining gate [Viable events] – 56.61% of viable events were FCV. (e) DNA doublet exclusion gate [FCV stained] – pulse geometry applied to the DNA fluorescence signal. (f) B-blast immunophenotyping gate [Blasts] – CD10CD19 polygon identifies the leukemic precursor B-cell population. (g, h) DNA content histograms for B-blast singlets (A) and T-cell diploid reference singlets (B). Panel A: X-CV = 4.82%, X-GMean = 234.25. Panel B: X-CV = 5.22%, X-GMean = 235.57. Calculated DNA Index = 0.994, indicating a diploid blast population. Both X-CV values are below the 6% technical acceptability threshold.

part of Sysmex, has a similar user-friendly interface for plant and animal genome sizing, with the ability to autoload up to 96 wells (CyFlow Ploidy Analyser, n.d.).

Kaluza: Kaluza from Beckman Coulter is great in clinical settings. It incorporates Michael H. Fox's cell cycle algorithm for low-event-rate acquisitions, such as 100 or more tumor events.

Table 9.3 Comparison of key features between different software analyzers.

Software	Debris/aggregate correction	Batch processing	Multiparameter gating	Platform/integration
ModFit LT	Auto advanced model	Yes	Via WinList	Win/Mac standalone
Multicycle AV	Rabinovitch/Bagwell models	Yes	FCS Express	FCS Express (free)
CyView (Sysmex)	Automated QC overlays	Autoloader	Basic scatter	Instrument-specific
Kaluza	Fox algorithm	Yes	Full	Beckman systems
MuPETFlow	Peak-based subtraction	Multifile	Limited	Web GUI (free)

The software processes Navios data, categorizes ploidy by DI, and correlates it with markers like CD38/CD138 in myeloma. FlowJo and Cytospec provide cytometry catalogs with ploidy plugins, though they are less specialized (Kaluza Analysis Software, n.d.).

MuPETFlow: MuPETFlow is a Shiny-based graphical user interface (GUI) that automates multifile ploidy estimation, peak detection, linear regression on standards, and the export of figures and tables. It has been benchmarked on yeast and plant data and outperforms competitors in handling multiple peaks and batch processing, making it ideal for large datasets without the need for scripting (Gómez-Muñoz & Fischer, 2025) (Table 9.3).

Decoding DNA histogram: DNA histogram portrays the distribution of DNA content in the various cycles of cell division in a visually understandable manner with the help of flow cytometry by measuring the peak DNA content in each cycle of cell division. A physiological/ routine histogram of normal native nonneoplastic cells demonstrates two peaks corresponding to the G0/G1 phase and G2/M phase, with a shoulder/plateau region in between corresponding to the S-phase of the cell cycle. The reasoning behind each peak and shoulder/plateau in the DNA histogram has already been discussed earlier in this Chapter. The aneuploid cancer cells have a deviation from this normal pattern, and it has been illustrated through the Fig. 9.3.

The DNA ploidy analysis of the CSF and body fluids helps identify aneuploid clones by identifying the alteration/abnormal presence of any peaks in the routine histogram, which is basically due to abnormal DNA content of the neoplastic cells, due to uncontrolled cellular proliferation. In routine, we can also encounter low-grade malignancies, which can possibly present with the normal pattern of DNA histogram and reactive cellular population, which could have increased DNA content mimicking the aneuploidy status. Hence, before reporting an aneuploid clone, the reporting pathologist should check for the three important metrics of the DNA ploidy analysis, which are DI, CV, and SPF, to exclude the variations in the peaks that could arise from improper processing, degeneration of the cells, presence of doublets, and aggregates.

Key metrics for diagnosis:

DI: DI is the measure of the DNA content of the cell. When performing flow cytometry of body fluids, we can often obtain a sample with neoplastic cells admixed among native nonneoplastic cells, such as lymphocytes and mesothelial cells. The DNA content of the G0/G1 peak of the

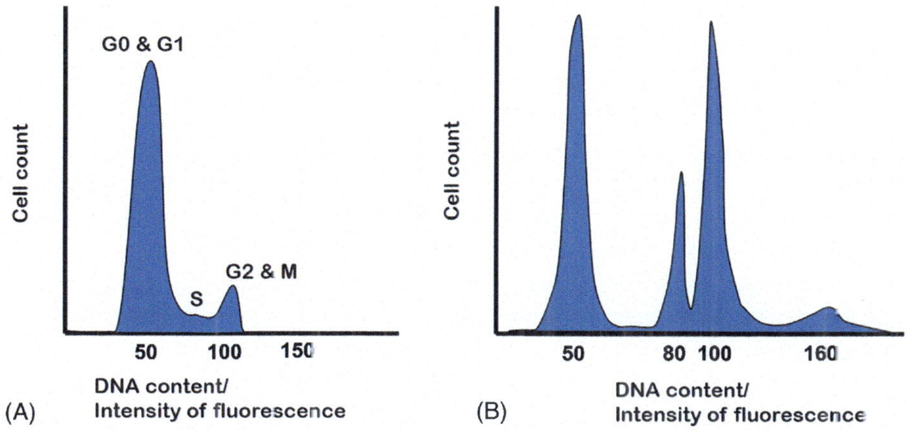

FIGURE 9.3 DNA histogram depicting in picture "A" on the left side with a normal pattern of tall G0 & G1, followed by a plateau-like area of "S" phase, followed by a smaller G2&M peak. In contrast, picture "B" demonstrates deviation from the normal pattern of a high-grade malignancy with the presence of multiple aneuploid peaks.

nonneoplastic native cells acts as an internal control (due to its euploid nature) for the DI, and its DI is fixed at 1. DI of any abnormal or aneuploid population is defined as the ratio of the mean fluorescence peak of the G0/G1 peak of the aneuploid or abnormal cellular population to the mean fluorescence of the internal control. The stromal cells act as a reference control in solid tissue analysis. DI less than or more than one is indicative of aneuploidy (less than one is referred to as hypoploidy and more than one is referred to as hyperploid) (Darzynkiewicz et al., 2010).

SPF: SPF refers to the percentage of cells present in the S-phase of the histogram; it is an indirect measure of cells that are actively performing DNA synthesis. SPF provides an overall assessment of the proliferative nature of the cells. Higher SPF indicates a larger number of cells in the cell cycle, which often correlates with aggressive clinical behavior and high histological grades. It is often expressed as a percentage of total cells or of the total of G0/G1, S, and G2/M gated populations in the DNA histogram. In diagnostic pathology, an elevated tumor SPF is frequently correlated with high-grade morphological features, rapid clinical progression, and aggressive biological behavior. It is essential to recognize that a higher SPF alone cannot be used to predict malignancy. Often, low-grade malignancies demonstrate lower SPF values, and the reactive cellular proliferation, especially reactive mesothelial proliferation, can show a higher SPF fraction. Hence, SPF value must be interpreted always with DI and Aneuploidy peak for accurate assessment of proliferative nature of the cells (Carloni et al., 2017).

Navigating complex histograms and strategies for interpretation:

Near diploid clones: Some abnormal clones may have a DI closer to 1.0 (around 0.95 to 1.05); they overlap with the normal diploid peak and pose a diagnostic challenge in distinguishing them. These scenarios happen with the low-grade malignancies. To accurately identify the near diploid clones and their proper distinction from the normal diploid peaks, adopting the following strategies can help: (1) Ensuring the normal diploid peak is defined clearly with a low CV, (2) Comparison of the histograms of the suspected near diploid clones with the benign control tissue. (3) By using bivariate gating (side scatter/fluorochrome intensity versus amount of the DNA) to separate the suspect population of cells before plotting the DNA histogram, and (4) performing immunophenotyping of the suspected tumor cells and then plotting the DNA histogram of the suspected tumor cells (Plander et al., 2003).

Multiple aneuploid clones: Due to the heterogeneous nature of the tumor, some patients can have two or more aneuploid clones with different DNA indices (e.g., 1.6 and 1.8 and so on). In these circumstances, we must label each clone separately, calculating the separate SPF and maintaining a lower CV. Also, it is advisable to consider the evolution of subclones in the tumor (for instance, the tumor that has disseminated into the body fluids and CSF can have the presence of new subclones). The presence of multiple aneuploid clones in a sample usually indicates aggressive tumor behavior with a high tumor mutational burden (Houtsma et al., 2021).

Presence of significant cellular debris: Body fluid samples are characterized by high amounts of cellular debris, mucin, apoptotic fragments, cellular clusters and clumps, which can distort the pattern of the DNA histogram, sub-G0 events (indicating less than 2N DNA content) and false 4N shoulder areas (Darzynkiewicz & Zhao, 2011). These pitfalls can be overcome by gentle and adequate centrifugation and meticulous sample processing. Apoptotic fragments commonly present a sub-G0 event and should not be categorized as a hypodiploid clonal population. In the histogram, identify and gate out the sub-G0 events (often apoptotic debris); they should not be mistaken for hypodiploid clones. In the case of the "4N" peak appearing with a wider CV, it is important to consider the possibility of cellular aggregates rather than a true "4N" hyperdiploid clone than a true tetraploid clone (Gabali, 2022).

Multiparameter flow cytometry: Conventional cytomorphological interpretation remains the initial first line investigation for the body fluids and CSF. In patients with ambiguous cytomorphological interpretation due to morphological overlap between malignant and reactive mesothelial cells, and paucicellular samples, combining DNA ploidy analysis with immunophenotyping using multiparameter flow cytometry helps confirm malignancy and guides in the identification of the lineage (Ferreira-Facio et al., 2013).

The CD45 is used as a lineage marker for hematolymphoid malignancies, and Pancytokeratin or EpCAM is used as a lineage identification marker for carcinomas. Multiparameter flow cytometry supplements traditional morphological and immunocytochemical methods (Puri et al., 2020). Aneuploid peak with CD45 expression and Pancytokeratin/EpCAM nonexpression indicates hematolymphoid malignancy. Aneuploid peak with CD45 nonexpression and Pancytokeratin/EpCAM expression indicates epithelial malignancy. Aneuploid peak with nonexpression of CD45 and Pancytokeratin will indicate a rare nonepithelial and nonhaematolymphoid neoplasms like mesothelioma and other neoplasms in melanomas, which commonly disseminate into the fluids and CSF apart from epithelial and hematolymphoid neoplasms. It is also important to keep in mind that antigen loss due to improper processing and the presence of significantly fewer numbers of the neoplastic cells in the effusion and CSF can also result in aneuploid peak with CD45 and Pancytokeratin/EpCAM nonexpression. In this type of scenario, it is essential to repeat the effusion tapping, perform cell blocks in these samples, and perform immunohistochemistry to give the accurate diagnosis.

Gating approach in multiparametric flow cytometry for simultaneous analysis of lineage and DNA ploidy: Contemporary MFC utilizes sequential gating approaches that initially recognize cell populations based on light scatter characteristics (size and complexity), then by immunophenotypic characteristics (lineage markers), and ultimately evaluate DNA content within these identified populations. This technique enables accurate identification of the particular cell population that displays aneuploidy (Gabali, 2022).

For instance, in multiple myeloma, this method can recognize aneuploid CD38-positive plasma cells and also pick up additional CD19-positive B-lymphocytes (which are aneuploid) and part of the neoplastic clone (Puri et al., 2020).

This combined method is helpful in distinguishing between reactive and malignant cellular populations in body fluids, as florid mesothelial hyperplasia can mimic adenocarcinoma, and a marked increase in benign lymphocytes can mimic lymphoma. The combination offers more evidence of malignancy when aneuploidy is identified in an immunophenotypically abnormal cellular population. Multiparameter flow cytometry, which links ploidy analysis with immunophenotypic analysis, is a quick and evidence-based analysis for diagnosing morphologically difficult cellular populations based on morphology to diagnostic information.

9.3 Image cytometry: Quantitative analysis with morphological context

9.3.1 Image cytometry

Image cytometry (also referred to as slide-based cytometry and/or laser scanning cytometry) derives data from the stoichiometric dye stained whole-slide images and converts them into digital pixels, extracting data regarding overall cell and nuclear size, shape, and nuclear DNA content (Piotr Pozarowski et al., 2013). This process converts the histological slides into meaningful biological data. Static image cytometry differs from flow cytometry in that it allows the evaluation of fixed tissue and archived biopsy material, enabling the assessment of morphological patterns, heterogeneity within the tumor population, and preservation of spatial resolution, along with nuclear details (Piotr Pozarowski et al., 2013).

Slide-based analysis and measuring DNA content via integrated optical density:

The process of static image cytometry begins with staining of either histological sections or cytological smears with a stoichiometric dye. These sections are then fed into the whole-slide image analyzer, which, using its digital camera, converts the sections into digital pixels. Next, using automated software, the cells of interest are selected by a process called segmentation (Tárnok, 2006). In segmentation, the software selects the outlines of the cell nucleus from the background structures. This is the most crucial step in the entire process, as segmentation identifies our area of interest; any errors in this step result in false interpretations. The automated software then analyses the amount (quantity) of DNA content in the nucleus, which is segmented, and calculates the IOD (Tárnok, 2006).

IOD is the measure of the total amount of DNA present in the entire cell. Now, let us understand the concept of IOD more clearly. After the segmentation process, each nucleus, which is stained with the dye, has absorbed a certain amount of dye and is further divided into smaller pixels. The optical density (OD) of the dye is measured using a densitometric microscope system, and the OD measured is representative of the light absorption. The light absorbed is proportional to the amount of dye stained, which in turn is proportional to the amount of DNA present. Hence, the IOD is the total of the OD values for every pixel within the nucleus (Tárnok, 2006).

When the IOD is estimated throughout many cells in the tissue section, this data can be plotted with a histogram, which reveals the DNA distribution across the cells. This visual representation will easily identify cells in the G0/G1, S-phase, and G2/M phases with varying 2N or 4N DNA ploidy status. Also, unlike flow cytometry, the morphology of the cells is available during image cytometry, hence we can either compare and contrast the DNA content of cells of our interest with background cells if required (if neoplastic cells are cells of interest, background inflammatory cells can be easily compared, since morphological evaluation is possible with image cytometry). Thus, image cytometry is a powerful tool that visually links morphology with the evaluation of genetic material.

A recent study investigated the diagnostic utility of DNA image cytometry (DNA-ICM) for identifying malignant pleural effusions by analyzing 830 cases alongside exfoliated cell smears (ECS) and cell block (CB) techniques. While individual sensitivities for DNA-ICM, ECS, and CB were high at 96.28%, 94.93%, and 95.95%, respectively, combining all three methods significantly improved diagnostic performance, reaching a peak sensitivity of 99.32%. Notably, DNA-ICM identified 22 positive cases that were missed by both ECS and CB, with subsequent clinical diagnosis confirming malignancy in four of those instances. The study concludes that integrating DNA-ICM ploidy analysis effectively minimizes missed pathological detections and increases identification rates, while the addition of cell blocks and immunochemistry remains essential for accurately determining the primary tumor site (Ma, 2023).

The role of the Feulgen stain for stoichiometric DNA measurement: The most important aim of the image cytometry is accurate estimation of amount of the nuclear DNA is the primary goal of image cytometry. The Feulgen stain is the gold standard procedure for stoichiometric measurement due to its specific DNA-binding, linearity, and ability to preserve morphological details in the tissue sections for a longer duration of time. It is a stoichiometric stain, which means the amount of dye bound to the nucleus is proportional to the amount of DNA present. Among all dyes available in the market, the Feulgen stain only fulfills the stoichiometric principle, making it the ideal and gold standard for DNA image cytometry (Biesterfeld et al., 2011).

Principle of the Feulgen stain: The Feulgen reaction is a two-step process, which includes hydrolysis by a milder normality weak acid followed by treatment with the Schiff reagent. Acid hydrolysis with milder hydrochloric acid targets the selective hydrolysis of adenine and guanine (which are purine bases) and exposes aldehyde groups in the DNA backbone, which is then treated with Schiff reagent, which provides Magenta color to the DNA in the sections (Biesterfeld et al., 2011).

Additionally, the Feulgen stain follows the Beer–Lambert law in brightfield absorbance. The per-pixel OD and summed IOD provide a linear proxy for nuclear DNA content over a broad range, enabling direct comparison to an internal control.

Image cytometry can compute integrated optical densities from transmitted images of the light after Feulgen staining from the sample and classify nuclei into diploid, aneuploid, tetraploid, and different phases of the cell cycle compartment after normalization to reference cells on the same slide. The normalization of the sample to the internal controls reduces inter-measurement reproducibility.

For accurate stoichiometry, one should focus on exact timings of hydrolysis; a weak or milder acid should be used for exact timings. Either a low or high degree of hydrolysis damages nuclear chromatin and reduces aldehyde formation. So individual laboratories should standardize the temperature of the reaction, the molarity of the acid, and the timings to achieve low coefficients of variation in the diploid peak. Best-practice guidance emphasizes consistent time for fixation, controlled use of Feulgen stain, background correction, and inclusion of internal controls (lymphocytes), which helps in better Feulgen staining of the DNA (Hardie et al., 2002).

9.3.2 Direct morphological correlation–the primary benefit

Morphology-based gating: In flow cytometry, the pathologist gates the cells based on the fluorescent intensities. However, in image cytometry, the pathologist can visually select the intended cells of interest, facilitating a morphology-guided gating process (Basiji et al., 2007). For example, one can choose the large pleomorphic cells with irregular nuclear contours having very prominent nucleoli in the sections, which are the morphological equivalents of malignancy, and avoid the background stromal and inflammatory cells. This eliminates the diagnostic dilemma of analyzing ground cells when the rare abnormal cells are diluted.

Data correlation and validation: After selecting the morphologically suspicious population of the malignant cells, the software delivers all the quantitative information for that chosen group. Crucially, selecting on any data point in a histogram or scatter plot automatically shows the original cell image. This enables visual confirmation that a cell with aberrant DNA content (e.g., an aneuploid peak) has the malignant cell morphology as suspected. This visual confirmation provides diagnostic confidence that other techniques cannot provide (Minderman et al., 2012).

Solid tissue analysis: This method is ideal for analyzing tissues, where preserving cell spatial and topographical context is necessary for understanding and evaluating various architectural patterns and tumor-associated inflammation (Piotr Pozarowski et al., 2013).

Reduction of ambiguity in the diagnosis: Image cytometry tackles a significant limitation of flow cytometry, which is the lack of morphological analysis, which can lead to uncertainty in data interpretation, especially in difficult samples (Piotr Pozarowski et al., 2013).

9.3.3 Specific advantages of image cytometry in body fluid and cerebrospinal fluid analysis

Hypocellular specimens: Hypocellular and sparsely cellular specimens require meticulous and accurate evaluation because the number of diagnostic cells may be extremely low. In CSF cytology, the presence or absence of just a few abnormal cells can decide whether the patient is upstaged for therapy in cases of hematolymphoid malignancies. In flow cytometric analysis, cells can be lost during processing, the fluidics system can accidentally overlook the analysis of rare cells, and, more importantly, cells are not evaluated morphologically.

The image cytometry overcomes these disadvantages by analyzing intact and fixed cells directly on tissue sections and cytological smears. Image cytometry achieves maximum cell retention by preserving even rare populations that remain in the field of view, particularly important when total cell counts are low. Also, it facilitates high-resolution capture of individual cells and allows for both quantitative measurements (DNA content, nuclear features, and morphological assessment). Also, unlike flow cytometry, image cytometry allows repeated re-examination of cells when needed, providing an opportunity for repeat auditing. Multiple parameters, such as DNA ploidy, nuclear pattern, chromatin nature, and cell size, can still be accurately evaluated even with a small total number of cells (Lazcano et al., 2000).

Confident distinction with benign reactive cells with atypical morphology: The most common pitfall in fluid cytology is morphological overlap between malignant cells and benign reactive populations such as mesothelial cells, histiocytes/macrophages, or degenerative debris, which is even more pronounced, especially when morphology is distorted by processing (Shidham & Layfield, 2021). Misinterpretation of these mimics can lead to false positives and unnecessary anxiety for the patient and wrong clinical intervention. Image cytometry prevents this mishappening by combining quantitative assessment of the DNA content with the morphological context.

a. Morphology-based discrimination, Reactive cells often show enlarged nuclei, prominent nucleoli with slight nuclear irregularity, a feature that can mimic malignancy in low-quality cellular preparations. With image cytometry, the cells are examined under a high-resolution imaging method, which details nuclear details like nuclear membrane contours, nucleolar prominence, cytoplasmic borders, and chromatin pattern. Macrophages often show vacuolation, and mesothelial "windows" can be easily identified. This helps distinguish the reactive atypia from malignancy (Shidham, 2021b).
b. Quantitative exclusion using DNA and nuclear metrics. In many instances, the benign reactive cells may look abnormal, but they retain their diploid DNA content. Image cytometry can measure DNA ploidy, SPF, nuclear texture, chromatin uniformity, and nuclear area and shape measurements. These measurements allow a clear distinction between the benign/reactive populations with diploid, nonproliferative profiles and the malignant cells, which show aneuploidy, increased SPF, and abnormal chromatin texture. Thus, even when morphology is overlapping, quantitative DNA measurements help clarify the distinction between benign and malignant cells.
c. Proper and accurate removal of debris and noncellular artifacts. Debris, proteinaceous debris, apoptotic fragments, and red blood cells are common contaminants in body fluids. Their presence may distort flow cytometry-based analysis and result in false positive analysis. Image cytometry eliminates these pitfalls, since the debris does not show identifiable nuclear boundaries and is automatically excluded by the segmentation process (Dimitriadis et al., 2024). During image cytometry, only intact nucleated cells are selected for analysis, and the

morphological review ensures noncellular material does not enter the analysis. This results in an accurate dataset and representation with trustworthy analysis.

Thus, image cytometry allows maximum cellular preservation and morphological interpretation, which is best for CSF and other body fluids with low cellularity. It allows for the analysis of every cell, providing an accurate assessment of nuclear details, and excludes reactive mimics and debris. It decreases diagnostic difficulties. As a result, patients and clinicians receive clearer, more reliable information when evaluating suspected malignancy in low cellularity specimens (Kouriefs & Gordon, 2001).

9.4 Clinical applications and method comparison

Cytometric technologies have become vital in modern diagnostic pathology, particularly for evaluating bodily fluids and CSF. Flow cytometry and image cytometry offer quantitative evaluations of cellular populations, but their techniques, strengths, and most effective clinical applications differ greatly.

When paired with clinical and morphological results, these methods help resolve difficult cytomorphology, confirm malignancy, and enhance patient care with reliability. The sections that follow explain their roles in three significant diagnostic areas: serous effusions, CSF analysis, and practical decision-making when selecting between the two modalities.

9.4.1 Diagnosing malignancy in serous effusions

In many laboratories around the world, cytomorphological interpretation remains the first line of investigation in body fluids and CSF. However, due to morphological overlap, low-grade nuclear atypia, degenerative changes, and reactive atypia can lead to confusion, making it difficult to render a confident diagnosis solely on cytomorphological examination (Shidham, 2021a). Flow cytometry and image cytometry are modern diagnostic tools that help provide confident, accurate diagnoses, supplementing cytomorphological examination by using multiparameter immunophenotyping analysis and DNA ploidy analysis.

Flow cytometry provides the advantage of multiparameter immunophenotyping and quick evaluation. By using fluorochrome-tagged antibodies against epithelial antigens, flow cytometry enables the screening of a large population of cells to identify abnormal epithelial antigen expression, thus helping in the diagnosis of carcinomas. These antibodies include Pancytokeratin, BerEp4, and EpCAM. This method is particularly useful when there is cytomorphological overlap between the reactive atypia and malignancy. Additionally, flow cytometry can identify a minor population of carcinoma cells that express epithelial markers among the reactive mesothelial cells in some patients with early malignant effusions, which tend to have a smaller number of carcinoma cells.

In contrast, image cytometry combines morphological analysis with ploidy analysis. While flow cytometry converts cells into mean fluorescent intensities, image cytometry converts cells into high-resolution digital images along with the evaluation of DNA ploidy. This combined approach reduces the likelihood of false positives, which can occur in flow cytometry due to the presence of benign atypical cells. Additionally, using image cytometry, we can analyze cell clusters and groups, which are crucial in the diagnosis of certain malignancies that exfoliate in groups and clusters. This combination of both methodologies offers an accurate diagnosis. Fluorescence serves as a screening tool for carcinomas by rapidly identifying cells that express an aberrant epithelial antigen expression, whereas image cytometry does morphological correlation of these abnormal cells that express epithelial markers (Ma et al., 2023).

9.4.2 Detecting CNS malignancy in cerebrospinal fluid

Cytological examination of the CSF sample is one of the most challenging tasks for cytopathologists due to its paucicellular nature, the rapid ability to degenerate if not transported to the laboratory, and the presence of small, subtle clonal abnormalities that have a significant impact on the patient's treatment. Flow cytometry and image cytometry help overcome these challenges and add significant value in patient management.

Flow cytometry is routinely used in the evaluation of central nervous system involvement by leukemia and lymphoma. It has its own advantages, including the ability to find small clones of B & T cells, extensive multiparameter immunophenotyping for accurate categorization, and a quick turnaround time. However, its performance in providing accurate diagnosis depends on the presence of adequate cellular numbers, good sample preparation, and preservation. Flow cytometry can give erroneous results if the sample is paucicellular, contains many debris, degenerated, and fragmented cells. These limitations can be overcome by image cytometry, which can be extremely helpful in evaluating low cellular counts, excluding debris, and morphological interpretation. The advantages of the image cytometry over cytometry in evaluation of CSF samples are its ability to diagnose even a sparse and occasional number of cells in a paucicellular population, evaluation of morphological parameters like nuclear chromatin, nuclear irregularities and amount of DNA, evidence-based identification of the aneuploid cells and excellent ability to exclude debris from the evaluation(De Graaf et al., 2011; Thastrup et al., 2022).

While evaluating leptomeningeal spread from the solid malignancies from breast, lungs, and metastatic melanomas, image cytometry excellently identifies aneuploid population of cells even with the atypical and indeterminate cytomorphology. In flow cytometry, image cytometry enhances flow cytometry by providing an accurate and visual estimation of DNA amount evaluation complemented by morphology, thus helping to differentiate leukemic blasts from reactive hematopoietic cells (Dunn et al., 2010). Thus, in CSF analysis, image cytometry provides excellent and valuable results when the sample is paucicellular, requiring mandatory morphological evaluation and assessment of rare neoplastic clones.

9.4.3 Flow versus image cytometry: A practical comparison

When to choose flow cytometry: Flow cytometry is still the best technique for high-throughput, quantitative investigation of cellular immunophenotypes across large cell populations. Its main advantages include.

Faster turnaround time and cellular acquisition: The advantage of analyzing a large number of cells in a shorter duration of time makes flow cytometry an ideal option for evaluating lymphomas and leukemia, especially cytological samples from lymph nodes, peripheral blood, and bone marrow aspirates.

Excellent detection capacity and precision: Flow cytometry measures mean fluorescent intensities of surface, cytoplasmic, and nuclear antigens and has a precise ability to detect subtle abnormalities in expression through gating. This property makes flow cytometry an ideal candidate for evaluating and classifying hematolymphoid malignancies, as well as assessing minimal residual disease.

Physical cell-sorting abilities: Flow cytometry helps isolate specific abnormal cellular populations through immunophenotyping, which can be further utilized for molecular analysis, cell culture, and functional analysis. This makes flow cytometry ideal for settings where genetic evaluation of abnormal cells is required.

In summary, flow cytometry is best for the evaluation of cytological samples with high cellularity, minimal debris, and a quick turnaround time, as well as when multiparameter immunophenotypic classification and quantitative population statistics are required.

When to choose image cytometry: Image cytometry is especially beneficial for morphological correlation and analysis of samples with a smaller cell population. The selective advantages of image cytometry are as follows:

Direct morphological visualization: Image cytometry combines direct morphological interpretation with DNA ploidy analysis. This makes image analysis an ideal method for distinguishing malignant cells and benign atypical cells with high precision. Also, for the evaluation of carcinomas and mesotheliomas, where morphological details such as cohesion in case of carcinomas and papillary architecture in case of mesothelioma can be better appreciated by image flow cytometry.

Remarkable performance with paucicellular samples: With image cytometry, each cell can be fully characterized. By obtaining images of all analyzed cells, the method ensures that the most information is extracted from limited samples.

Ability to exclude debris: In flow cytometry, debris, cellular clumps, and benign cells like macrophages and mesothelial cells interfere with the analysis of malignant cells. This limitation can be easily prevented by image cytometry, in which the cell of interest can be easily selected, and debris can be excluded from analysis.

To summarize, image flow cytometry is particularly useful for paucicellular samples, samples with debris, when morphological correlation and aneuploid detection are necessary.

Table 9.4 summarizes clinical scenarios and suggested techniques to reach a diagnosis.

The diagnostic ability of body fluids and CSF can be significantly increased by combining flow cytometry and image cytometry. Instead of viewing both techniques as competitive, the pathology laboratory must utilize both of them as supporting techniques with different and distinctive purposes to categorize challenges faced in the samples. For good quality results,

- Laboratories must employ a layered testing approach depending on the initial specimen assessment. Samples with high cellularity can be triaged for flow cytometry to allow for

Table 9.4 Suggested techniques for various clinical scenarios.

Scenario	Best method	Reason
Suspicious highly cellular samples for Lymphoma and Leukemia	Flow cytometry	High throughout and Multiparameter immunophenotyping
Suspicious carcinoma in a paucicellular sample	Image cytometry	Morphological evaluation combined with aneuploidy evaluation
Minimal residual disease detection	Flow cytometry	Quantitative precision
Central Nervous System (CNS) involvement by leukemia	Both methods on combination	Flow cytometry for sensitivity and image cytometry for morphological assessment
Mesothelioma versus adenocarcinoma differentiation	Image cytometry	For architectural assessment combined with assessment of markers and aneuploidy

quick and multiparameter immunophenotyping, whereas low cellularity samples and morphologically ambiguous samples must be sent to image cytometry to derive useful information from limited cellular material.

- When cytological morphological evaluation and flow cytometry produce discrepant results, or when flow cytometry detects abnormal cell populations needing morphological evaluation, image cytometry can be a valuable tool for solving the problem.
- Create integrated processes that utilize the statistical power of flow cytometry with the context that is offered by image cytometry. This method can be particularly useful in CSF analysis, where some systems now allow sequential analysis of the same sample using both approaches.

9.5 Conclusion: An integrated approach to liquid cytopathology

Cytology has evolved from being a branch focused solely on morphological evaluation under the microscope to an advanced field utilizing image cytometry and multiparameter immunophenotyping. This collaborative approach enables the cytologist to think beyond the morphological evaluation, toward a meaningful understanding of cellular changes. Among the recent advancements shaping this field, flow cytometry and static image cytometry work together as complementary technologies which improve diagnosis. DNA ploidy analysis, which can be performed using either approach, has enhanced cytology's diagnostic capabilities by providing a measurable indicator of genomic instability that directly reflects tumor biology.

In developing a new diagnostic workflow for cytology, integrating DNA ploidy analysis with routine cytomorphological examination adds a new dimension and an era that enhances accurate diagnosis. By measuring the DNA content of cells, DNA ploidy analysis also helps link the morphological examination with molecular details, thus enhancing both sensitivity and specificity. This incorporation of technologies allows pathologists to identify subtle morphological changes and helps confident stratification of preneoplastic lesions and malignancies.

This current method of diagnosis encourages clinicians, pathologists, and molecular pathologists to work together for the development of standardized procedures for sample processing, data analysis, and result interpretation. This cooperation minimizes variability and improves consistency, making diagnoses more reliable. Additionally, recent developments in artificial intelligence for image cytometry present excellent opportunities in streamlining detailed measurement processes of cellular details. These advancements help streamline work, improve turnaround time, and enhance diagnosis in cytology, ultimately improving patient care by combining expert knowledge with advanced technology.

References

Bakloushinskaya, I. (2022). Chromosome changes in soma and germ line: Heritability and evolutionary outcome. *Genes, 13*(4), 602. doi:10.3390/genes13040602.

Bansal, C., Tiwari, V., Singh, U., Srivastava, A., & Misra, J. (2011). Cell cannibalism: A cytological study in effusion samples. *Journal of Cytology, 28*(2), 57–60. doi:10.4103/0970-9371.80736.

Bansbach, C. E., & Cortez, D. (2011). Defining genome maintenance pathways using functional genomic approaches. *Critical Reviews in Biochemistry and Molecular Biology, 46*(4), 327–341. doi:10.3109/10409238.2011.588938.

Basiji, D. A., Ortyn, W. E., Liang, L., Venkatachalam, V., & Morrissey, P. (2007). Cellular image analysis and imaging by flow cytometry. *Clinics in Laboratory Medicine, 27*(3), 653–670. doi:10.1016/j.cll.2007.05.008.

Biesterfeld, S., Beckers, S., Cadenas, M. D. C. V., & Schramm, M. (2011). Feulgen staining remains the gold standard for precise DNA image cytometry. *Anticancer Research, 31*(1), 53–58.

Breusegem, S. Y., Clegg, R. M., & Loontiens, F. G. (2002). Base-sequence specificity of Hoechst 33258 and DAPI binding to five (A/T)4 DNA sites with kinetic evidence for more than one high-affinity Hoechst 33258-AATT complex. *Journal of Molecular Biology, 315*(5), 1049–1061. doi:10.1006/jmbi.2001.5301.

Canonico, B., Betti, M., Luchetti, F., Battistelli, M., Falcieri, E., Ferri, P., Zamai, L., Barnett, D., & Papa, S. (2010). Flow cytometric profiles, biomolecular and morphological aspects of transfixed leukocytes and red cells. *Cytometry Part B - Clinical Cytometry, 78*(4), 267–278. doi:10.1002/cyto.b.20510.

Carloni, S., Gallerani, G., Tesei, A., Scarpi, E., Verdecchia, G. M., Virzì, S., Fabbri, F., & Arienti, C. (2017). DNA ploidy and S-phase fraction analysis in peritoneal carcinomatosis from ovarian cancer: Correlation with clinical pathological factors and response to chemotherapy. *OncoTargets and Therapy, 10*, 4657–4664. doi:10.2147/OTT.S141117.

Cossarizza, A., Chang, H. D., Radbruch, A., Abrignani, S., Addo, R., Akdis, M., Andrä, I., Andreata, F. Annunziato, F., Arranz, E., Bacher, P., Bari, S., Barnaba, V., Barros-Martins, J., Baumjohann, D., Beccaria, C. G., Bernardo, D., Boardman, D. A., Borger, J., … Yang, J. (2021). Guidelines for the use of flow cytometry and cell sorting in immunological studies (third edition). *European Journal of Immunology, 51*(12), 2708–3145. doi:10.1002/eji.202170126.

Cossarizza, A., Chang, H. D., Radbruch, A., Akdis, M., Andrä, I., Annunziato, F., Bacher, P., Barnaba, V., Battistini, L., Bauer, W. M., Baumgart, S., Becher, B., Beisker, W., Berek, C., Blanco, A., Borsellino, G., Boulais, P. E., Brinkman, R. R., Büscher, M., … Zimmermann, J. (2017). Guidelines for the use of flow cytometry and cell sorting in immunological studies. *European Journal of Immunology, 47*(10), 1584–1797. doi:10.1002/eji.201646632.

Crissman, H. A., & Tobey, R. A. (1990). Specific staining of DNA with the fluorescent antibiotics, mithramycin, chromomycin, and olivomycin. *Methods in Cell Biology, 33*(C), 97–103. doi:10.1016/S0091-679X(08)60515-4.

Cusick, E. L., Milton, J. I., & Ewen, S. W. B. (1990). The resolution of aneuploid DNA stem lines by flow cytometry: Limitations imposed by the coefficient of variation and the percentage of aneuploid nuclei. *Analytical Cellular Pathology, 2*(3), 139–148.

CyFlow Ploidy Analyser. (n.d.). Sysmex. https://www.sysmex-ap.com/product/cyflow-ploidy-analyser/.

Darzynkiewicz, Z., Halicka, H. D., & Zhao, H. (2010). Analysis of cellular DNA content by flow and laser scanning cytometry. *Advances in Experimental Medicine and Biology, 675*, 137–147. doi:10.1007/978-1-4419-6199-0_9.

Darzynkiewicz, Z., Traganos, F., Zhao, H., Halicka, H.D., & Li, J. (2011). Cytometry of DNA replication and RNA synthesis: Historical perspective and recent advances based on "click chemistry." Cytometry Part A, 79A(5), 328–337. https://pubmed.ncbi.nlm.nih.gov/20687474/

Davies, G. E., & Stark, G. R. (1970). Use of dimethyl suberimidate, a cross-linking reagent, in studying the subunit structure of oligomeric proteins. *Proceedings of the National Academy of Sciences of the United States of America, 66*(3), 651–656. doi:10.1073/pnas.66.3.651.

De Graaf, M. T., De Jongste, A. H. C., Kraan, J., Boonstra, J. G., Smitt, P. A. E. S., & Gratama, J. W. (2011). Flow cytometric characterization of cerebrospinal fluid cells. *Cytometry Part B - Clinical Cytometry, 80*(5), 271–281. doi:10.1002/cyto.b.20603.

Darzynkiewicz, Z., Huang, X., & Zhao, H. (2017). Analysis of cellular DNA content by flow cytometry. *Current Protocols in Cytometry, 82*(1), 7.5.1–7.5.20. https://pubmed.ncbi.nlm.nih.gov/28967991/.

Dimitriadis, S., Dova, L., Kotsianidis, I., Hatzimichael, E., Kapsali, E., & Markopoulos, G. S. (2024). Imaging flow cytometry: Development, present applications, and future challenges. *Methods and Protocols, 7*(2). doi:10.3390/mps7020028.

Dunn, J. M., MacKenzie, G. D., Oukrif, D., Mosse, C. A., Banks, M. R., Thorpe, S., Sasieni, P., Bown, S. G., Novelli, M. R., Rabinovitch, P. S., & Lovat, L. B. (2010). Image cytometry accurately detects DNA ploidy abnormalities and predicts late relapse to high-grade dysplasia and adenocarcinoma in Barrett's oesophagus following photodynamic therapy. *British Journal of Cancer, 102*(11), 1608–1617. doi:10.1038/sj.bjc.6605688.

Fahlberg, M. D., Forward, S., Assita, E. R., Mazzola, M., Kiem, A., Handley, M., Yun, S. H., & Kwok, S. J. J. (2024). Overcoming fixation and permeabilization challenges in flow cytometry by optical barcoding and multi-pass acquisition. *BioRxiv*. doi:10.1101/2024.08.13.607771.

Ferreira-Facio, C. S., Milito, C., Botafogo, V., Fontana, M., Thiago, L. S., Oliveira, E., da Rocha-Filho, A. S., Werneck, F., Forny, D. N., Dekermacher, S., de Azambuja, A. P., Ferman, S. E., de Faria, P. A. S., Land, M. G. P., Orfao, A., & Costa, E. S. (2013). Contribution of multiparameter flow cytometry immunophenotyping to the diagnostic screening and classification of pediatric cancer. *PLoS One, 8*(3). doi:10.1371/journal.pone.0055534Brazil.

Forment, J. V., & Jackson, S. P. (2015). A flow cytometry-based method to simplify the analysis and quantification of protein association to chromatin in mammalian cells. *Nature Protocols, 10*(9), 1297–1307. doi:10.1038/nprot.2015.066.

Fuller, M. Y., & Thrall, M. J. (2016). Utility of flow cytometry analysis for pleural and peritoneal fluids. *Journal of the American Society of Cytopathology, 5*(6), 339–344. doi:10.1016/j.jasc.2016.07.001.

Gabali, A. (2022). Flow cytometry, molecular analysis, and other special techniques (in serous fluid cytopathology). *Cytojournal, 19*, 18. doi:10.25259/cmas_02_13_2021.

Giam, M., & Rancati, G. (2015). Aneuploidy and chromosomal instability in cancer: A jackpot to chaos. *Cell Division, 10*(1). doi:10.1186/s13008-015-0009-7.

Gómez-Muñoz, C., & Fischer, G. (2025). MuPETFlow: Multiple ploidy estimation tool from flow cytometry data. *BMC Genomics [Electronic Resource], 26*(1). doi:10.1186/s12864-025-11470-8.

Hardie, D. C., Gregory, T. R., & Hebert, P. D. N. (2002). From pixels to picograms: A beginners' guide to genome quantification by Feulgen image analysis densitometry. *Journal of Histochemistry and Cytochemistry, 50*(6), 735–749. doi:10.1177/002215540205000601.

Hjerpe, A., Demir, E., Abd-Own, S., & Dobra, K. (2023). Utility of BerEp4/calretinin and desmin/epithelial membrane antigen (EMA) dual immunocytochemical staining in effusion cytology. *Cancer Medicine, 12*(5), 5334–5340. doi:10.1002/cam4.5353.

Holtfreter, H. B., & Cohen, N. (1990). Fixation-Associated quantitative variations of DNA fluorescence observed in flow cytometric analysis of hemopoietic cells from adult diploid frogs. *Cytometry, 11*(6), 676–685. doi:10.1002/cyto.990110604.

Houtsma, R., Hogeling, S. M., & Schuringa, J. J. (2021). CombiFlow: Flow cytometry-based identification and characterization of genetically and functionally distinct AML subclones. *STAR Protocols, 2*(4). doi:10.1016/j.xpro.2021.100864.

Huang, Q., Yu, C., Zhang, X., & Goyal, R. K. (2008). Comparison of DNA histograms by standard flow cytometry and image cytometry on sections in Barrett's adenocarcinoma. *BMC Clinical Pathology, 8*(1). doi:10.1186/1472-6890-8-5.

Hume, S., Dianov, G. L., & Ramadan, K. (2020). A unified model for the G1/S cell cycle transition. *Nucleic Acids Research, 48*(22), 12483–12501. doi:10.1093/nar/gkaa1002.

Kaluza Analysis Software. (n.d.). https://www.mybeckman.co/resources/product-applications/cell-based-therapeutics/automated-cellular-analysis/kaluza-analysis-software.

Karaman, S., & Detmar, M. (2014). Mechanisms of lymphatic metastasis. *Journal of Clinical Investigation, 124*(3), 922–928. doi:10.1172/jci71606.

Katherine-Rittenbach. (n.d.). Where can I get more information regarding DNA analysis using the multicycle AV? De Novo Software. https://denovosoftware.com/faq/kb-where-can-i-get-more-information-regarding-dna-analysis-using-the-multicycle-av/.

Kim, K. H., & Sederstrom, J. M. (2015). Assaying cell cycle status using flow cytometry. *Current Protocols in Molecular Biology, 2015*, 28.6.1. doi:10.1002/0471142727.mb2806s111.

Kouriefs, C., & Gordon, G. J. (2001). Diagnostic accuracy of dna image cytometry and urinary cytology with cells from voided urine in the detection of bladder cancer. *Urology, 58*(3), 499. doi:10.1016/s0090-4295(01)01162-1.

Kraan, J., Gratama, J. W., Haioun, C., Orfao, A., Plonquet, A., Porwit, A., Quijano, S., Stetler-Stevenson, M., Subira, D., & Wilson, W. (2008). Flow cytometric immunophenotyping of cerebrospinal fluid. *Current Protocols in Cytometry*, (45), 6.25.16. doi:10.1002/0471142956.cy0625s45.

Lazcano, O., Chen, L. M., Tsai, C., Li, C. Y. Katzmann, J. A., Sebo, T. J., Kimlinger, T. K., & Baker, J. (2000). Image analysis and flow cytometric DNA studies of benign and malignant body cavity fluids: Reappraisal of the role of current methods in the differential diagnosis of reactive versus malignant conditions. *Modern Pathology, 13*(7), 788–796. doi:10.1038/modpathol.3880137.

Ma, Z., Li, P., Gai, X., Li, X., Sun, B., Wang, T., Jiang, P., Wang, H., & Zhang, J. (2023). DNA image cytometry ploidy analysis technique improves the detection rate of pleural effusion cytology. *Diagnostic Cytopathology, 51*(3), 159–165. doi:10.1002/dc.25077.

Mahajan, S., Nambirajan, A., Gupta, I., Gupta, N., Gupta, P., & Jain, D. (2022). Malignant pleural effusion cell blocks are reliable resources for PD-L1 analysis in advanced lung adenocarcinomas: A concordance study with matched histologic samples. *Journal of the American Society of Cytopathology, 11*(5), 253–263. doi:10.1016/j.jasc.2022.04.004.

McKinnon, K. M. (2018). Flow cytometry: An overview. *Current Protocols in Immunology, 2018*, 5.1.1. doi:10.1002/cpim.40.

Minderman, H., Humphrey, K., Arcadi, J. K., Wierzbicki, A., Maguire, O., Wang, E. S., Block, A. W., Sait, S. N. J., George, T. C., & Wallace, P. K. (2012). Image cytometry-based detection of aneuploidy by fluorescence in situ hybridization in suspension. *Cytometry Part A, 81*(9), 776–784. doi:10.1002/cyto.a.22101.

Mishra, S., Awasthi, N. P., Husain, N., Anand, A., Pradeep, Y., Ansari, R., & Saxena, S. (2017). Flow cytometric analysis of DNA ploidy in liquid-based cytology for cervical pre-cancer and cancer. *Asian Pacific Journal of Cancer Prevention, 18*(6), 1595–1601. doi:10.22034/APJCP.2017.18.6.1595.

ModFit LTTM. (n.d.). Retrieved March 23, 2026, from https://www.vsh.com/products/mflt/.

Plander, M., Brockhoff, G., Barlage, S., Schwarz, S., Rothe, G., & Knuechel, R. (2003). Optimization of three- and four-color multiparameter DNA analysis in lymphoma specimens. *Cytometry Part A, 54A*(1), 66–74. doi:10.1002/cyto.a.10051.

Pollice, A. A., Mccoy, J. P., Shackney, S. E., Smith, C. A., Agarwal, J., Burnolt, D. R., Janocko, L. E., Hornicek, F. J., Singh, S. G., & Hartsock, R. J. (1992). Sequential paraformaldehyde and methanol fixation for simultaneous flow cytometric analysis of DNA, cell surface proteins, and intracellular proteins. *Cytometry, 13*(4), 432–444. doi:10.1002/cyto.990130414.

Pozarowski, P., & Darzynkiewicz, Z. (2004). Analysis of cell cycle by flow cytometry. *Methods in Molecular Biology (Clifton, N.J.), 281*, 301–311. doi:10.1385/1-59259-811-0:301.

Pozarowski, P., Holden, E., & Darzynkiewicz, Z. (2013). Laser scanning cytometry: Principles and applications-an update. *Methods in Molecular Biology (Clifton, N.J.), 931*. doi:10.1007/978-1-62703-056-4_11.

Puri, M., Sen, R., Gupta, M., Kalra, R., Bhargava, S., & Puri, K. (2020). DNA ploidy analysis and cell block immunohistochemistry in the diagnosis of malignant effusions. *Acta Cytologica, 64*(3), 256–264. doi:10.1159/000501790.

Replogle, J. M., Zhou, W., Amaro, A. E., McFarland, J. M., Villalobos-Ortiz, M., Ryan, J., Letai, A., Yilmaz, O., Sheltzer, J., Lippard, S. J., Ben-David, U., & Amon, A. (2020). Aneuploidy increases resistance to chemotherapeutics by antagonizing cell division. *Proceedings of the National Academy of Sciences of the United States of America, 117*(48), 30566–30576. doi:10.1073/pnas.2009506117.

Roederer, M. (2002). Compensation in flow cytometry. *Current Protocols in Cytometry, 22*(1). doi:10.1002/0471142956.cy0114s22.

Schmid, I., Uittenbogaart, C. H., & Giorgi, J. V. (1991). A gentle fixation and permeabilization method for combined cell surface and intracellular staining with improved precision in DNA quantification. *Cytometry, 12*(3), 279–285. doi:10.1002/cyto.990120312.

Shen, Y., Vignali, P., & Wang, R. (2017). Rapid profiling cell cycle by flow cytometry using concurrent staining of DNA and mitotic markers. *Bio-Protocol, 7*(16). doi:10.21769/BioProtoc.2517.

Shidham, V. B. (2021a). Diagnostic pitfalls in effusion fluid cytology. *CytoJournal, 18*(1). doi:10.25259/CMAS_02_04_2021.

Shidham, V. B. (2021b). The panorama of different faces of mesothelial cells. *CytoJournal, 18*(1). doi:10.25259/CMAS_02_02_2021.

Shidham, V. B., & Layfield, L. J. (2021). Approach to diagnostic cytopathology of serous effusions. *CytoJournal, 18*(1). doi:10.25259/CMAS_02_03_2021.

Suthanthiraraj, P. P. A., Shreve, A. P., & Graves, S. W. (2024). Essential fluidics for a flow cytometer. *Current Protocols, 4*(10). doi:10.1002/cpz1.1124.

Tárnok (2006). Slide-based cytometry for cytomics—A minireview. *Cytometry. Part A: The Journal of the International Society for Analytical Cytology, 69*(7), 555–562. doi:10.1002/cyto.a.20317.

Thastrup, M., Marquart, H. V., & Schmiegelow, K. (2022). Flow cytometric detection of malignant blasts in cerebrospinal fluid: A biomarker of central nervous system involvement in childhood acute lymphoblastic leukemia. *Biomolecules, 12*(6). doi:10.3390/biom12060813.

Viryasova, G. M., Golenkina, E. A., Tatarskii, V. V., Galkin, I. I., Sud'ina, G. F., & Soshnikova, N. V. (2019). An optimized permeabilization step for flow cytometry analysis of nuclear proteins in myeloid differentiation of blood cells into neutrophils. *MethodsX, 6*, 360–367. doi:10.1016/j.mex.2019.02.011.

Yao, Y., Peng, M., Shen, Q., Hu, Q., Gong, H., Li, Q., Zheng, Z., Xu, B., Li, Y., & Dong, Y. (2019). Detecting EGFR mutations and ALK / ROS1 rearrangements in non-small cell lung cancer using malignant pleural effusion samples. *Thoracic Cancer, 10*(2), 193–202. doi:10.1111/1759-7714.12932.

Wersto, R. P., Chrest, F. J., Leary, J. F., Morris, C., Stetler-Stevenson, M. A., & Gabrielson, E. (2001). Doublet discrimination in DNA cell-cycle analysis. *Cytometry, 46*(5), 296–306. https://pubmed.ncbi.nlm.nih.gov/11746105/.

10

Flow cytometry for pediatric small round cell tumors

Jasmita Dass

DEPARTMENT OF HEMATOLOGY, ALL INDIA INSTITUTE OF MEDICAL SCIENCES, NEW DELHI, INDIA

Cancers in the pediatric and adolescent age groups are a serious health issue. About 300,000 new cases of pediatric cancers occur every year, with the number of yearly deaths of ~80,000 worldwide (Ward et al., 2019). The majority of pediatric cancers are hematolymphoid, comprising leukemias and lymphomas, which together account for ~40%–41% of all pediatric cancers. The next most commonly occurring pediatric cancers are of the central nervous system (CNS), including the spinal cord tumors, and constitute ~25%–26% of all pediatric cancers. Among the renal and adrenal tumors, Wilm's tumor and neuroblastoma account for ~6% each (Agarwal, 2023; Apps et al., 2024; Ferreira-Facio et al., 2021). The survival of pediatric cancers has improved in high-income countries (HIC) and exceeds 80%. However, the incidence of pediatric cancers in low and middle-income countries (LMIC) is high, and the survival rates still lag behind those in the HIC. Contributory factors to this phenomenon are (1) delayed diagnosis with a prolonged prediagnosis interval causing the patient to present at an advanced stage, (2) wrong diagnosis, (3) under-diagnosis and under-reporting, (4) higher treatment toxicity partly stemming from poorer nutritional status, (5) treatment abandonment, (6) deficiency of health care facilities, (7) drug shortages, etc. (Khan et al., 2025). In addition, sarcomas in the pediatric population carry an exceptionally poor prognosis with a 3-year event-free survival of only 30%–40% despite intensive treatment strategies. The survival drops further to 20% when negative prognostic factors are present (Oberlin et al., 2008).

Delays in diagnosis are a contributory factor in poor outcomes in many pediatric tumors (Brasme et al., 2012) (7). The World Health Organization (WHO) has a Global Initiative for Childhood Cancers (GICC), which aims to improve survival rates for childhood cancers to ≥60% by the year 2030 (Ortiz et al., 2024) [8]. To achieve this target and address the knowledge gap in health care workers for childhood cancers, the International Society of Pediatric Oncology (SIOP), in collaboration with the International Pediatric Association (IPA) organized the "Childhood Cancer Early Diagnosis and Appropriate Referral (CEDAR)" in 2022, under which they conducted webinars to address six common childhood cancers and these include unfavorable site of the primary tumor, ≥3 metastatic sites, or the presence of blood or bone marrow involvement (Khan et al., 2025).

Advancing Body Fluid Diagnostics with Flow Cytometry: A Comprehensive Guide to Effusions and CSF.
DOI: https://doi.org/10.1016/B978-0-443-27426-8.00012-0

Table 10.1 Panel of Immunohistochemistry markers applied for the diagnosis of pediatric small round cell tumors.

Myogenic	Neural	Germ cell	Mesenchymal	Hematopoietic
MyoD1 (nuclear)	NSE	Alpha-fetoprotein	Vimentin	CD45
Myogenin (nuclear)	NB-84	Placental alkaline phosphatase	Smooth muscle actin	CD20/CD19/PAX-5 (nuclear)- B-cell differentiation
Desmin	TrkA	Cytokeratin	–	CD3, CD5, CD4, and CD8 (T-cell differentiation)
Myoglobin	NFTP	–	–	MPO, CD68, and CD163 (myeloid)
–	Synaptophysin	–	–	CD34, TdT, and CD99: Immaturity
–	CD56	–	–	–
–	CD99	–	–	–

Clinical manifestations, imaging, and laboratory findings provide an initial guide to the diagnosis and classification of pediatric tumors. A directed biopsy from the site of involvement is mandatory for establishing the diagnosis. These biopsies are subjected to basic histopathology, immunohistochemistry (IHC), and molecular pathology to arrive at a correct diagnosis. Using this approach, most pediatric cancers are grouped under "small round cell tumors (SRCTs)" arising from embryonic tissues (Triche et al., 2006). An extensive IHC panel is required to classify these SRCTs correctly (Parham, 2010). Some of the markers are listed in Table 10.1.

Even with extensive IHC panels, some cases are difficult to classify. This occurs due to overlapping morphological and immunohistochemical features, paucicellular small biopsies, and intrinsic tumor variations (Magro et al., 2015). This has led to expanding IHC panels and inclusion of markers like WT1 and Cyclin D1 (Salvatorelli et al., 2021), SATB2, ERG, FLI-1, BCOR, and many others to further differentiate various SRCTs (Dehner et al., 2024).

Flow cytometry (FCM) is a widely used technique for classifying, prognosticating, and monitoring hematolymphoid malignancies. It is amenable to assessing various solid organ malignancies and is being adapted to pediatric cancers as well. It can phenotype millions of cells and simultaneously determine the expression of multiple markers per cell, with a short turnaround time (Ferreira-Facio et al., 2013). However, the use of FCM for pediatric cancer assessment is limited by the need for single-cell suspensions and a limited panel of markers, as most centers do not routinely assess these cases (Ferreira-Facio et al., 2021, 2013). Moreover, because FCM uses single-cell suspensions, the tissue structure cannot be assessed (Ferreira-Facio et al., 2021). The availability of markers also somewhat restricted the wider application of FCM to characterize pediatric SRCTs (Ferreira-Facio et al., 2013). FCM has primarily been used to assess deoxyribonucleic acid (DNA) ploidy in frozen and paraffin sections in combination with one or two phenotypic markers. The tissues have been converted to single-cell suspensions using enzymatic dissociation (Corver et al., 2005; Mora et al., 2007). As most FCM labs already assess bone marrow and peripheral blood, this technique has also been extended to identify metastatic pediatric cancers in these samples (Sugimoto et al., 1988). In rare-event analysis, some users have also tried to identify CTCs in pediatric cancers such as Ewing's sarcoma (DuBois et al., 2010).

The uses of FCM in pediatric SRCTs are: (1) Ploidy analysis, (2) Correct diagnosis of the tumor in the tissue, (3) Detecting metastases, (4) Measurable residual disease analysis, and (5) Studying the interactions between the stroma and the tumor. Other uses in basic research are a study of cell survival and repair, signal transduction, studying cellular differentiation, etc.

10.1 Processing protocols

The samples that can be submitted for FCM in pediatric SRCTs include tumor tissue, either fresh or formalin-fixed, as well as liquid samples such as peripheral blood, bone marrow, and effusions. Since FCM requires single-cell suspensions, the liquid samples remain easy to process. The challenge is always with the tissue samples that need disaggregation to yield single-cell suspensions. The tumor cells must be dislodged from the fibrous stroma, recovered intact, and all subpopulations should be represented. The disaggregation protocols use either mechanical or enzymatic methods. Both methods may be combined. The enzymes that aid disaggregation, such as trypsin, pronase, DNase, and collagenase, may be added to the solutions. These enzymes, though they aid disaggregation and improve cell counts achieved, may affect the surface expression of markers and require optimization. After disaggregation, to remove debris and cell aggregates, the samples require filtration with 70–100 μm strainers or meshes. The steps for surface staining and intracellular staining are similar to the regular FCM staining protocols (Chang & Hedley, 2012; Ferreira-Facio et al., 2013).

Viability determination is important when disaggregated tissue samples are processed. Most laboratories use the forward and side scatter (FSC and SSC, respectively) to determine viability on peripheral blood and bone marrow samples. However, this may prove to be inadequate when assessing solid tumors. Hence, viability assessment may require the use of dyes such as DAPI, PI, or DRAQ5 (Chang & Hedley, 2012; Ferreira-Facio et al., 2021, 2013). Delayed sample transport to the lab impacts the viability of cells when performing FCM. Hence, the samples should be transported to the lab immediately after collection, as delays of >12–24 h compromise viability more significantly than those from peripheral blood, bone marrow, and effusion. However, despite low viability, the samples should still be processed as they are precious, and a diagnosis may still be achieved (Ferreira-Facio et al., 2021).

10.2 Ploidy assessment in pediatric cancers

Ploidy assessment of round cell tumors was one of the first assays performed using FCM. The assay uses dyes that bind stoichiometrically to the DNA and emit fluorescence. FCM, even in the 1990s, had an advantage over other techniques for ploidy determination as it could count a higher number of cells, and its less labor-intensive nature than the static cytophotometric study. To assess the ploidy of any tumor cell, the DNA content of the G0-G1 phase of the tumor cells is compared to the DNA content of the G0-G1 phase of the normal cells. Tumor cells with an altered G0-G1 content compared with normal cells are considered aneuploid. Events with DNA content below the G1 peak should be excluded (Chang & Hedley, 2012). The DNA index (DI) is calculated as a ratio of the mean channel number of the tumor G0-G1 peak to the mean channel number of the G0-G1 peak of the normal cells. The level of resolution achieved is at a DI of 1.04, meaning that the DNA content of tumor cells and that of the normal cells should at least have a 4% difference. This assay also measures the proportion of cells in S-phase and in the G2-M phase of the cell cycle. Incidentally, ploidy can be assessed on both fresh and formalin-fixed tissue samples. However, ploidy on paraffin embedded tissue does pose problems of poorer resolution of G0-G1 peaks when compared to fresh tissue, they have a higher subcellular debris due to the process of enzymatic digestion and this requires careful analysis to avoid calculations of a falsely high S-phase, and external standards cannot be used for G0-G1 profiling due to fixation induced fluorophore binding variations requiring a comparison with normal cells present in the tumor itself (Merkel & McGuire, 1990). Ploidy analysis can also be aided with the addition of antibodies that bind to the different types of cell types in the sample to enable the determination of the normal cell populations present within the tumor (Corver et al., 2005).

Ploidy has been assessed in pediatric SRCTs using FCM. The assessment of DNA ploidy in neuroblastoma showed that there was a ploidy heterogeneity in nearly all neuroblastomas. It emerged as a powerful prognostic variable in multivariate analysis, with the presence of a 2n/4n peak alone associated with worse overall and progression-free survival. In contrast, triploid (3n) peaks were related to low-risk, nonprogressive neuroblastomas (Mora et al., 2007). Even frozen tissues are amenable to FCM for ploidy determination, enabling a good analysis of ploidy in >90% pediatric malignant solid tumors (Smyth et al., 1993) (21). In rhabdomyosarcomas, aneuploidy is the most frequently observed pattern, with many patients exhibiting multiple aneuploid peaks. Using the DI values, the tumors can be determined as aneuploid with values ranging from 1.10 to 1.80, and near tetraploidy is defined as values ranging from 1.80 to 2.20. The patients with multiple aneuploid peaks and with aneuploidy tend to have a high risk of treatment failure (Jung et al., 1994).

10.3 Dissecting the phenotype of small round blue cells with flow cytometry

Among the pediatric cancers, neuroblastoma has been assessed most systematically using FCM. The first assessment dates back to 1988, when Komada et al. (1998) discovered that neuroblastoma cells are typically CD56+/CD9+/CD45− and demonstrated that FCM can detect disease at levels of 1 cell per $10^{(4\text{-}5)}$, enabling the detection of occult disease. This has been confirmed by multiple investigators, who have shown that neuroblastoma can be detected in bone marrow using FCM (Warzynski et al., 2002). CD56 is expressed by natural-killer T-cells (NK-cells), but the expression on neuroblastoma tends to be brighter than that on NK-cells. After this discovery, investigators used intracellular FCM to detect neuron-specific enolase (NSE), an enzyme expressed in neuroblastoma cells. In addition, disialoganglioside (GD2) was investigated as a surface marker for neuroblastoma. An additional marker positive in neuroblastoma and available in many labs doing measurable residual disease analysis for acute lymphoblastic leukemia is CD81. The addition of GD2 to CD45 and CD56 increased the assay's specificity for detecting neuroblastoma. GD2 is additionally expressed on other tumors of neuroectodermal origin, such as melanoma, small cell carcinoma, and glioma. The detection limit of this 3-color assay was determined as 0.002% (Warzynski et al., 2002).

The CD56+/CD45− phenotype is the most frequent phenotype associated with pediatric round blue cell tumors and was discovered as early as 2002 (Bryson et al., 2002). More antibodies were developed against different targets, including cytokeratin and epithelial cell adhesion molecule (EpCAM), for epithelial malignancies. Antibodies to myogenin and CD99 were also added (Bozzi et al., 2008; Chang et al., 2003) (26, 27)

10.4 Detecting metastases in the bone marrow

Neuroblastoma was the first pediatric small round cell tumor in which extensive work on FCM to detect disseminated disease in bone marrow is available. Moreover, this is one tumor for which FCM was used for MRD.

Ewing's sarcoma is another pediatric SRCT that frequently disseminates to the bone marrow. Even with localized disease, patients with Ewing's sarcoma need systemic therapy, as they develop metastases if only the local lesion is treated. This indicates that micrometastases are present even in patients with localized lesions. The characteristic EWS: ETS fusion can be detected in peripheral blood using reverse transcriptase polymerase chain reaction (RT-PCR). The malignant cells are characteristically positive for CD99 and negative for CD45, enabling their detection in peripheral blood and bone marrow using FCM. CD99 is also frequently expressed by immature lymphocytes and blasts in precursor T-cell acute lymphoblastic leukemia. However, these are usually CD45 positive. In addition, CD90 is also frequently expressed on Ewing's sarcoma cells. Using the approach of detecting CD99bright/CD45-negative cells, the sensitivity of detection in peripheral blood and bone marrow was demonstrated to be 1 in 500,000 peripheral blood mononuclear cells. In healthy controls, cells with this CD99bright/CD45-negative phenotype occur at very low frequency with a median value of 0.0001% (range: 0%–0.00046%). The addition of exclusion markers for monocytes, blasts, and gating live cells only enhances the diagnostic accuracy (DuBois et al., 2010) (28).

In the first series on the detection of marrow metastases in cases of rhabdomyosarcoma, all seven cases expressed the CD45−/CD56+ phenotype and showed strong nuclear myogenin expression, detected with the myf4 antibody. In addition, half the cases expressed CD90, and a few also expressed CD57 (Bozzi et al., 2008) (27)

10.5 Detecting disease in the cerebrospinal fluid

Among the pediatric brain tumors, medulloblastoma is the most common, accounting for 20% of all pediatric brain tumors. It is followed by diffuse intrinsic pontine glioma (DIPG), ependymoma, two high-grade gliomas, which account for 12% and 8% of all pediatric brain tumors, respectively (Hargrave & Zacharoulis, 2007). Due to the presence of an effective blood-brain barrier, the cerebrospinal fluid (CSF), which lies in close proximity to the pediatric brain tumors, becomes a good source to detect circulating tumor DNA (ctDNA) and for even molecular diagnosis (Buzova et al., 2023).

Incidentally, imaging FCM (IFCM) has been used to detect histones in the CSF using antibodies marked to these histones using the ImageStream MkII (Luminex Corporation). The most frequently detected histones were the H2A, its variants macroH2A1.1/macroH2A1/2, and the H3K27M, which is the histone H3 protein carrying the pathognomonic mutation for DIPG (Buzova et al., 2023).

10.6 Screening tubes to identify pediatric small round cell tumors

The early efforts concentrated on identifying hematopoietic neoplasms and also the pediatric round cell tumors. After excluding T-lymphocytes gated as surface CD3/cytoplasmic CD3-positive cells, B-lymphocytes gated as CD19-positive cells, and neutrophils as myeloperoxidase (MPO)-positive cells, they focused on the CD45-negative population. This population is the area of interest when screening for pediatric round cell tumors. They had further markers to identify the different pediatric tumors. The markers included in the panel are shown in Table 10.2 (Ferreira-Facio et al., 2013).

Tissue samples collected in the surgical units are usually split into two: one for morphology, and the second for FCM. The first block undergoes standard tissue processing and is fixed in formalin. The second block is kept in cold phosphate-buffered saline (PBS) at 4°C (14). Alternate media for preserving the tissue, such as the RPMI medium, may also be used instead of PBS. The tissue must be converted into a single-cell suspension. For this, the sample tissue can be placed in petri dishes containing PBS with 1% bovine serum albumin (BSA). It is minced into small pieces, which are then mechanically disintegrated using two needles. After the mentioned procedure, this solution in the petri dish containing disintegrated cells in PBS-BSA is filtered through a filter/syringe with a 100 μm pore size. This process eliminates cell debris and clumps, which can interfere with FCM analysis and also may clog the flow cytometer. The filtrate is then centrifuged at 540 g for 10 min and is resuspended in PBS-BSA to achieve a cell concentration of 10^6 cells/mL. Liquid samples such as effusion fluids, bone marrow, and peripheral blood can be processed directly. The antibodies, as per the tubes shown in Table 10.2, were added to 50 μL solution, followed by cell lysis. After lysis, the cells were centrifuged, and the cell pellet was washed with PBS and suspended in 0.5mL PBS for acquisition. If cell permeabilization is required, the cells after surface staining are fixed, a permeabilizing agent is added, followed by the nuclear/cytoplasmic antibodies, and the cells are then acquired. The study showed a high concordance of 96% between FCM and conventional histology and immunohistochemistry to identify both reactive and malignant samples. The two discordant samples were from a case of Hodgkin lymphoma (HL) and anaplastic lymphoma each. All the reactive/inflammatory cases

Table 10.2 The first comprehensive approach to identifying multiple hematopoietic and nonhematopoietic tumors.

Tube	PacB	PacO	FITC	PE	PerCPCy5.5	PE-Cy7	APC	APC-H7
Orientation panel for screening								
1	cyCD3 (UCHT1)	CD45 (HI30)	MPO (MPO-7)	cyCD79a (HM57)	CD34 (8G12)	CD19 (J3119)	CD7 (124-1D1)	sCD3 (SK7)
2	CD20 (2H7)	CD45 (HI30)	CD8/sIgλ (lymohoclonal)	CD56/Igκ (lymohoclonal)	CD19/CD4 (lymphoclonal)	CD56 (N901/NKH1)	CD3	–
Characterization panel if orientation panel positive for pediatric tumor								
1	–	CD45 (HI30)	CD57 (HNK1)	CD90	CD34 (8G12)	CD56 (N901/NKH1)	EpCAM (EBA-1)	–
2	CD20 (2H7)	CD45 (HI30)	CD99 (TÜ12)	CD81 (JS-81)	CD9 (M-L13)	CD56 (N901/NKH1)	CD117 (104D2)	–
3	–	CD45 (HI30)	CD58 (1C3)	CD38 (HB-7)	–	CD56 (N901/NKH1)	CD10 (HI10A)	–
4	–	CD45 (HI30)	–	CD271 (C40-1457)	–	CD56 (N901/NKH1)	–	–
Tubes 5-8 containing one respective marker with rabbit anti-IgG in FITC								
5-8	–	–	Rabbit anti-IgG (tube 5-8)	nuMYOD (Moab5.8)	nuMyogenin (F5D)	GD2 (14.G2a)	cyDesmina (RD301)	–

From Ferreira-Facio, C. S., Milito, C., Botafogo, V., Fontana, M., Thiago, L. S., Oliveira, E., da Rocha-Filho, A. S., Werneck, F., Forny, D. N., Dekermacher, S., de Azambuja, A. P., Ferman, S. E., de Faria, P. A. S., Land, M. G. P., Orfao, A., & Costa, E. S. (2013). Contribution of multiparameter flow cytometry immunophenotyping to the diagnostic screening and classification of pediatric cancer. PLOS ONE, 8(3), Article e55534. https://doi.org/10.1371/journal.pone.0055534.

Table 10.3 Marker profile for the pediatric tumors using the comprehensive panel. The markers in bold represent the most informative markers for the respective entities.

Tumor	Phenotypic profile	Remarks
Neuroblastoma	$CD45^-/CD56^+/CD81^{br}/CD9^+/$ **$GD2^{++}$**	$CD58^{+/++}$ and CD57 were heterogeneous; CD90 expression was present in all but one case. CD271 was rarely expressed B-, T-, myeloid markers were negative ~60% Ganglioneuroblastomas showed a bright expression of CD56, CD81, and CD27 with a higher forward scatter (FSC) and side scatter (SSC)
PNET	$CD45^-/CD56^{br}/$**$CD99^+$**$/CD81^+/CD90^+/CD9^+/CD127^{br}$	nuMYOD and nuMyogenin were negative. B-, T-, myeloid markers were negative. GD2 predominantly negative
Rhabdomyosarcomas	$CD45^-/CD56^{dim}/$**$nuMYOD1^{br}$**$/CD90^+/$**$nuMyogenin^{br}$**	B-, T-, myeloid markers were negative GD2 negative Some cases expressed CD81, CD9, and CD58 Embryonal rhabdomyosarcoma expressed bright CD99 CD271 is expressed (3)
Hemangiopericytoma	$CD45^-/$**$CD34^{br}$**	B-, T-, myeloid markers were negative
Adrenal carcinoma	$CD45^-/$**$EpCAM^{br}$**$/CD90^{het}$	B-, T-, myeloid markers were negative
Nasopharyngeal carcinoma	$CD45^-/$**$EpCAM^{br}$**$/CD56^+/CD90^+$	–
Germ cell tumor	$CD45^-/CD56^+/$**$CD10^+/NG2^+$**	B-, T-, myeloid markers were negative CD38 was negative
Wilm's tumor	Two distinct cell populations: Population 1: $CD45^-/CD56^+/CD58^+/CD99^-/GD2^+/nuMYOD1^-/nuMyogenin^-/CD10^-/NG2^-/CD90^-/EpCAM^-/CD57^-$ Population 2: $CD45^-/CD56^+/CD58^+/CD99^-/GD2^+/nuMYOD1^-/nuMyogenin^-/CD10^-/NG2^-/CD90^+/EpCAM^+/CD57^+$	–

were correctly identified, and there was no false positive diagnosis (Ferreira-Facio et al., 2013). The phenotypes detected in the tumors are shown in Table 10.3 and Fig. 10.1.

Neuroblastomas showed the highest expression of CD56, followed by peripheral neuroectodermal tumor (PNET). However, CD56 was positive in many pediatric cancers, limiting its utility as a single marker for any pediatric tumor. GD2 was positive only in neuroblastoma

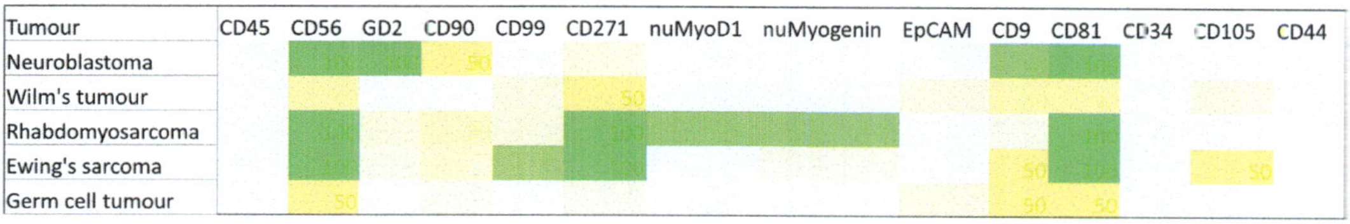

FIGURE 10.1 Immunophenotype of pediatric small round cell tumors. Diagrammatic representation of the immunophenotypic profiles of the various pediatric small round cell tumors. (*Gray* indicates absent expression; *green* indicates expression in over half the cases, with intensity increasing with increasing expression; *yellow* indicates less than half the cases expressing the tumor, with intensity increasing with increasing expression).

and enabled its distinction from PNET with CD271, which was positive in PNET but negative in neuroblastoma and possibly correlates with the mesenchymal origin of PNET. Interestingly, bone marrow metastases of PNET are CD271-negative and possibly associated with a more aggressive behavior and an immature phenotype. MYOD1 and myogenin were most specific for rhabdomyosarcoma. Wilm's tumor is an interesting case with two distinct populations of malignant cells that can be discriminated using CD90, EpCAM, and CD57 (Ferreira-Facio et al., 2013).

This laid the field for the development of a more structured orientation tube for solid tumors. The EuroFlow Consortium designed a tube to identify solid tumors in an 8-color format. This tube is an 8-color, 12-marker combination for screening of solid tumors and is referred to as the "solid tumor orientation tube (STOT)." These markers include hematopoietic cell markers, such as CD45, which is negative on nonhematolymphoid tumors, and multiple markers to identify mature and immature lymphoid cells, including surface CD3, cytoplasmic CD3, and CD19. The tube has EpCAM for epithelial tumors, CD56 and GD2 for neuroblastomas, nuclear myogenin and CD56 for rhabdomyosarcomas, and CD99 and CD56 for Ewing's sarcoma/PNET. This tube went through three design phases, the third of which was split into two, with minor modifications to clones of a few markers, such as CD99, and a change in the manufacturer for CD45. Myogenin is a nuclear marker and requires permeabilization. The inclusion of hematopoietic markers also enables the diagnosis of leukemias and lymphomas that may mimic pediatric round cell tumors. The markers and their fluorochrome conjugates included in the STOT are shown in Table 10.4. In addition, a special HL tube was also used to identify HL and anaplastic large cell lymphoma (ALCL) (Ferreira-Facio et al., 2021).

This STOT enables the identification of normal cell populations in addition to tumoral populations. The normal cell populations detected are lymphocytes, including T-lymphocytes and their CD4+ and CD8+ subsets, B-lymphocytes, monocytes, and neutrophils. Once the screening tube shows a nonhematopoietic tumor, a specific nonhematopoietic solid tumor (NHST) classification panel should be applied. The STOT identified all the cases of neuroblastoma, rhabdomyosarcoma, Ewing's sarcoma, and Wilms' tumor. However, though it could identify a tumoral population in other relatively uncommon pediatric cancers, the specific diagnosis required the application of additional markers. The most frequent phenotype observed among malignant cells was CD45−/CD56+, whereas benign tumors showed abnormal CD45−/CD56− populations. As expected, the tube also performed well in identifying leukemia/lymphoma

Table 10.4 Euroflow consortium designed solid tumor orientation tube (STOT).

BV421	PO	FITC	PE	PerCPCy5.5	PE-Cy7	APC	APC-H7
cyCD3 (UCHT1)	CD45 (HI30)	CD99 (3B2TA8)	nuMyogenin (F5D)	EpCAM (EBA-1)	CD56 (N901)	GD2 (14.G2a)	sCD3 (SK7)
CD27 (C40-1457)	–	CD8 (UCH-T4)	–	CD4 (SK3)	–	–	CD19 (SJ25C1)

Modified from de Sá Ferreira-Facio, C., Botafogo, V., Ferrão, P. M., Canellas, M. C., Milito, C. B., Romano, S., Lopes, D. V., Teixeira, L. C., Oliveira, E., Bruno-Riscarolli, E., Mello, F. V., Siqueira, P. F. R., Moura, P., Macedo, F. N., Forny, D. N., Simião, L., Pureza, A. L., Land, M. G. P., Pedreira, C. E., . . . da Costa, E. S. (2021). Flow cytometry immunophenotyping for diagnostic orientation and classification of pediatric cancer based on the EuroFlow Solid Tumor Orientation Tube (STOT). Cancers, 13(19), Article 4945. https://doi.org/10.3390/cancers13194945.

samples requiring further characterization. The tube performs poorly to detect HL and mature B-cell lymphoma (Ferreira-Facio et al., 2021). It is, however, important to identify normal CD45-negative populations, such as mesenchymal stromal cells and endothelial cells, and to avoid misclassifying them as tumor cells. The mesenchymal stromal cells are typically $CD45^{-}/CD34^{-}/CD81^{++}/CD90^{++}/CD271^{++}/CD10^{++}$ (Lv, Tuan, Cheung, & Leung, 2014), and endothelial cells are $CD81^{++}/CD10^{-}/CD56^{-}/CD271^{+}/CD34^{+}$ (Mund et al., 2012) The protocol for processing samples using the STOT is available from the webpage https://euroflow.org/protocols/ (Standardized EuroFlow Protocols, 2026). Using the STOT, a multicentric study from Brazil reported a 33% detection rate of bone marrow infiltration and showed 97.8% concordance with histopathology diagnosis. Incidentally, due to the higher sensitivity of FCM, some cases of neuroblastoma were only positive in FCM (Fingolo et al., 2024).

10.7 Use of imaging flow cytometers for the detection of pediatric small round cell tumors

The imaging flow cytometers combine the principles of fluorescence microscopy with multiparameter analysis and FCM, with the analysis of a large number of cells in a short period of time. This technique of IFCM is not widely available yet for diagnostic use, with only a few studies available exploring its utility in pediatric SRCTs (Merugu et al., 2020). IFCM has also been used to identify different tumor subpopulations and apoptotic cells after in vitro experiments (Ivanov et al., 2015).

10.8 Circulating tumor cell detection

CTCs and disseminated tumor cells (DTCs) can be detected in peripheral blood and bone marrow, respectively. These CTCs and DTCs may play a role in explaining intratumoral heterogeneity, as it is often difficult to obtain multiple biopsies from the tumor. Moreover, these CTCs may predict tumor evolution and aid disease monitoring. This also highlights the importance of liquid biopsy. Methods for detecting CTCs and DTCs include flow cytometric cell sorting, microfluidic devices with antibody-coated microspots (CTC-chip) for separation, ultrasound, magnetic bead enrichment, etc (Merugu et al., 2020). The interest in detecting CTCs is not new. In the early 2000s, the CellSearch system, an immunomagnetic bead-based preparation system for detecting CTCs in epithelial malignancies, was used to detect CD45−/cytokeratin+ cells within a fixed volume of peripheral blood (Allard et al., 2004; Cristofanilli et al., 2004).

Epithelial-mesenchymal transition (EMT) and its reverse, the mesenchymal-epithelial transition (MET), are extremely important for metastases of tumor cells (Agnoletto et al., 2021). EpCAM is the most frequent marker for epithelial lineage and is a transmembrane protein. Its expression is dynamic, and high levels of EpCAM expression are seen in stem cells from colorectal cancer, tumor-causing breast cancer cells, and the human embryonic stem cells (hESCs). It is believed to be a critical protein in the EMT and MET processes. EpCAM binds to E-cadherin, and in the initial phase of EMT, its expression is reduced on the cell surfaces, reducing cell adhesion of the tumor cells. It is later acquired again when the metastases are well-established (Tombolan et al., 2020). Incidentally, Ewing's sarcoma/ PNET tumors exhibit epithelial markers, and sarcomas exhibiting both epithelial and mesenchymal markers can have a "metastable phenotype" and are characterized by properties of proliferation, motility, and stemness. Ewing's sarcoma cells also transiently downregulate the expression of the key EWSR1-FLI1 transcript and contribute to the process of EMT (Sannino et al., 2017).

This CellSearch system was also used to detect CTCs in pediatric sarcomas using the phenotype CD45−/DAPI+/EpCAM+/Cytokeratin (8, 18, and 19)+. FCM was used to detect the EpCAM expression in cell lines of different pediatric sarcomas. EpCAM was expressed in all the alveolar rhabdomyosarcoma cell lines but was weak in embryonal rhabdomyosarcoma. The Ewing's sarcoma cell line showed an absence of EpCAM expression. The CellSearch system could detect CTCs at ≥1/7.5 mL peripheral blood in 63% patients. The detection rate at ≥2 CTCs/7.5 mL peripheral blood was 45%. Incidentally, in 57% patients, the CTCs also expressed desmin, a marker typically of the mesenchymal lineage. At the molecular level, patients exhibiting a high EpCAM expression at disease onset had a poorer overall survival. The approach, despite detecting sarcomas, does not rely on vimentin and N-Cadherin as they are expressed on normal peripheral blood leucocytes (Tombolan et al., 2020).

Investigators have used IFCM to detect CTCs and DTCs in peripheral blood and bone marrow from neuroblastoma patients using the GD2 antibody. After depleting red cells from the peripheral blood and bone marrow samples, the cells were permeabilized, and antibodies to GD2, neural cell adhesion molecule (NCAM), and CD45 were added with DAPI as the nuclear counterstain. After staining, the samples were acquired on the Amnis IFCM, and in the peripheral blood, a CD45 vs GD2 plot was used to identify CTCs; imaging data provided the confirmation, and IDEAS software was used for the bone marrow. The identification of CTCs was performed using brightfield morphology, the phenotype of GD2+/CD45− on nuclear cells. The addition of imaging aided confirmation of the suspected CTCs, gated on the GD2/CD45 scatter plots. Simultaneous assessment of ploidy could also be done as DAPI is a dye that stoichiometrically binds to the cell DNA. A DI of >1.25 was considered hyperdiploid (Merugu et al., 2020).

10.9 Flow cytometry to identify markers of chemoresistance and assessment for potential treatment approaches

In medulloblastoma, CD114+ cells were isolated from the patient tumors, cell lines, and patient-derived xenograft tumors using FCM. This CD114+ population in medulloblastoma expands when the patient is treated with granulocyte-colony stimulating factor (G-CSF), exhibits slower growth, and is also resistant to chemotherapy. The expression of downstream genes differs from that in the CD114− medulloblastoma cells, with higher expression of *CSF3R, NRP-1, TWIST-1*, and *MYCN*. This also impacts the treatment options for prolonged neutropenia, as the treatment with G-CSF gives a survival advantage to the CD114+ population of medulloblatomas (Paul et al., 2020). CD114 is also a marker of chemoresistance in melanomas, in which at least a subset of CD114-expressing cells exhibits resistance to etoposide, temozolomide, and nifurtimox (Zhang et al., 2015).

FCM can also be used to perform in vitro experiments to assess the chemosensitivity of malignant cells in pediatric round cell tumors. In a study on neuroblastomas, upon in vitro exposure of the cell lines, peripheral blood, and bone marrow samples containing CTCs to an MDM-2 inhibitor, the cells were labeled with fluorochrome-tagged antibodies to GD2, p53, and p21, along with DAPI, and run on an imaging flow cytometer. The neuroblastoma patients showed a higher p53 and p21 expression compared to controls in the CTCs (Merugu et al., 2020).

FCM is also amenable to assessing the expression of PD-L1, which can be targeted by checkpoint inhibitors. A study in medulloblastoma showed that, though the tumors per se show a low expression of PD-L1 as assessed by FCM. However, its expression could be increased in vitro by exposing the cells to Th1 cytokines and radiation (Martin et al., 2018).

10.10 Flow cytometry to identify cancer stem cells

FCM is highly effective for detecting and isolating cancer stem cells (CSCs) for downstream analysis. In medulloblastomas, CD133 and CD15 have been identified as markers of CSCs (Singh et al., 2003, 2016). Brain tumor stem cell population expressing CD133 was isolated as early as 2003, when it was also noted that medulloblastomas have a significantly higher proportion of cells that express CD133 than gliomas (Singh et al., 2003). CD133-enriched medulloblastomas have a poorer clinical outcome and a higher metastatic rate. They also exhibit selective STAT3 pathway activation, and c-MYC promoted tumorigenesis (Garg et al., 2017). CD15+ stem cells from medulloblastoma were also resistant to multiple chemotherapeutic agents, including cisplatin and temozolomide. These tumor-propagating CD15+ cells in medulloblastoma were found to use the PI-3 kinase signaling pathway, indicating a possible combination of a PI-3 kinase inhibitor with chemotherapy for their treatment (Singh et al., 2016). CD114, or G-CSF receptor, is also a possible marker of CSCs in neuroblastomas, as cells expressing CD114 exhibited self-renewal and further differentiation. The cell cycle, microRNA, and gene expression profiles of CD114+ populations in neuroblastoma resemble those of the pluripotent stem cells and correspond to the dedifferentiated neural crest population (Hsu et al., 2013).

Even in retinoblastomas, CSCs can be detected using FCM as FSC^{lo}/SSC^{lo} cells that express CD44, considered as a stem cell marker, while lacking CXCR4, and CD90 were negative. CD133 was uncharacteristically negative on this cell population, as for retinoblastomas, CD133 is a marker associated with cell differentiation, along with NSE. The progenitor nature can be confirmed by assessing the expression of PROX1 and Syntaxin 1A on the sorted cell population (Balla et al., 2009).

10.11 Conclusions

FCM is an excellent tool for the diagnosis and subclassification of pediatric SRCTs. This is especially true when looking for marrow involvement and to identify if fluids are showing metastases. The solid biopsies can also be subjected to FCM, but require conversion to a single-cell suspension, for which either mechanical or enzyme treatments are required to convert the solid tissue to a single-cell suspension. Standardized panels are available for the detection and classification of SRCTs, such as the STOT, and further classification tubes by the EuroFlow consortium. FCM enables the detection of ploidy and viability in the samples. There is an emerging role of IFCM to detect disease in the CNS and to identify intracellular markers if required. It also enables the detection of CTCs, CSCs, and the study of EMT. However, it is still fairly underutilized for the pediatric SRCTs.

References

Agarwal, S. (2023). Pediatric cancers: Insights and novel therapeutic approaches. *Cancers, 15*(14), 3537. doi:10.3390/cancers15143537.

Agnoletto, C., Caruso, C., & Garofalo, C. (2021). Heterogeneous circulating tumor cells in sarcoma: Implication for clinical practice. *Cancers, 13*(9), 2189. doi:10.3390/cancers13092189.

Allard, W. J., Matera, J., Miller, M. C., Repollet, M., Connelly, M. C., Rao, C., Tibbe, A. G. J., Uhr, J. W., & Terstappen, L. W. M. M. (2004). Tumor cells circulate in the peripheral blood of all major carcinomas but not in healthy subjects or patients with nonmalignant diseases. *Clinical Cancer Research, 10*(20), 6897–6904. doi:10.1158/1078-0432.CCR-04-0378.

Apps, J., Ritzmann, T. A., Liu, J. F., Shanmugavadivel, D., Halsey, C., Jones, K. P., Atun, R., Oliver, K., Vedhara, K., Ball-Gamble, A., Ranasinghe, N., Polanco, A., Adamski, J., Green, A. L., & Walker, D. A. (2024). A review calling for research directed at early detection of childhood cancers: The clinical, scientific, and economic arguments for population screening and surveillance. *EJC Paediatric Oncology, 4*. doi:10.1016/j.ejcped.2024.100191.

Balla, M. M. S., Vemuganti, G. K., Kannabiran, C., Honavar, S. G., & Murthy, R. (2009). Phenotypic characterization of retinoblastoma for the presence of putative cancer stem-like cell markers by flow cytometry. *Investigative Ophthalmology and Visual Science, 50*(4), 1506–1514. doi:10.1167/iovs.08-2356India.

Bozzi, F., Collini, P., Aiello, A., Barzanò, E., Gambirasio, F., Podda, M., Meazza, C., Ferrari, A., & Luksch, R. (2008). Flow cytometric phenotype of rhabdomyosarcoma bone marrow metastatic cells and its implication in differential diagnosis with neuroblastoma. *Anticancer Research, 28*(3 A), 1565–1569.

Brasme, J. F., Morfouace, M., Grill, J., Martinot, A., Amalberti, R., Bons-Letouzey, C., & Chalumeau, M. (2012). Delays in diagnosis of paediatric cancers: A systematic review and comparison with expert testimony in lawsuits. *The Lancet Oncology, 13*(10), e445. doi:10.1016/S1470-2045(12)70361-3.

Bryson, G. J, Lear, D., Williamson, R., & Wong, R. CW. (2002). Detection of the CD56+/CD45-immunophenotype by flow cytometry in neuroendocrine malignancies. *Journal of Clinical Pathology, 55*(7), 535–537. doi:10.1136/jcp.55.7.535.

Buzova, D., Frohlich, J., Zapletalova, D., Raffaele, M., Lo Re, O., Tsoneva, D. K., Sterba, J., Cerveny, J., & Vinciguerra, M. (2023). Detection of cell-free histones in the cerebrospinal fluid of pediatric central nervous system malignancies by imaging flow cytometry. *Frontiers in Molecular Biosciences, 10*. doi:10.3389/fmolb.2023.1254699.

Chang, A., Benda, M. P., Wood, B. L., & Kussick, S. J. (2003). Lineage-specific identification of nonhematopoietic neoplasms by flow cytometry. *American Journal of Clinical Pathology, 119*(5), 643–655. doi:10.1309/FU3F-DKYN-8AU0-891N.

Chang, Q., & Hedley, D. (2012). Emerging applications of flow cytometry in solid tumor biology. *Methods, 57*(3), 359–367. doi:10.1016/j.ymeth.2012.03.027.

Corver, W. E., ter Haar, N. T., Dreef, E. J., Miranda, N. F. C. C., Prins, F. A., Jordanova, E. S., Cornelisse, C. J., & Fleuren, G. J. (2005). High-resolution multi-parameter DNA flow cytometry enables detection of tumour and stromal cell subpopulations in paraffin-embedded tissues. *Journal of Pathology, 206*(2), 233–241. doi:10.1002/path.1765.

Cristofanilli, M., Budd, G. T., Ellis, M. J., Stopeck, A., Matera, J., Miller, M. C., Reuben, J. M., Doyle, G. V., Allard, W. J., Terstappen, L. W. M. M., & Hayes, D. F. (2004). Circulating tumor cells, disease progression, and survival in metastatic breast cancer. *New England Journal of Medicine, 351*(8), 781–791. doi:10.1056/NEJMoa040766.

Dehner, C. A., Lazar, A. J., & Chrisinger, J. S. A. (2024). Updates on WHO classification for small round cell tumors: Ewing sarcoma vs. everything else. *Human Pathology, 147*, 101–113. doi:10.1016/j.humpath.2024.01.007.

de Sá Ferreira-Facio, C., Botafogo, V., Ferrão, P. M., Canellas, M. C, Milito, C. B., Romano, S., Lopes, D. V., Teixeira, L. C., Oliveira, E., Bruno-Riscarolli, E., Mello, F. V., Siqueira, P. F. R., Moura, P., Macedo, F. N., Forny, D. N., Simião, L., Pureza, A. L., Land, M. G. P., Pedreira, C. E., … Sobral da Costa, E. (2021). Flow cytometry immunophenotyping for diagnostic orientation and classification of pediatric cancer based on the EuroFlow solid tumor orientation tube (STOT). *Cancers, 13*(19), 4945. doi:10.3390/cancers13194945.

DuBois, S. G., Epling, C. L., Teague, J., Matthay, K. K., & Sinclair, E. (2010). Flow cytometric detection of Ewing sarcoma cells in peripheral blood and bone marrow. *Pediatric Blood & Cancer, 54*(1), 13–18. doi:10.1002/pbc.22245.

Ferreira-Facio, C. S., Milito, C., Botafogo, V., Fontana, M., Thiago, L. S., Oliveira, E., da Rocha-Filho, A. S., Werneck, F., Forny, D. N., Dekermacher, S., de Azambuja, A. P., Ferman, S. E., de Faria, P. A. S., Land, M. G. P., Orfao, A., & Costa, E. S. (2013). Contribution of multiparameter flow cytometry immunophenotyping to the diagnostic screening and classification of pediatric cancer. *PLoS ONE, 8*(3). doi:10.1371/journal.pone.0055534.

Fingolo, A. R. F., Riscarolli, E. B, Abraham, C. A, Lima, P. M, Xavier, G., Oliveira, E., Siqueira, P. FR , Torres, R. C, Lisboa, V. C, Camargo, R., Pontes, R. M, Facio, C. SF., & Costa, E. S (2024). Citometria de fluxo multiparamétrica na avaliação diagnóstica e seguimento de células tumorais metastáticas em tumores sólidos pediátricos - tubo stot ("solid tumor orientation tube"). *Hematology, Transfusion and Cell Therapy, 46*, S196. doi:10.1016/j.htct.2024.09.328.

Garg, N., Bakhshinyan, D., Venugopal, C., Mahendram, S., Rosa, D. A., Vijayakumar, T., Manoranjan, B., Hallett, R., Mcfarlane, N., Delaney, K. H., Kwiecien, J. M., Arpin, C. C., Lai, P. S., Gómez-Biagi, R. F, Ali, A. M., De Araujo, E. D., Ajani, O. A., Hassell, J. A., Gunning, P. T., … Singh, S. K. (2017). CD133+ brain tumor-initiating cells are dependent on STAT3 signaling to drive medulloblastoma recurrence. *Oncogene, 36*(5), 606–617. doi:10.1038/onc.2016.235.

Hargrave, D. R., & Zacharoulis, S. (2007). Pediatric CNS tumors: Current treatment and future directions. *Expert Review of Neurotherapeutics, 7*(8), 1029–1042. doi:10.1586/14737175.7.8.1029.

Hsu, D. M., Agarwal, S., Benham, A., Coarfa, C., Trahan, D. N., Chen, Z., Stowers, P. N., Courtney, A. N., Lakoma, A., Barbieri, E., Metelitsa, L. S., Gunaratne, P., Kim, E. S., & Shohet, J. M. (2013). G-CSF receptor positive neuroblastoma subpopulations are enriched in chemotherapy-resistant or relapsed tumors and are highly tumorigenic. *Cancer Research, 73*(13), 4134–4146. doi:10.1158/0008-5472.CAN-12-4056.

Ivanov, D. P., Parker, T. L., Walker, D. A., Alexander, C., Ashford, M. B., Gellert, P. R., & Garnett, M. C. (2015). In vitro co-culture model of medulloblastoma and human neural stem cells for drug delivery assessment. *Journal of Biotechnology, 205*, 3–13. doi:10.1016/j.jbiotec.2015.01.002.

Jung, W. H., Jung, S. H., Yoo, C. J., Kim, Y. J., Park, C., & Kim, B. S. (1994). Flow cytometric analysis of DNA ploidy in childhood rhabdomyosarcoma. *Yonsei Medical Journal, 35*(1), 34–42. doi:10.3349/ymj.1994.35.1.34.

Khan, M. S, Sampor, C., Farmer, M., Ranasinghe, N., Sniderman, E., Prasad, M., Raj, T. S, Dalvi, R., Kambugu, J. B, Schoot, R. A., Moreira, D. C., Peckham-Gregory, E. C, Challinor, J., Weber, L. S, Pulungan, A., Chantada, G., Wollaert, S., Atanasova, A., Kebudi, R., … Pritchard-Jones, K. (2025). Childhood cancer early diagnosis and appropriate referral: An educational initiative of the international society of paediatric oncology (SIOP), in collaboration with international pediatric association. *Pediatric Blood & Cancer, 72*(7). doi:10.1002/pbc.31702.

Komada, Y., Zhang, X. L., Zhou, Y. W., Inaba, H., Deguchi, T., Azuma, E., & Sakurai, M. (1998). Flow cytometric analysis of peripheral blood and bone marrow for tumor cells in patients with neuroblastoma. *Cancer, 82*(3), 591–599. doi:10.1002/(SICI)1097-0142(19980201)82:3⟨591::AID-CNCR23⟩3.0.CO;2-W.

Lv, F-J., Tuan, R. S., Cheung, K. M. C., & Leung, V. Y. L. (2014). Concise Review: The surface markers and identity of human mesenchymal stem cells. *Stem Cells, 32*(6), 1408–1419. doi:10.1002/stem.1681.

Magro, G., Longo, F. R., Angelico, G., Spadola, S., Amore, F. F., & Salvatorelli, L. (2015). Immunohistochemistry as potential diagnostic pitfall in the most common solid tumors of children and adolescents. *Acta Histochemica, 117*(4-5), 397–414. doi:10.1016/j.acthis.2015.03.011.

Martin, A. M., Nirschl, C. J., Polanczyk, M. J., Bell, W. R., Nirschl, T. R., Harris-Bookman, S., Phallen, J., Hicks, J., Martinez, D., Ogurtsova, A., Xu, H., Sullivan, L. M., Meeker, A. K., Raabe, E. H., Cohen, K. J., Eberhart, C. G., Burger, P. C., Santi, M., Taube, J. M., … Lim, M. (2018). PD-L1 expression in medulloblastoma: An evaluation by subgroup. *Oncotarget, 9*(27), 19177–19191. doi:10.18632/oncotarget.24951.

Merkel, D. E., & McGuire, W. L. (1990). Ploidy, proliferative activity and prognosis. DNA flow cytometry of solid tumors. *Cancer, 65*(5), 1194–1205. doi:10.1002/1097-0142(19900301)65:5⟨1194::AID-CNCR2820650528⟩3.0.CO;2-M.

Merugu, S., Chen, L., Gavens, E., Gabra, H., Brougham, M., Makin, G., Ng, A., Murphy, D., Gabriel, A. S., Robinson, M. L., Wright, J. H., Burchill, S. A., Humphreys, A., Bown, N., Jamieson, D., & Tweddle, D. A. (2020). Detection of circulating and disseminated neuroblastoma cells using the imageStream flow cytometer for use as predictive and pharmacodynamic biomarkers. *Clinical Cancer Research, 26*(1), 122–134. doi:10.1158/1078-0432.CCR-19-0656.

Mora, J., Lavarino, C., Alaminos, M., Cheung, N.-K. V., Ríos, J. é, Torres, C., Illei, P., Juan, G., & Gerald, W. L. (2007). Comprehensive analysis of tumoral DNA content reveals clonal ploidy heterogeneity as a marker with prognostic significance in locoregional neuroblastoma. *Genes, Chromosomes and Cancer, 46*(4), 385–396. doi:10.1002/gcc.20421.

Mund, J. A., Estes, M. L., Yoder, M. C., Ingram, D. A., & Case, J. (2012). Flow cytometric identification and functional characterization of immature and mature circulating endothelial cells. *Arteriosclerosis, Thrombosis, and Vascular Biology, 32*(4), 1045–1053. doi:10.1161/ATVBAHA.111.244210.

Oberlin, O., Rey, A., Lyden, E., Bisogno, G., Stevens, M. C. G., Meyer, W. H., Carli, M., & Anderson, J. R. (2008). Prognostic factors in metastatic rhabdomyosarcomas: Results of a pooled analysis from United States and European Cooperative Groups. *Journal of Clinical Oncology, 26*(14), 2384–2389. doi:10.1200/JCO.2007.14.7207.

Ortiz, R., Vásquez, L., Giri, B., Kapambwe, S., Dille, I., Mahmoud, L., & Ilbawi, A. (2024). Developing and sustaining high-quality care for children with cancer: The WHO global initiative for childhood cancer. *Revista Panamericana De Salud Publica, 47,* e164. doi: 10.26633/RPSP.2023.164.

Parham, D. M. (2010). Modern diagnosis of small cell malignancies of children. *Surgical Pathology Clinics, 3*(3), 515–551. doi:10.1016/j.path.2010.06.002.

Paul, M. R., Huo, Y., Liu, A., Lesperance, J., Garancher, A., Wechsler-Reya, R. J., & Zage, P. E. (2020). Characterization of G-CSF receptor expression in medulloblastoma. *Neuro-Oncology Advances, 2*(1). doi:10.1093/noajnl/vdaa062.

Salvatorelli, L., Parenti, R., Broggi, G., Vecchio, G. M., Angelico, G., Puzzo, L., Di Cataldo, A., Di Benedetto, V., Alaggio, R., & Magro, G. (2021). Wt1 and cyclin d1 immunohistochemistry: A useful adjunct for diagnosis of pediatric small round blue cell tumors on small biopsies. *Diagnostics, 11*(12). doi:10.3390/diagnostics11122254.

Sannino, G., Marchetto, A., Kirchner, T., & Grünewald, T. G. P. (2017). Epithelial-to-mesenchymal and mesenchymal-to-epithelial transition in mesenchymal tumors: A paradox in sarcomas? *Cancer Research, 77*(17), 4556–4561. doi:10.1158/0008-5472.CAN-17-0032.

Singh, A. R., Joshi, S., Zulcic, M., Alcaraz, M., Garlich, J. R., Morales, G. A., Cho, Y. J., Bao, L., Levy, M. L., Newbury, R., Malicki, D., Messer, K., Crawford, J., Durden, D. L., & Castresana, J. S (2016). PI-3K inhibitors preferentially target CD15+ cancer stem cell population in SHH driven medulloblastoma. *PLOS ONE, 11*(3), e0150836. doi:10.1371/journal.pone.0150836.

Singh, S. K., Clarke, I. D., Terasaki, M., Bonn, V. E., Hawkins, C., Squire, J., & Dirks, P. B. (2003). Identification of a cancer stem cell in human brain tumors. *Cancer Research, 63*(18), 5821–5828.

Smyth, C. M., Helmer, M. A., Pozza, L. D., & Rowe, P. B. (1993). Flow cytometric DNA analyses of frozen samples from children's solid tumors. *Pathology, 25*(4), 388–393. doi:10.3109/00313029309090865.

Standardized EuroFlow Protocols [Internet]. 2026, https://euroflow.org/protocols/.

Sugimoto, T., Sawada, T., Matsumura, T. Kemshead, J. T., Horii, Y., Saida, T., Imashuku, S., & Takeuchi, Y. (1988). Diagnosis of neuroblastoma metastasis in bone marrow with a panel of monoclonal antibodies. *Medical and Pediatric Oncology, 16*(3), 190–196. doi:10.1002/mpo.2950160308.

Tombolan, L., Rossi, E., Zin, A., Santoro, L., Bonvini, P., Zamarchi, R., & Bisogno, G. (2020). Pediatric sarcomas display a variable EpCAM expression in a histology-dependent manner. *Translational Oncology, 13*(11), 100846. doi:10.1016/j.tranon.2020.100846.

Triche, T. J., Hicks, J., & Sorensen, P. H. E. (2006). Diagnostic pathology of pediatric malignancies. In P. A. Pizzo, & D. G. Poplack (Eds.), *Principles and practice of pediatric oncology* (pp. 185–235). Lippincott Williams and Wilkins.

Ward, Z. J., Yeh, J. M., Bhakta, N., Frazier, A. L., & Atun, R. (2019). Estimating the total incidence of global childhood cancer: A simulation-based analysis. *The Lancet Oncology, 20*(4), 483–493. doi: 10.1016/S1470-2045(18)30909-4

Warzynski, M. J., Graham, D. M., Axtell, R. A., Higgins, J. V., & Hammers, Y. A. (2002). Flow cytometric immunophenotyping test for staging/monitoring neuroblastoma patients. *Clinical Cytometry, 50*(6), 298–304. doi:10.1002/cyto.10159.

Zhang, L., Agarwal, S., Shohet, J. M., & Zage, P. E. (2015). CD114 expression mediates melanoma tumor cell growth and treatment resistance. *Anticancer Research, 35*(7), 3787–3792. http://ar.iiarjournals.org/content/35/7/3787.full.pdf+html.

11

Challenges in conventional flow cytometry

Ruchi Gupta

DEPARTMENT OF HEMATOLOGY, SANJAY GANDHI POST GRADUATE INSTITUTE OF MEDICAL SCIENCES, LUCKNOW, UTTAR PRADESH, INDIA

11.1 Background

11.1.1 Flow cytometry: A powerful technology

Since its invention in the late 1960s, flow cytometry (FCM) has revolutionized the field of research and diagnostics. It has proven to be a potent and versatile tool for the quantitative and qualitative analysis of cells on a single-cell level. This concept of analysis of single-cells in a fluidic stream was proposed way back in 1934 for the first time by Moldavan for cell counting in Montreal. Subsequently, in 1945, Coulter developed a mechanism that permitted counting events and measuring cell size by variation in their conductance. Coulter's cell counters evolved into cell analyzers, which were rapidly adopted by clinical laboratories for blood cell counting. This was the first foundation step wherein there was a flow of cells in a constant fluid stream, and these cells were intercepted with multiple lasers to analyze their properties (Picot et al., 2012). Eventually, in the 1968–1970s, multiple technical concepts were incorporated into a single system, and the first fluorescence-based device (ICP 11) was developed by Wolfgang Göhde, University of Münster, Germany. Kamentsky et al. (1965) founded the Bio/Physics Systems (Mahopac, NY, USA) and, in 1971, sold the *Cytograph* and *Cytofluorograph* systems, where, for the first time, Van Dilla et al. (1969) equipped the analyzers with a 633 nm He–Ne laser and 488 nm argon ion lasers. Excitation by a laser instead of a lamp allowed a better focusing of the beam on the cells, and also simultaneously allowed the measurement of green and red fluorescence emission, besides the cell scatter. Subsequently, in 1974, Becton–Dickinson (now BD Biosciences, San Jose, CA, USA) commercialized a second cell sorter version named *FACS II*, which measured FSC and 530 nm emission fluorescence. The original name of the FCM technology was pulse cytophotometry (*German: Impulszytophotometrie*), and it was only at the Conference of the American Engineering Foundation in Pensacola, Florida, in 1978, that the name was changed to FCM (Reinherz et al., 1979). Since then, there has been no going back, and the machines have undergone dramatic evolution, leading to exponential growth in the use of FCM experiments in different disciplines of science. The newer fluorochromes, fluorescent proteins, tandem dyes, polymer dyes, and availability of different lasers have made it possible to work with up to 30–50 parameters at once. Besides analyzing single-cells, surface markers, cytoplasmic proteins, and

Advancing Body Fluid Diagnostics with Flow Cytometry: A Comprehensive Guide to Effusions and CSF.
DOI: https://doi.org/10.1016/B978-0-443-27426-8.00003-X

movement of ions, the newer systems have been designed for single-cell sorting, combined FCM with microscopy, mass cytometry, and incorporating high-end analysis software (Picot et al., 2012).

Discussing the limitations of a technology that has made such a multitude of impacts in the field of research and diagnostics is difficult; this chapter focuses rather on the challenges in successfully implementing this powerful tool in routine practice and provides some troubleshooting clues. As far as the practical aspects of this technology are concerned, the three main realms include: the reagents used in different experiments, the instrumentation, and the analytical methods for large data sets. These are being discussed in the following subheadings: Preanalytical and analytical intricacies in flow cytometry, though eventually, they are interlinked with each other.

11.1.2 Preanalytical issues

The preanalytical issues can be related to the following: The type and quality of the sample, panel selection and designing, validation, and subsequently, sample processing.

11.2 Sample-related challenges

Practically any sample that can be converted into a single-cell suspension can be subjected to flow cytometric analysis. However, to obtain optimal results for interpretation, the quality of the sample, its viability, and the availability of the cell of interest are critical. In routine clinical practice, besides blood and bone marrow samples, body fluids are also frequently subjected to FCM analysis. Some of the basic sample-related problems encountered during FCM analysis include.

11.2.1 Quality of the sample

The first and foremost challenge is to ensure that a representative sample has been subjected to testing. Hemolyzed or partly clotted samples, particulate bone marrow aspirates, hemodiluted nonrepresentative samples, nonviable cells in fine needle aspirates, or paucity of the cell of interest in the sample can influence the sensitivity of flow cytometry. Peripheral blood contaminated specimens will show increased numbers of mature granulocytes, and this should be mentioned in the final report. Delay in transportation of samples like cerebrospinal fluid (CSF) or improper transport may lead to cell degeneration. Hence, inclusion of viability dyes like 7-AAD in the assay would be beneficial.

11.2.2 Need for fresh samples

While the fixed samples have a longer viability and can be subjected to immunophenotyping by immunohistochemistry, FCM generally requires fresh samples. In stored samples over a period, there is cell death/ loss of viability due to natural apoptosis of cells, thereby reducing the specificity of flow cytometric analysis and increasing nonspecific binding of antibodies, thereby complicating analysis. A variety of protocols have been described to ascertain the cell viability, like (1) cell morphology on smear prepared from the FCM sample, or (2) staining with viability dyes that selectively bind dead or dying cells.

11.2.3 Lacks morphological correlation

A hardcore pathologist relies heavily on the morphological details of a cell. While this benefit of immunophenotyping and simultaneous morphological appreciation is addressed by techniques like immunohistochemistry, most of the flow cytometers in clinical practice fail to offer this correlation with immunophenotyping. This has been addressed and overcome by newer technologies like imaging FCM (Rees et al., 2022); however, it is not in routine clinical use.

11.2.4 Loss of fragile cells/cells of interest

Loss of fragile cells/cells of interest in the sample due to physical damage or stress from the lysis solutions used during the red cell lysis. Fragile cells may degrade due to decompression when they move rapidly out of higher pressures in the flow cytometer, at least theoretically, and if in low numbers. The physical alterations of the flow cytometer can cause cells to disappear during analysis, especially when samples have low cell numbers.

11.2.5 Rare event analysis

This can be challenging if the sample collected is nonrepresentative or in low volumes.

11.3 Panel designing and antibody selection

Panel design in FCM is a critical step in the process of designing a successful experiment and requires a certain level of technical expertise and background immunology knowledge to ensure that the intensity of a marker analyzed on FCM is optimal. The purpose of the panel should be well defined, to draft an accurate panel of antibodies required for obtaining all necessary information (Herzenberg et al., 2006). The panel should address the following questions: The population of interest, the antibodies required to identify the cells, stability of the target protein, nature of the antibody (monoclonal vs. polyclonal, specificity and affinity), need for gating markers, intracellular staining, etc. A poorly designed panel and sample preparation may fail to provide optimal information, lead to overlapping emission spectra, cross laser excitation, and thereby lead to the loss of resolution of cell populations.

11.3.1 Knowledge of antigens and equipment

In-depth knowledge of the primary antigens (well characterized CD3, CD19 etc.), secondary antigens (exhibiting a continuum spectrum of expression patterns, like CD38, CD27, HLA-DR) and, the tertiary antigens (expressed in low density, CD25) is mandated to ensure the selection of

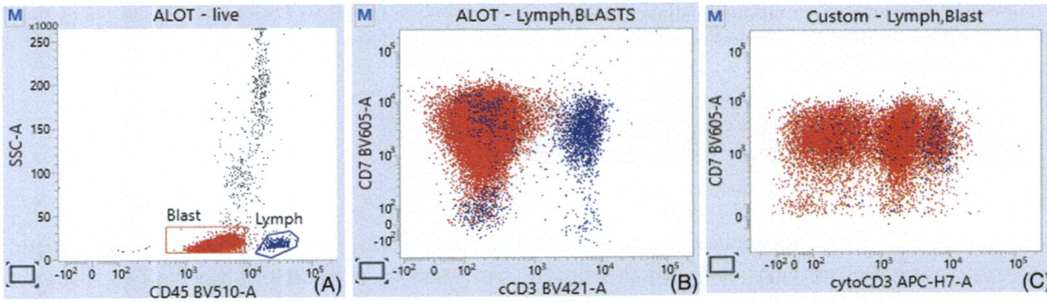

FIGURE 11.1 Flow cytometry dot plots in a case of T-ALL. (A and B) show blasts which are negative for CD45 and cytoplasmic CD3 (UCHT-1 clone); while (C) shows partial positivity for cytoplasmic CD3 using SK-7 clone.

antibody conjugates, besides the information about the instrument configuration that includes the type of lasers, laser power, filters, and detectors etc (Kalina et al., 2020).

11.3.2 Clone identification

Clone identification is a key point to ensure the robustness of the results generated by FCM (Cruz et al., 2018). A laboratory may need to keep multiple clones of a particular antigen, as some antibodies may not detect all antigens, particularly in malignant cells (Fig. 11.1). Some antibodies may show a difference in binding when obtained from different manufacturers even though they are conjugated to the same fluorochrome and prepared from the same clone. Once the antibody with the required clone is defined, an optimal fluorochrome must be selected for use with each antibody.

11.3.3 Fluorochrome selection

The pairing of antibodies with the optimal fluorochromes is one of the most critical aspects of panel design. The purpose of the appropriate selection of antibody-fluorochrome combinations is to minimize the spectral spread/spillover and maximize the resolution of the population(s) of interest. In general, weakly expressed antigens should be analyzed on a channel that does not receive significant fluorescence spillover from other channels, while a strongly expressed antigen is best tagged to fluorochrome that does not contribute to the signal on other channels (Kalina et al., 2020). If multiple antigens are present on the same cell (i.e., coexpressed antigens), they should not be on adjacent detectors, if possible. They should be spread across as many lasers as possible to minimize spillover (Fig. 11.2).

11.3.4 Tandem dyes-related issues

These dyes can complicate the process due to the spillover fluorescence of the donor fluorophore into many other fluorophore channels. Use of tandem dyes such as PE-Cy7, Per-CP-Cy5.5, or APC-Cy7 has increased the emission range from a single laser. However, they may undergo uncoupling when exposed to light or in cocktails, resulting in fluorescence emission at two

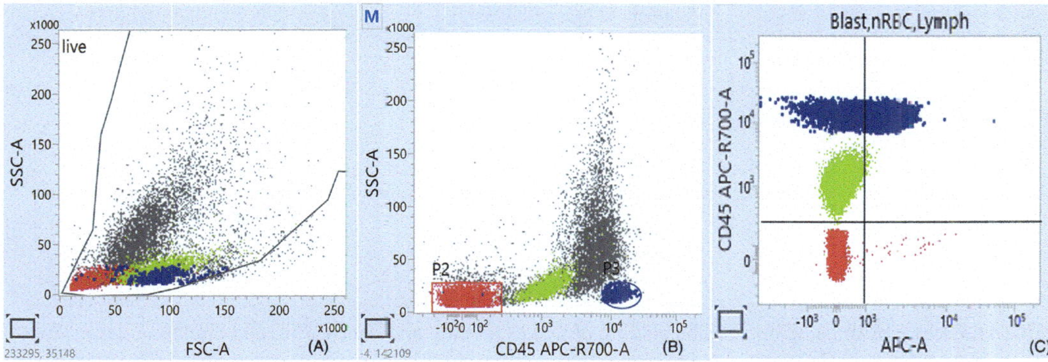

FIGURE 11.2 Dot plots depicting spillover of CD45 APCR-700 in APC. *Red*–nucleated RBC, *green*—blasts, and *blue*—lymphocytes.

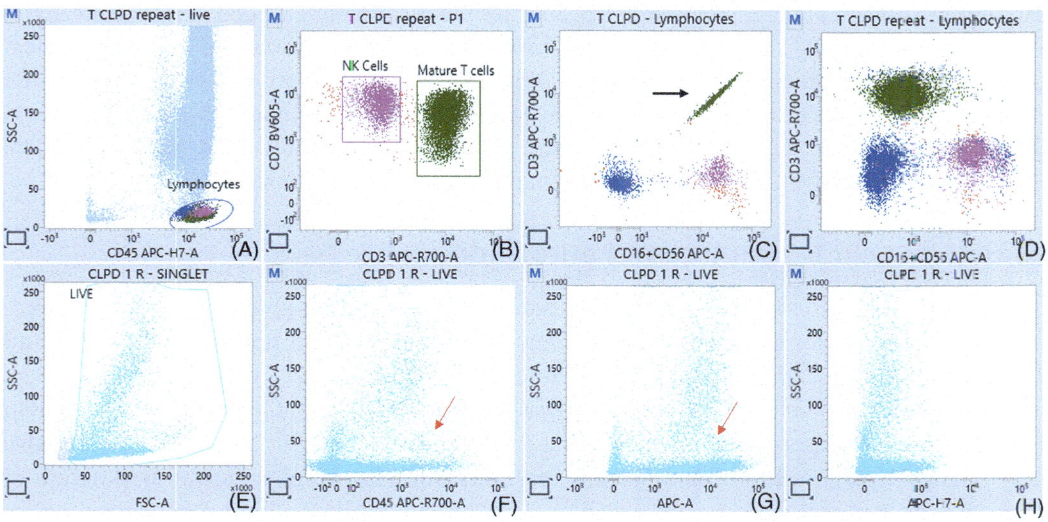

FIGURE 11.3 Flow cytometry dot plots depicting T-cell subset analysis when the staining was performed in cocktail. (A–D), lymphocytes are gated in the bright CD45 and low SSC region. Due to uncoupling of APCR700, the mature T lymphocytes are showing a skewed expression of CD3 as well as CD16/CD56 in (C), *black arrow*. However, when the staining was carried out individually, a distinct population of mature T-cell, negative for CDCD16/56, was appreciated. *SSC*, side scatter.

wavelengths (Fig. 11.3). When this occurs, the product reverts to the parent fluorochrome, that is, allophycocyanin (APC) or phycoerythrin (PE), and may lead to false data interpretation. Sometimes the brightness of tandem dyes may be reduced by fixation and permeabilization.

11.3.5 Type of antibodies

Certain antibodies, like those developed against highly O-glycosylated structures such as CD235a (glycophorin A), can behave in unpredictable ways; PE-conjugated forms sometimes

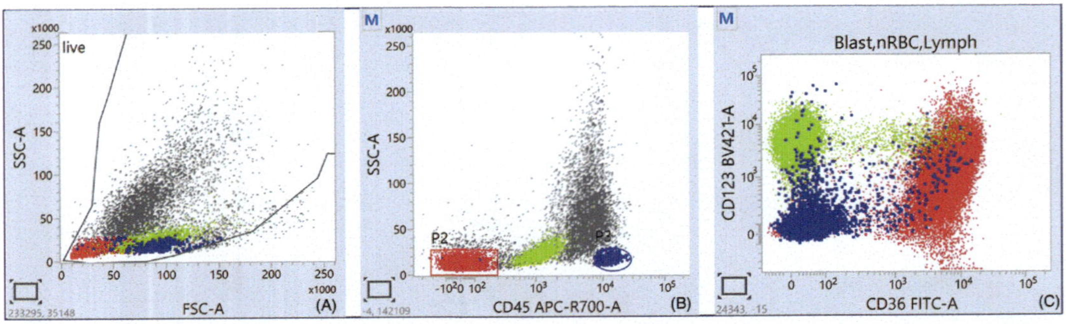

FIGURE 11.4 Dot plot showing nonspecific binding of CD123 to nRBC and debris. The red population depicts nRBC in a case of acute leukemia, where the blasts (*green*) are expressing CD123, while the partly lysed nRBCs have falsely taken up CD123.

cause significantly greater aggregation of RBCs than their negatively-charged FITC counterparts (Sutherland et al., 2018). A huge array of antibodies is available for evaluation of hematological neoplasms, and numerous panels have been validated for specific disorders like acute leukemia and myeloma; however, the list of antibodies targeting the nonhematological neoplasms is limited. The majority of the panels include epithelial cell adhesion molecule (EpCAM), CD326, CD117, CD56, and MOC-31 for the detection of malignant epithelial effusions.

11.3.6 Nonspecific staining and autofluorescence

All cells exhibit autofluorescence due to excitation of intracellular metabolic compounds; therefore, increasing detector sensitivity may enhance both specific and nonspecific background fluorescent signals. Certain cells, like the leukemic promyelocytes, have a high autofluorescence, which interferes with data interpretation unless proper unstained biological controls to help ensure optimal panel performance and overcome the challenges with autofluorescence false positives. High background and nonspecific antibody binding can obscure the population of interest and result in ambiguous or skewed data (Fig. 11.4). Even high antibody concentration can increase nonspecific binding; therefore, all antibodies should be properly titrated to maximize the signal-to-noise ratio. Nonspecific binding of antibodies can be minimized by preincubating cells with appropriate Fc blockers before staining or including 1%–5% bovine serum albumin (BSA) in phosphate-buffered saline (PBS), while autofluorescence can be checked using unstained controls.

11.3.7 Steric hindrance in cocktails

It is critical to demonstrate that equivalent staining characteristics are achieved when using the antibody in a cocktail (fully stained sample) against the same sample stained with a single antibody. The addition of multiple reagents to the same tube raises the potential for one reagent to interfere with the binding of one or more other reagents, especially when two or more antibodies are used to label antigens that are in close proximity to each other on the cell surface. This will lead to reduced binding of the sterically hindered antibody and eventually to a reduction in the associated fluorescence and lower-than-expected signal that could lead to an erroneous interpretation of FCM results (Kalina et al., 2020).

11.3.8 Antibody titration

To determine the optimal antibody concentration, staining with several dilutions of antibody must be performed. Titrations should always be performed using the same sample preparation process (whether cells are unfixed, fixed, or fixed and permeabilized), reagent systems, staining conditions, final staining volumes, and cell concentration to be used during the assay. The concentration that shows the best separation between negative versus positive cells and that exhibits a negligible signal on nontarget cells should be used (Bonilla et al., 2024).

11.4 Sample processing

Sample processing is one of the most critical steps in determining the overall outcome of a FCM experiment. The reagents and staining protocols should be stringent to ensure cell viability and preferably cell-specific staining. Certain rules or standard operating procedures are mandatory to ensure consistent results. Some of the issues to be addressed during sample processing are briefly listed later.

11.4.1 Improper staining protocol/procedure

In multicolor panels, antibodies added in insufficient quantities, false labeling, expired antibodies, and inadequate washing/permeabilization can be potential sources of error in data interpretation. Improper incubation of samples with antibodies can lead to the creation of shadow populations and consequently false interpretations (Fig. 11.5).

11.4.2 Cocktails problems

Antibody cocktails are immensely useful in the clinical flow laboratory to aid in the consistent performance of an assay, reduce the time of sample processing, minimize manual pipetting errors, and standardize the results. All the antibodies used for the cocktail need to be titrated to obtain the optimal separation of negative and positive expression using signal-to-noise ratio (S/N), stain index (SI), and median (or geometric mean) fluorescence intensity (MFI). Ready-to-use cocktails of antibodies like "*DURAClone Panels*" or "*BD Horizon Dri Multicolor*" cocktails are now commercially available, which provide customization of single or multicolor antibody conjugates and are available in liquid or room temperature-stable dry formats.

11.4.3 Rare event analysis

While carrying out bulk lysis, particularly in experiments designed for rare event analysis, discriminating the cells of interest from the background noise is vital. The limit of blank needs to be ascertained by running particle-free reagents, fluorescence minus one control (FMOs) to determine assay accuracy.

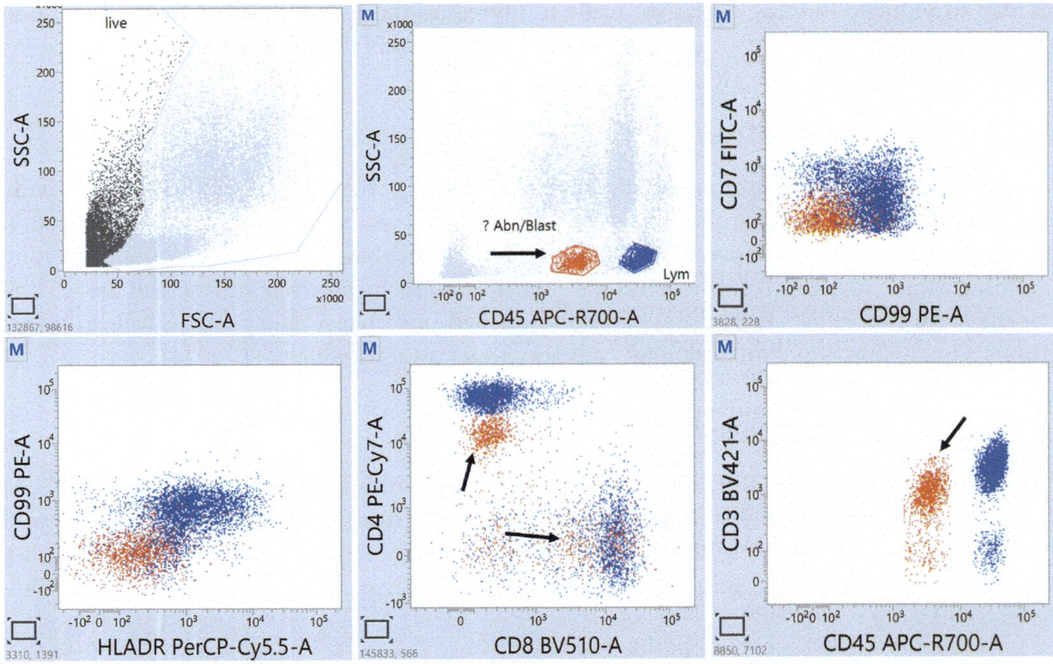

FIGURE 11.5 Dot plots of a hemorrhagic pleural fluid in a case of mixed phenotype acute leukemia. Suboptimal binding of antibodies to cells may result in the creation of a "shadow population", (arrow, *orange*). This population labeled as?Abnormal/ blasts was actually a shadow population of normal T-cells mimicking blasts but revealed dimmer expression of all antigens as compared to the true population (*blue*, mature lymphocytes).

11.4.4 Specialized experiments

Special experiments require specific processing, for example, while analyzing solid tumors, enzymes may be required to disrupt protein-protein interactions and the extracellular matrix that holds cells together. Their action depends on factors including pH, temperature, and cofactors, so care must be taken when choosing an enzyme, and conditions should be optimized to ensure that cell surface antigens of interest are not lost. Similarly, interrogation of certain cellular targets may require some form of modification; for example, while studying platelet activation and expression of annexin V, calcium supplementation is needed, or if the secreted molecules such as cytokines are to be evaluated, a protein transport inhibitor such as Monensin or Brefeldin A should be used to trap the targets inside the cells and avoid false negative results (Bedoya et al., 2013; Pasalic et al., 2017).

11.4.5 Sample fixation

While fixing samples, 1% Paraformaldehyde (PFA) is used frequently, which aids in preserving the stained cells for at least 1 week; but PFA may break down and release methanol and other breakdown products. This can lead to increased autofluorescence, altered cell morphology, potential for antibody epitope masking, and nonspecific binding due to PFA breakdown products (Vistica Sampino et al., 2022).

11.5 Analytical issues

The analytical challenges in flow cytometers are related to the software, data handling, and interpretation.

11.5.1 Software related

Various types of FCM analysis software is available, and these require a certain degree of expertise for data handling and refining it for analysis. Several advancements have been made to enhance the performance of flow cytometry.

11.5.1.1 Raw data handling

Even before considering the expertise required for data analysis, the collection of raw FCM data can be an issue. For example, when different flow cytometers are used within an organization, their outputs are typically encrypted in vendor-specific formats or nonfile-based formats; as a result, aligning data outputs from different machines and creating models to make data usable is challenging. However, this can be overcome by using other third-party software, which can automatically ingest files from any file-based flow cytometer into the cloud. The data is then processed and automatically made available for more in-depth analysis of FCM results.

11.5.1.2 Spectral overlap and compensation

FCM relies on the use of multiple fluorescent markers, which invariably have spectral overlap. This spectral overlap between fluorochromes can complicate data interpretation. Though now the flow cytometers have necessary inbuilt compensation algorithms, handling loss of data resolution due to spectral overlap when using multiple colors can still be challenging. Another issue with FCM data interpretation is that it only permits measurements above a defined threshold of signal intensity and may therefore not be able to detect cells that are expressing antigens below the detection limits of a flow cytometer. With the advent of newer techniques, many problems related to conventional FCM have been addressed. For example, spectral FCM enables the simultaneous detection of over 40 fluorophores at a time, and mass cytometry avoids technical issues of autofluorescence and spectral overlap by using isotopic labeling of antibodies and mass spectrometry-based detection (Siddiqui & Liválk, 2023).

11.5.1.3 Analysis software

With a growing number of available fluorescent reagents and lasers, output datasets have increased exponentially. By 2020, up to about 40 dimensions can be analyzed, which is likely to increase beyond 100 dimensions when spectral FCM becomes more rampant in clinical practice (Park et al., 2020). As a result, the number of pairwise plots that will need assessment will become too cumbersome to do so manually. As cytometers capture more complete spectral data, manual processes will become unmanageable.

11.5.2 Data interpretation

A big hurdle in using FCM is turning the plentiful data coming from the tests into actionable insights. A few notable challenges stand in the way.

11.5.2.1 Analysis "gates"

Gating populations of interest is central to FCM analysis. While hematopoietic elements occupy well-established regions, epithelial cells, being variable in size and cytoplasmic complexity, may be present in any region of the forward versus side scatter dot plot. Hence, all flow cytometric events should be included in the initial analysis, including the debris region (Gaur et al., 2023; Pillai & Dorfman, 2016). This may be particularly important while analyzing effusion samples for epithelial malignancies or blasts being completely negative for CD45 expression. Other factors that may make FCM data interpretation challenging are hypocellular samples, with paucity of malignant cells, antigenic switches/aberrant expression of antigens while analyzing MRD samples, or lack of EpCAM expression in malignant cells (Ramalingam et al., 2025) in effusion samples. Epithelial-mesenchymal transitions (EMT) are another important issue that one should be cautious about while analyzing effusion samples for epithelial malignancies. EMT is indicative of the migratory and invasive behavior of a tumor. During the EMT, there is a modification in the expression of adhesion molecules of a cell, and thus, the expression of EpCAM may be downgraded.

11.5.2.2 Data presentation

Traditional 2D dot plots displaying fluorescence intensities associated with markers of interest may be insufficient in high-dimensional cases. The 2D FCM dot plots are insufficient for the multidimensional data. Visualization of the data processed through algorithmic analyses will gradually become an essential aspect of analysis workflows. These algorithms are aimed at processing data containing multiple parameters and reducing it to two dimensions that can be easily interpreted. These software algorithms are categorized into supervised and unsupervised learning for identifying cell populations and classifying entire cytometry samples (Cheung et al., 2021). Principal component analysis (PCA) is a well-established dimensionality reduction method; however, newer algorithms such as t-stochastic neighborhood embedding (t-SNE) are being developed to improve and generate a visual map of the data where similar points are clustered together. The FCM critical assessment of population (FlowCAP) identification methods consortium initiated the effort to objectively evaluate the new computational methods, and some of the highly ranked software included: viSNE, t-SNE, SPADE1, PhenoGraph, FlowSOM, and Citrus (Cheung et al., 2021). A common attribute of these software packages is the availability of a graphical user interface (GUI) that increases ease of use and appearance.

11.6 Conclusion

Although FCM is a revolutionary tool in research and diagnostics, successful operation of FCM requires a significant level of technical expertise; the challenges must be acknowledged and addressed. There is an unsaid need to understand these constraints and design experiments in a manner that ensures the reliability of the findings. Lack of expertise in any field, starting from experiment design to data analysis, can result in poor-quality data and erroneous conclusions. However, with the availability of more robust techniques of sample preparation, evolving cytometry hardware technologies, and data analysis software, many limitations of conventional FCM are gradually being addressed; thereby simultaneously creating the need to make oneself familiar with the advanced FCM pipelines and their applications in routine diagnostics.

References

Bedoya, S. K., Wilson, T. D., Collins, E. L., Lau, K., & Larkin, J. (2013). Isolation and Th17 differentiation of Naïve CD4 T lymphocytes. *Journal of Visualized Experiments*, (79). doi:10.3791/50765.

Bonilla, D. L., Paul, A., Gil-Pulido, J., Park, L. M., & Jaimes, M. C. (2024). The power of reagent titration in flow cytometry. *Cells, 13*(20). doi:10.3390/cells13201677.

Cheung, M., Campbell, J. J., Whitby, L., Thomas, R. J., Braybrook, J., & Petzing, J. (2021). Current trends in flow cytometry automated data analysis software. *Cytometry Part A, 99*(10), 1007–1021. doi:10.1002/cyto.a.24320.

Cruz, N. M., Sugita, M., Ewing-Crystal, N., Lam, L., Galetto, R., Gouble, A., Smith, J., Hassane, D. C., Roboz, G. J., & Guzman, M. L. (2018). Selection and characterization of antibody clones are critical for accurate flow cytometry-based monitoring of CD123 in acute myeloid leukemia. *Leukemia and Lymphoma, 59*(4), 978–982. doi:10.1080/10428194.2017.1361023.

Gaur, G., Awasthi, N. P., Gupta, A., Agarwal, A., Sachan, R., Malhotra, K. P., Shukla, S., Singh, A. K., Singh, P., & Husain, N. (2023). Diagnostic accuracy of flow cytometry in detecting malignant epithelial cells in serous effusions. *Journal of the American Society of Cytopathology, 12*(6), 423–435. doi:10.1016/j.jasc.2023.09.003.

Herzenberg, L. A., Tung, J., Moore, W. A., Herzenberg, L. A., & Parks, D. R. (2006). Interpreting flow cytometry data: A guide for the perplexed. *Nature Immunology, 7*(7), 681–685. doi:10.1038/ni0706-681

Kalina, T., Lundsten, K., & Engel, P. (2020). Relevance of antibody validation for flow cytometry. *Cytometry Part A, 97*(2), 126–136. doi:10.1002/cyto.a.23895.

Kamentsky, L. A., Melamed, M. R., & Derman, H. (1965). Spectrophotometer: New instrument for ultrarapid cell analysis. *Science, 150*(3696), 630–631. doi: 10.1126/science.150.3696.630.

Park, L. M., Lannigan, J., & Jaimes, M. C. (2020). OMIP-069: Forty-color full spectrum flow cytometry panel for deep immunophenotyping of major cell subsets in human peripheral blood. *Cytometry Part A, 97*(10), 1044–1051. doi:10.1002/cyto.a.24213.

Pasalic, L., Pennings, G. J., Connor, D., Campbell, H., Kritharides, L., & Chen, V. M. (2017). Flow cytometry protocols for assessment of platelet function in whole blood. *Methods in Molecular Biology, 1646*, 369–389. doi:10.1007/978-1-4939-7196-1_28.

Picot, J., Guerin, C. L., Le Van Kim, C., & Boulanger, C. M. (2012). Flow cytometry: Retrospective, fundamentals and recent instrumentation. *Cytotechnology, 64*(2), 109–130. doi:10.1007/s10616-011-9415-0.

Pillai, V., & Dorfman, D. M. (2016). Flow cytometry of nonhematopoietic neoplasms. *Acta Cytologica, 60*(4) 336–343. doi:10.1159/000448371.

Ramalingam, T. R., Mani, S., Narla, S., Lakshmanan, A., Muthu, A., Nk, H. R., Mohanraj, S., Ramachandran, E. D., Subramanyam, A., & Vaidhyanathan, L. (2025). Simple and rapid flow cytometry assay for the detection of malignant epithelial cells in body fluids. *Cytopathology, 36*(1), 23–30. doi:10.1111/cyt.13408.

Rees, P., Summers, H. D., Filby, A., Carpenter, A. E., & Doan, M. (2022). Imaging flow cytometry. *Nature Reviews Methods Primers, 2*(1). doi:10.1038/s43586-022-00167-x.

Reinherz, E. L, Kung, P. C, Goldstein, G., & Schlossman, S. F (1979). Separation of functional subsets of human T cells by a monoclonal antibody. *Proceedings of the National Academy of Sciences, 76*(8), 4061–4065. doi:10.1073/pnas.76.8.4061.

Siddiqui, S., & Liválk, F. (2023). Principles of advanced flow cytometry: A practical guide. *Methods in Molecular Biology, 2580*, 89–114. doi:10.1007/978-1-0716-2740-2_5.

Sutherland, D. R., Illingworth, A., Marinov, I., Ortiz, F., Andreasen, J., Payne, D., Wallace, P. K., & Keeney, M. (2018). ICCS/ESCCA consensus guidelines to detect GPI-deficient cells in paroxysmal nocturnal hemoglobinuria (PNH) and related disorders part 2 – reagent selection and assay optimization for high-sensitivity testing. *Cytometry Part B - Clinical Cytometry, 94*(1), 23–48. doi:10.1002/cyto.b.21610.

Van Dilla, M. A., Trujillo, T. T., Mullaney, P. F., & Coulter, J. R. (1969). Cell microfluorometry: A method for rapid fluorescence measurement. *Science, 163*(3872), 1213–1214. doi:10.1126/science.163.3872.1213.

Vistica Sampino, E., Morgan, J., Chorzalska, A., Nguyen, L., Yu, C., Rodriguez, A., Pardo, M., Bonal, D., Liang, O., Kim, M., De Vito, R., Lulla, R. R., & Dubielecka, P. M. (2022). Comparative flow cytometry-based immunophenotyping analysis of peripheral blood leukocytes before and after fixation with paraformaldehyde. *Journal of Immunological Methods, 511*. doi:10.1016/j.jim.2022.113379.

12

Beyond flow cytometry: Emerging cellular assays

Firoz Ahmad[a] and Sowmya Sharma[b]

[a]MOLECULAR GENOMICS, UNIPATH SPECIALITY LABORATORIES LIMITED, AHMEDABAD, GUJARAT, INDIA [b]CLIN LABS BELLA VISTA, ANATOMIC PATHOLOGY, BELLAVISTA, NSW, AUSTRALIA

12.1 Precision theranostics in patients with malignant effusions

12.1.1 Background

Metastatic spread is responsible for the vast majority of cancer fatalities; it represents the most critical challenge in oncology (He et al., 2026). Given that cancer survival outcomes exhibit greater international disparity, the etiological factors driving these variations remain largely unknown and may be attributable to differing approaches to diagnosis and treatment (McPhail et al., 2022).

The dissemination of malignant tumors to serous cavities (pleural, peritoneal, and pericardial) frequently results in the pathological accumulation of fluid, a condition termed malignant serous effusion (MSE). This process is commonly observed in cancers such as those of the lung, breast, ovary, and stomach. The pathogenesis of MSE is multifactorial, involving direct tumor infiltration of serous membranes, increased vascular permeability, obstruction of lymphatic drainage, and the secretion of inflammatory factors by tumor cells (He et al., 2026). The clinical manifestations of MSE are directly related to the affected cavity and can include dyspnea, chest pain, and cough (pleural effusion); abdominal distension, pain, and anorexia (ascites); or cardiac compression leading to dyspnea and hemodynamic instability (pericardial effusion) (He et al., 2026). The presence of an effusion is a significant clinical event, as it is associated with a poor prognosis and a marked decline in patient quality of life (QoL). A critical step in management is pathological confirmation of the effusion to definitively distinguish a malignant etiology from nonmalignant causes (He et al., 2026). The link between pleural metastasis and poor prognosis is strongly associated with the development of effusion, which occurs in approximately half of patients. This negative outcome is particularly tied to "wet" rather than "dry" disease. While the majority of these effusions are exudative, the recent increase in identifying eosinophilic subtypes highlights a shift toward more precise classification of malignant involvement. However, the data—showing that 42% to 77% of effusions are exudative—also underscores a key diagnostic

Advancing Body Fluid Diagnostics with Flow Cytometry: A Comprehensive Guide to Effusions and CSF.
DOI: https://doi.org/10.1016/B978-0-443-27426-8.00002-8

challenge: even in known cancer patients, a significant proportion of exudative effusions may stem from causes other than the malignancy itself (Sharma et al., 2025).

Conventional therapeutic strategies, including fluid drainage and systemic chemotherapy, often fail to provide durable control of MSE. This therapeutic resistance is attributed to the inability of these modalities to effectively target the underlying immunosuppressive tumor microenvironment and the inherent heterogeneity of the tumor cells, which together drive the recurrence of the effusion. Consequently, the development of novel targeted immunotherapies designed to address these specific biological barriers represents an urgent and critical area of research (He et al., 2026).

All cancers are fundamentally genomic diseases, characterized by a cancer cell genome harboring tens of thousands of passenger mutations, a smaller set of 5–10 driver alterations, and 1–2 actionable targets. Somatic mutations therefore have diagnostic, prognostic, and predictive value (Khranovska et al., 2024). We are aware that the application of next-generation sequencing (NGS) to cancer cells is paramount for guiding personalized therapeutic strategies. While tissue biopsy remains the established gold standard for obtaining tumor material for NGS, the procurement of sufficient tissue for successful analysis frequently presents a significant clinical challenge. In such cases, malignant effusions represent an alternative and valuable source of tumor cells for genomic profiling (Gupta et al., 2025).

Lately, malignant pleural effusion (MPE), which is primarily characterized by the presence of malignant cells in pleural fluid on histological or cytological examination during thoracotomy, thoracoscopy, or autopsy, is being used for the detection of mutations (Awadallah et al., 2019; Guo et al., 2019). In 2005, MPE was first analyzed to detect epidermal growth factor receptor (EGFR) mutation as a liquid biopsy specimens. Over time, with the advent of PCR or NGS, it has become more feasible and can detect exon 19 deletion, L858R, and T790M. MPE supernatant is an alternative liquid biopsy specimen for comprehensive mutational profiling (Guo et al., 2019). Analysis of MPE supernatant from 154 lung adenocarcinoma patients demonstrated that it harbors significantly higher tumor DNA fractions than matched plasma or cell pellets, is comparable to tissue samples, and yields superior detection of copy number variations (Guo et al., 2019).

12.2 Which next-generation sequencing tests would be useful?

Several NGS tests are valuable for analyzing malignant effusions, primarily to guide treatment decisions by identifying actionable genetic alterations. Research has firmly established that tumor cells and DNA from malignant effusions are suitable for and highly concordant with traditional tissue biopsies for molecular profiling. The choice of test depends on the clinical question, whether it is identifying targets for therapy, diagnosing the cancer type, or detecting infections (Gupta et al., 2025).

12.3 Actionable genes in malignant effusions

The following table encapsulates the data from a multicenter retrospective study that enrolled 95 patients with advanced lung cancer who experienced disease progression after at least one line

Table 12.1 A clinically referenced table with a complete summary of patient characteristics, genomic alterations, resistance mechanisms, and survival outcomes in lung cancer patients undergoing pleural effusion cfDNA NGS (n = 95).

Category	Parameter	Finding
Demographics and cohort characteristics		
–	Total patients (N)	95
–	Age and median (range)	66 years (36–88)
–	Gender	Male: 38 (40%); female: 57 (60%)
–	Smoking status	Never smoker: 70 (73.7%); former/current: 25 (26.3%)
–	Histology	Adenocarcinoma: 92 (96.8%); squamous: 3 (3.2%)
–	Prior EGFR TKI exposure	First-/second-generation: 27/95 (28.4%); first-line osimertinib: 14/95 (14.7%); second-line osimertinib:15/95 (15.8%)
–	PE cytology (%)	>5%:13 (13.6%); <5%:34 (35.8%); negative:7 (7.4); not available:41 (43.2)
Baseline routine molecular detection		
–	EGFR mutation	EGFR exon 19 deletion (n = 36, 37.9%), EGFR L858R mutation (n = 22, 23.2%), uncommon or compound EGFR mutations (n = 4, 4.2%), EGFR exon 20 insertion (n = 2, 2.1%),
–	ALK rearrangements	ALK fusion (n = 3, 3.2%),
–	ROS1 rearrangements	ROS1 fusion (n = 3, 3.2%)
–	No mutation	25 (26.3%)
–	Overall detection rate of driver mutation	73.7% (70/95)
cfDNA NGS enhanced detection		
–	cfDNA NGS driver mutation detection rate	85.3 patients (81%)
–	Net increase in detection	+11 patients (+11.6%)

cfDNA, cell-free DNA; *NGS*, next-generation sequencing; *EGFR*, epidermal growth factor receptor; *TKI*, tyrosine kinase inhibitor; *Ex19del*, exon 19 deletion; *L858R*, leucine-to-arginine substitution at codon 858; T790M.
From Chang, S.-C., Huang, C.-Y., Lai, Y.-C., Hsieh, M.-S., Ho, C.-C., Yang, C.-Y., Shih, J.-Y., Wu, S.-G., Hu, P.-W., Hsu, C.-L., Chen, C.-Y., & Liao, W.-Y. (2026). Clinical utility of pleural effusion supernatant cell-free DNA genotyping in previously treated patients with advanced non-small-cell lung cancer and disease progression: a multicenter retrospective study. ESMO Open, 11( ). https://doi.org/10.1016/j.esmoop.2025.106023.

of systemic therapy. Comprehensive NGS of cell-free DNA (cfDNA) derived from pleural effusion samples was performed to characterize the genomic landscape of treatment resistance. The complete dataset, encompassing patient demographics, baseline molecular profiles, resistance mechanisms to EGFR tyrosine kinase inhibitors (TKIs), and prognostic implications of identified alterations, is presented in Table 12.1.

Out of the 95 total patients, 27 were treated with first- or second-generation EGFR TKIs. Of these, 40.7% (11 patients) developed the T790M resistance mutation, detected using cfDNA from pleural effusion supernatant. In these 11 patients, the median progression-free survival (PFS) following osimertinib treatment was not reached after a median follow-up of 6 months. Other common genetic alterations after progression on first-/second-generation EGFR TKIs included

PIK3CA-activating missense mutations and *MET* copy number gain (CNG). Acquired EGFR C797S, *MET* and *ERBB2* amplifications, and small-cell transformation were also identified in these patients. Thus, it is evident that newly identified alterations were detected in 11 patients who were wild-type by routine molecular assay on tissue biopsies at baseline, primarily due to the methodological limitations of these assays relative to the comprehensive testing of pleural effusion samples at follow-up. Additionally, this method identifies potential resistance mechanisms, underscoring the value of posttreatment genomic profiling as a convenient and reliable clinical approach. These findings support the feasibility of pleural effusion cfDNA NGS to complement routine testing at progression. These findings warrant validation in larger, prospective studies with prespecified survival endpoints (Chang et al., 2026). Nonetheless, MPE supernatant is positioned as an optimal alternative for comprehensive mutation profiling. Although circulating tumor DNA analysis from blood enables less invasive and expedited genotyping, malignant effusions yield additional genomic information, such as emerging alterations and PDL1 status (PD-L1 remains the sole NCCN-recommended biomarker for guiding immunotherapy treatment strategies in metastatic NSCLC (Qi et al., 2024). The complementary use of both methods enhances diagnostic precision for tailored therapy, underscoring the applicability of precision oncology in community-based NGS application, which directly addresses the critical diagnostic gap when conventional tissue biopsy is unattainable (Guo et al., 2019; Gupta et al., 2025).

As per current guidelines, patients with metastatic or advanced cancer should undergo multi-gene panel-based genomic sequencing in a certified laboratory when results guide treatment decisions, including for site-agnostic biomarkers. In this context, malignant effusion represents a valuable alternative specimen for comprehensive DNA/RNA NGS. Its direct proximity to the tumor provides a richer source of tumor-derived DNA than plasma, and cfDNA analysis from effusion supernatant effectively overcomes the challenge of low cellularity that often limits traditional cytology specimens. This approach enables both initial genotyping and monitoring of acquired resistance to guide subsequent therapy (Chakravarty et al., 2022; Gibbs et al., 2023; Gibbs et al., 2025).

12.4 Comparison of sample preparation methods for next-generation sequencing testing

Cell blocks cannot be assessed for adequacy at collection, and even with dedicated needle rinses, sufficient material for testing is not guaranteed. Liquid-based cytology (LBC) shares these limitations—real-time assessment is not possible, and neoplastic cellularity varies. However, LBC offers practical advantages: easy collection and transport, excellent preservation, and removal of blood and debris, making it attractive for NGS. Most institutions using LBC for NGS perform preanalytical visualization of tumor cells to reduce false negatives. When tissue is limited, or cell blocks are insufficient, direct smears provide a valuable alternative. DNA from smears avoids formalin-induced damage and retains intact nuclei. However, using smears for NGS requires slide sacrifice, which may raise medicolegal concerns in some jurisdictions, and demands additional in-house validation (Chen et al., 2015). Routine NGS utilization is recommended by ESCAT and European Society for Medical Oncology (ESMO) guidelines for advanced cancers including nonsquamous NSCLC, prostate, ovarian, and cholangiocarcinoma. The technique's inherent sensitivity permits reliable detection of actionable mutations in circulating tumor DNA from plasma and alternative body fluid specimens (Khranovska et al., 2024). NGS of MPEs performs comparably to tissue biopsies, yields superior DNA quantities, and detects more alterations than blood-based liquid biopsies. Supernatant analysis may outperform cellular preparations. In patients with inaccessible tumors or paucicellular biopsies, MPE-based NGS offers a viable alternative for guiding therapy and identifying resistance. Evidence in nonlung malignancies remains limited. Beyond diagnosis, MPE processing provides critical prognostic and theranostic insights (Grigoriadou et al., 2020). Early genomic profiling relied on separate tests to detect individual alterations predicting response to a single drug in a specific cancer type. Traditional methods like Sanger sequencing, pyrosequencing, and allele-specific PCR remain in use today for DNA analysis. However, their low throughput restricts them to detecting only the most common mutations in target genes (Khranovska et al., 2024).

12.5 The "wet" laboratory

For genomic sequencing results to inform clinical management, testing must be performed in a laboratory with appropriate certification. When making clinical decisions, two things matter: first, whether the specific genetic change actually affects protein function, and second, whether there's solid evidence that targeting this change with a particular drug works. If multiple targeted therapies are approved for different biomarkers in a given cancer, a multigene panel test should be used rather than single-gene testing (Chakravarty et al., 2022).

High-quality specimens in sufficient quantity are essential for accurate NGS results. Standardization is therefore critical. Collection sites should implement standard operating procedures requiring harvested fluid specimens to be either transported quickly or refrigerated at 2–8°C (Shin et al., 2021). DNA and RNA quality from FFPE samples depends on tissue age, ischemia time, fixation conditions, storage, and extraction methods. Even with optimal handling, molecular profiling fails in 5%–30% of cases due to poor quality or insufficient material, typically requiring 20–150 ng of nucleic acid input. If an FFPE sample fails quality control, tissue re-biopsy is recommended; when not feasible, liquid biopsy NGS should be pursued (Stenzinger et al., 2024).

12.6 Collection of effusion sample and cell-free DNA extraction

Following cytological confirmation of malignancy, approximately 40 mL of pleural effusion is collected from each patient and immediately transferred into cell-free DNA BCT tubes (Streck, USA) (Chang et al., 2024). Although each laboratory follows its own protocol and kit, we are presenting a standard protocol using ascites (ASC), peritoneal effusion, and pleural effusion. ASC and PE samples are centrifuged at 1600 × g and then 16,000 × g (10 min each). Supernatants

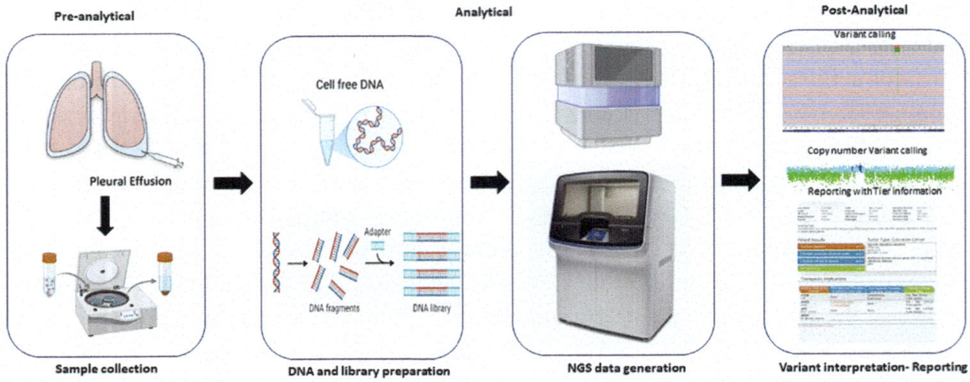

FIGURE 12.1 Application of the next-generation sequencing technology to identify the variants in the pleural effusion. Overview of the next-generation sequencing (NGS) workflow using pleural effusion samples. The process comprises three key stages: preanalytical (sample collection of pleural effusion), analytical (cell-free DNA extraction, adapter ligation, and DNA library preparation followed by NGS data generation), and postanalytical (variant calling including copy number variants and clinical reporting with tiered information).

are used for cfDNA extraction, and sediments for genomic DNA isolation. cfDNA is extracted from supernatant and plasma using the QIAamp Circulating Nucleic Acid Kit (Qiagen, Valencia, CA) or any other commercially available kits. Later, genomic DNA from tumor tissues and sediments is purified with the QIAamp DNA Mini Kit (Qiagen, Valencia, CA). White blood cell DNA, serving as germline control, is isolated using the DNeasy Blood Kit (Qiagen, Valencia, CA) (Wu et al., 2022). The extracted cfDNA is quantified and assessed for fragment size distribution using the 4200 TapeStation system (Agilent Technologies, Santa Clara, CA, USA) with the high sensitivity D1000 ScreenTape assay. Purified cfDNA is stored at −20°C or -80°C until further use (Chang et al., 2024).

12.7 Next-generation sequencing

The two main NGS platforms used in molecular oncology clinical laboratories—Thermofisher and Illumina-based NGS platform which employ multiplex PCR-based and/or hybridization capture-based methods, respectively, to generate DNA libraries. Both incorporate sequential quality control checkpoints at critical steps: preanalytical qualification, DNA quantification, library preparation, clonal amplification, sequencing, and postanalytical data analysis (Fig. 12.1) (Chen et al., 2015; Yan et al., 2025). Other NGS platforms are also commercially available and are being used as per the individual laboratory's choice.

12.8 Preanalytical considerations

Rigorous preanalytical sample processing is crucial, as inadequate specimen preparation—especially poor choice of DNA extraction material—can cause false negatives, especially when mutations are limited to a small cell fraction. Such diagnostic errors directly affect therapeutic decisions and patient management. Therefore, integrating a deep understanding of disease biology with the strengths and limits of molecular methods is essential to achieve reliable, clinically actionable results (Suver Stević et al., 2025).

Another crucial factor is the preanalytic quality of the specimen being tested. The DNA integrity number (DIN) is a scale from 1 (completely degraded) to 10 (completely intact) for assessing gDNA quality; samples with a DIN > 3 are considered suitable for NGS analysis. DNA input requirements for NGS typically range from 50 to 300 ng depending on the platform and panel size; lower input amounts can compromise variant detection sensitivity. Given that a single nucleated cell yields approximately 6 pg of DNA, roughly 2000 cells are required to obtain 10 ng of DNA (Noda et al., 2023).

For hematologic specimens, tumor content is estimated indirectly using tests like white blood cell differentials or flow cytometry. For solid tumors, a pathologist must review the sample to confirm the diagnosis, ensure there is enough viable tumor tissue, and optionally perform macrodissection to enrich for cancer cells. However, estimating tumor percentage solely by microscopic examination is often inaccurate, as normal cells can be easily overlooked and interobserver variability is high. A more reliable approach is to correlate with sequencing data—specifically, mutant allele fractions—which provide a precise measure of tumor purity and improve the accuracy of downstream analyses (Jennings et al., 2017).

In FFPE samples, formalin fixation causes DNA fragmentation and cross-linking, reducing sensitivity for variant detection. It also induces artifactual C→T changes that must be filtered out to avoid false-positive mutations (Stenzinger et al., 2024). ctDNA NGS follows the same workflow as tissue biopsy, with ESMO consensus recommendations supporting its use for genotyping advanced cancers (Stenzinger et al., 2024).

12.9 Analytical

Pathologists play critical roles in advising clinicians on when to perform NGS, selecting suitable tests, and optimizing outcomes. Whenever possible, NGS should be performed early—ideally at first diagnosis and initiation of first-line systemic therapy for advanced disease—to enable timely second-line treatment with actionable genomic alterations if first-line therapy fails. Key considerations for test selection include gene panel size, sample type, and turnaround time. With appropriate processing, both DNA and RNA can be extracted from a single biopsy for NGS. However, some panels are designed for specific tumor types and may not yield comprehensive results if used for other tumors (Stenzinger et al., 2024), while whole-genome sequencing captures the full spectrum of variants but involves higher cost, complexity, and turnaround time. Standardized analysis pipelines and benchmarking initiatives are essential to enable reliable, consistent use of both approaches in routine clinical practice (Bewicke-Copley et al., 2019).

NGS workflow is initiated with the random fragmentation of genomic DNA to create libraries that represent the target nucleic acids. Three main approaches are used for library preparation. The first clonally amplified libraries rely on PCR; adapter-ligated fragments are amplified by emulsion PCR (ePCR) or bridge PCR (bPCR), depending on the platform (e.g., ePCR for SOLiD, Roche/454, and Polonator; bPCR for Illumina). Raw sample concentration is typically kept below 20 ng/μL (e.g., ~15 ng/μL for Illumina, 5–10 ng/μL for Roche/454). The second approach, single-molecule libraries, avoids PCR entirely, reducing amplification bias and sequencing errors. It requires more starting material (<1 μg), allows longer DNA fragments (tens of kilobases), and supports real-time sequencing methods. The third approach uses circularized single-stranded DNA, which undergoes rolling circle replication to generate tandem copies; reads are then grouped into families to substantially lower error rates. Library preparation fundamentally determines data quality, and the choice of method depends on the application. For quantitative analyses such as transcriptomics or gene expression profiling, the single-molecule approach is preferable to avoid amplification bias. For qualitative analyses like mutation detection or methylation studies, amplified libraries help capture complete genomic regions, though care must be taken with AT-rich and GC-rich bias. The circle-template method is particularly suited for cancer profiling, rare-variant calling, microbial diversity studies, and immunogenetics, and its utility continues to expand (Li et al., 2016).

Next is sequencing, which involves synthesizing new DNA strands using single-stranded libraries as templates. Nucleotides are incorporated in a sequence-specific order, and each incorporation generates a signal that is digitally recorded. Different platforms detect this signal through distinct mechanisms: Illumina and SOLiD use fluorescence upon nucleotide incorporation; Ion Torrent detects pH changes resulting from hydrogen ion release during DNA synthesis; and Roche measures light produced by a pyrophosphate-driven enzymatic cascade (Khranovska et al., 2024).

12.10 Postanalytical

After confirming sequence read quality, analysis proceeds through a bioinformatics pipeline tailored to the sequencing method and clinical question (Fig. 12.2). Base-calling generates FastQ files, which are aligned to a reference genome. Variants are identified, annotated against public databases, and compiled into a variant table. For diagnostic interpretation, variants are classified for clinical significance using pathogenicity databases and prediction tools that assess impact on protein function or splicing (Fig. 12.2). Final confirmation of pathogenicity may require functional studies or cosegregation analysis in families. Similar principles apply to somatic cancer mutations (Yan et al., 2025), liquid biopsy using ascitic fluid and pleural effusion supernatants for genomic profiling in gastrointestinal and lung cancers (Kamps et al., 2017; Wu et al., 2022).

NGS has enabled more detailed characterization of numerous cancers, facilitating the discovery of new subtypes, biomarkers, and therapeutic targets—efforts that culminated in the completion of The Cancer Genome Atlas (TCGA) (Khranovska et al., 2024). Ongoing efforts—including public databases like NCBI's ClinVar, the Clinical Genome Resource (ClinGen), and the Human Variome Project—are systematically collecting and curating human genetic variants and alleles that influence health. These and similar initiatives are expected to accelerate the integration of NGS into routine medical practice (Gargis et al., 2015). Best practice standards and guidelines

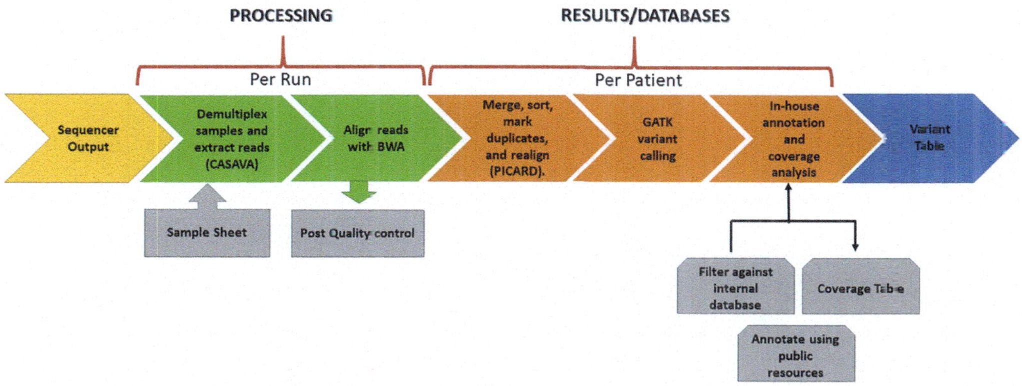

FIGURE 12.2 Bioinformatics Pipeline for variant prioritization and selection. This figure details the bioinformatics workflow for NGS data processing: per-run outputs are demultiplexed and quality-controlled, followed by per-patient alignment, variant calling, and annotation against internal and public databases to generate a final variant table.

for somatic variant curation are now available. Cancer hotspots (https://www.cancerhotspots.org) identifies recurrently mutated positions from ~25,000 tumors, with good solid tumor coverage but limited for pediatric/rare cancers. COSMIC (https://cancer.sanger.ac.uk/cosmic) provides somatic variant frequencies, though data are unadjusted for mutability and coverage varies (Horak et al., 2022). GenomOncology Pathology Workbench is an example of a one-stop-shop platform for somatic NGS testing. It includes customized QC metrics, licensed databases for reference, annotated classification of variants aligned with the Association of Molecular Pathologists/American Society of Clinical Oncology/College of American Pathologists (AMP/ACSO/CAP), and supports generation of comprehensive and customized clinical reporting (Schilter et al., 2023).

12.11 Next-generation sequencing for RNA

While circulating tumor cells (CTCs) are not feasible for routine use due to rarity and technical challenges, cfRNA has emerged as a promising adjunctive biomarker for NSCLC. It demonstrates superior sensitivity for detecting fusions and splice variants, boosting detection rates by up to 28.6% when combined with cfDNA—particularly valuable in cases with low tumor burden. However, widespread clinical adoption awaits standardized protocols and improved extraction techniques to overcome current issues of degradation and reliability (Wang & Shih, 2025).

RNA sequencing techniques are categorized by library preparation method into short-read sequencing, long-read cDNA sequencing, and long-read direct RNA sequencing. Short-read and long-read cDNA approaches share similar core steps—RNA extraction, fragmentation, reverse transcription to cDNA, adapter ligation, and amplification—but differ in sample input requirements and computational analysis. Short-read cDNA sequencing generates short read archive

Table 12.2 Major molecular approaches applied to effusions, their clinical utility, illustrative findings, and practical considerations.

Category	Key targets/approach	Clinical utility	Specific findings	Key considerations and challenges
Liquid biopsy components				
Circulating tumor DNA (ctDNA)	NGS or PCR on ctDNA from supernatant	High-throughput mutation profiling with superior sensitivity and specificity; captures tumor heterogeneity; useful when tissue biopsy is insufficient	Mutant allele fraction (MAF)was significantly higher in pleural effusion supernatant than in plasma ctDNA: Pleural fluid MAF: 28.5% versus plasma MAF: 1.3% ($p < 0.0001$) Wu et al. (2022) Pleural fluid had 5- to 46-fold higher driver mutation abundance than plasma (*KRAS*, *EGFR* L858R, and *EGFR* T790M) compared to plasma de Kock et al. (2021) median MAF in pleural fluid: 15.8% (sensitizing *EGFR*) versus plasma 0.8%; for T790M: 8.69% versus 0.16% ($p < 0.05$) Pérez-Barrios et al. (2021)	Superior to cell pellet for NGS in some studies; preanalytical processing critical de Kock et al. (2021); Pérez-Barrios et al. (2021); Wu et al. (2022)

Effusion-derived tumor cells (including CTCs and CTMs)	Enumeration and molecular characterization of tumor cells from effusion fluid	Provides phenotypic and molecular data; captures tumor diversity	Prognostic value established for blood CTCs; explored in effusions for disease monitoring	Challenges include heterogeneity, rarity, and assay variability Bisht et al. (2025)
Extracellular vesicles (EVs)	Isolation of nanoscale membranous particles (including exosomes, microvesicles) carrying proteins, lipids, DNA, and RNA	Noninvasive genomic profiling; complements or replaces tissue biopsy when cell blocks are scant; can capture tumor heterogeneity	EV-DNA from pleural effusion detects actionable mutations (e.g., EGFR and KRAS) with higher variant allele frequency than plasma cfDNA in some studies	Not tumor-specific; need for standardized isolation; safety and scalability of cellular sources; deeper understanding of EV structure and turnover required for therapeutic use Carreca et al. (2024)
Exosomes	Isolation of nano-sized exosomes (30–150 nm) carrying RNAs and proteins; analysis using standardized methods, high-throughput omics, and biosensors	Early diagnosis, prognosis, personalized therapy, and treatment monitoring in lung cancer; complements existing screening	Exosomal components (e.g., RNAs, surface proteins) are detectable in serum and plasma as biomarkers; same principles apply to effusions	Heterogeneity; coisolation from nontumor cells obscures tumor signals; variable isolation methods cause inconsistent results; adherence to MISEV2023 guidelines; validation in large cohorts needed Liu et al. (2026)

Table 12.3 Advanced molecular profiling.

Technique	Description	Main applications	Advantages	Limitations
Single-cell RNA-seq (scRNA-seq)	Transcriptomic analysis at individual cell resolution	Reveals cellular heterogeneity within effusion; identifies rare cell populations	Characterizes tumor microenvironment composition; identifies cell-type-specific drug resistance markers	High cost; technically demanding; requires fresh samples; complex data analysis Whitfield et al. (2023)
Multiomics approaches	Integration of genomic, transcriptomic, epigenomic, proteomic, and metabolomic data to analyze the complex molecular landscape underlying cancer development, progression, and treatment response	Identifying key cancer-driving changes; revealing tumor diversity; finding biomarkers for early detection, prognosis, and treatment response; discovering new therapeutic targets; developing innovative treatments; supporting clinical decisions	Provides deep insight into how different biological components interact in cancer cells; supports personalized treatment; reveals new pathways and targets; advances precision medicine	Challenges include integrating, standardizing, and interpreting large datasets; the need for better analytical tools; and ethical concerns about patient data privacy Chakraborty et al. (2024)

Other techniques

Technique	Description	Main applications	Advantages	Limitations
Microfluidic technology/lab-on-a-chip	Lab-on-a-chip technologies by manipulating fluid on a micron scale (1–1000 μm) for isolation and analysis	Diagnostics; molecular diagnostic devices; point-of-care tools; sample handling; sample preparation; sensing; total analysis systems	High capacity; time-saving; low reagent consumption; precision; can integrate complicated procedures on a compact chip	Most lab-on-a-chip tools remain at TRL 3–4 with low commercial uptake. Assistive devices are bulky and hard to miniaturize because they were designed for other uses. There is a lack of standards and poor coordination between academia and industry, making the field more "chip-in-a-lab" than true lab-on-a-chip. Standardized modular building blocks for mass production are being developed to bridge the gap Bargahi et al. (2022)
Spatial transcriptomics	Technologies that enable the measurement and mapping of gene expression within the spatial context of tissue architecture	Cancer research; studying tumor heterogeneity; characterizing the tumor microenvironment; investigating tumor-immune interactions	Preserves spatial context; reveals interactions between cancer cells and surrounding stroma/immune cells; can be applied to clinical tissue sections such as biopsies	Emerging technology; challenges in data integration, resolution, and throughput; limited standardization; complex data analysis Yu et al. (2022)

(SRA) data comprising the majority of mRNA sequences, whereas long-read cDNA sequencing enables characterization of full-length transcript isoforms (Ergin et al., 2022).

Raw sequencing data (FastQ files) undergo quality control using tools like Trimmomatic to remove low-quality reads and adapters. Clean reads are then aligned to a reference genome with aligners such as STAR or BWA, followed by transcript assembly using StringTie or Trinity. Finally, expression is quantified at gene/transcript/exon levels (FeatureCounts, RSEM, and DEXSeq), with differential expression analyzed using DESeq2 or edgeR (Hong et al., 2020).

12.12 Advanced molecular approaches in effusions

12.12.1 Cell free and particle-based analytes in effusions

Lately, liquid biopsy has emerged as a promising alternative to invasive diagnostic methods. Key liquid biopsy candidates—including CTCs, circulating cell-free tumor DNA (cfDNA), tumor-educated platelets (TEP), and extracellular vesicles (EVs)—carry cancer-derived biomolecules. They enable the identification of driver mutations, detection of acquired resistance, monitoring of refractory disease, and assessment of prognosis. Effusions (pleural, peritoneal, and pericardial) represent an especially rich source of these analytes, often yielding higher concentrations of tumor-derived material than blood (Chang et al., 2024). The following Table 12.2 summarizes the major molecular approaches applied to effusions, their clinical utility, illustrative findings, and practical considerations.

12.12.2 Emerging technologies for deep molecular characterization of effusions

While new technologies have made routine detection of solid tumors possible, more advanced approaches are still needed to detect cancers before symptoms arise, monitor disease progression, and assess treatment response. Such advancements would improve prognosis, enhance QoL, reduce healthcare costs, and increase recovery rates for cancer patients (Bargahi et al., 2022).

The following Table 12.3 summarizes the main applications of advanced molecular profiling, its advantages and limitations.

12.13 Conclusion

In clinical scenarios where tissue biopsy is not feasible—because it is unsafe, contraindicated, or declined by the patient; because the available tumor tissue is insufficient in quantity or quality for accurate molecular diagnosis; or because significant delay is expected in obtaining tissue—malignant effusion offers a valuable alternative. It has emerged as a key liquid biopsy substrate, providing direct genomic insight into advanced malignancies. Its proximity to the tumor yields a richer source of tumor-derived DNA than plasma, effectively overcoming the low cellularity that often limits traditional tissue biopsies. By enabling comprehensive genomic profiling through sensitive cfDNA analysis, this approach supports both driver mutation detection at diagnosis and real-time monitoring of treatment response and acquired resistance. As NGS panels continue to advance, integrating malignant effusion testing into routine practice reduces the need for repeat invasive procedures and ensures that therapeutic decisions are guided by accurate and current tumor profiles, ultimately supporting more personalized treatment strategies.

References

Awadallah, S. F, Bowling, M. R., Sharma, N., & Mohan, A. (2019). Malignant pleural effusion and cancer of unknown primary site: A review of literature. *Annals of Translational Medicine, 7*(15), 353. doi:10.21037/atm.2019.06.33.

Bargahi, N., Ghasemali, S., Jahandar-Lashaki, S., & Nazari, A. (2022). Recent advances for cancer detection and treatment by microfluidic technology, review and update. *Biological Procedures Online, 24*(1). doi:10.1186/s12575-022-00166-y.

Bewicke-Copley, F., Kumar, E. A, Palladino, G., Korfi, K., & Wang, J. (2019). Applications and analysis of targeted genomic sequencing in cancer studies. *Computational and Structural Biotechnology Journal, 17*, 1348–1359. doi:10.1016/j.csbj.2019.10.004.

Bisht, D., Kumar, V., Gupta, S., & Sachan, M. (2025). Navigating circulating tumour cells and tumour microemboli: Where we stand in bridging their molecular dynamics with clinical translation. *Biochimica et Biophysica Acta (BBA) - Reviews on Cancer, 1880*(4), 189384. doi:10.1016/j.bbcan.2025.189384.

Carreca, A. P, Tinnirello, R., Miceli, V., Galvano, A., Gristina, V., Incorvaia, L., Pampalone, M., Taverna, S., & Iannolo, G. (2024). Extracellular vesicles in lung cancer: Implementation in diagnosis and therapeutic perspectives. *Cancers, 16*(11), 1967. doi:10.3390/cancers16111967.

Chakraborty, S., Sharma, G., Karmakar, S., & Banerjee, S. (2024). Multi-OMICS approaches in cancer biology: New era in cancer therapy. *Biochimica et Biophysica Acta (BBA) - Molecular Basis of Disease, 1870*(5), 167120. doi:10.1016/j.bbadis.2024.167120.

Chakravarty, D., Johnson, A., Sklar, J., Lindeman, N. I., Moore, K., Ganesan, S., Lovly, C. M., Perlmutter, J., Gray, S. W., Hwang, J., Lieu, C., André, F., Azad, N., Borad, M., Tafe, L., Messersmith, H., Robson, M., & Meric-Bernstam, F. (2022). Somatic genomic testing in patients with metastatic or advanced cancer: ASCO provisional clinical opinion. *J Clin Oncol, 40*(11), 1231–1258. doi:10.1200/JCO.21.02767.

Chang, S.-C., Huang, C.-Y., Lai, Y.-C., Hsieh, M.-S., Ho, C.-C., Yang, C.-Y., Shih, J.-Y., Wu, S.-G., Hu, P.-W., Hsu, C.-L., Chen, C.-Y., & Liao, W.-Y. (2026). Clinical utility of pleural effusion supernatant cell-free DNA genotyping in previously treated patients with advanced non-small-cell lung cancer and disease progression: A multicenter retrospective study. *ESMO Open, 11*(1), 106023. doi:10.1016/j.esmoop.2025.106023.

Chang, S. C., Wei, Y. F., Chen, C. Y., Lai, Y. C., Hu, P. W., Hung, J. C., & Chang, C. Y. (2024). Profiling Cell-Free DNA from malignant pleural effusion for oncogenic driver mutations in patients with treatment-naive stage IV adenocarcinoma: A multicenter prospective study. *Molecular Diagnosis and Therapy, 28*(6), 803–810. https://www.springer.com/journal/40291. doi:10.1007/s40291-024-00736-8.

Chen, H., Luthra, R., Goswami, R. S., Singh, R. R., & Roy-Chowdhuri, S. (2015). Analysis of pre-analytic factors affecting the success of clinical next-generation sequencing of solid organ malignancies. *Cancers, 7*(3), 1699–1715. http://www.mdpi.com/2072-6694/7/3/0859/pdf. doi:10.3390/cancers7030859.

de Kock, R., Knoops, C., Baselmans, M., van den Borne, B., Brunsveld, L., Scharnhorst, V., & Deiman, B. (2021). Sensitive cell-free tumor DNA analysis in supernatant pleural effusions supports therapy selection and disease monitoring of lung cancer patients. *Cancer Treatment and Research Communications, 29*, 100449. doi:10.1016/j.ctarc.2021.100449.

Ergin, S., Kherad, N., & Alagoz, M. (2022). RNA sequencing and its applications in cancer and rare diseases. *Molecular Biology Reports, 49*(3), 2325–2333. doi:10.1007/s11033-021-06963-0.

Gargis, A. S., Kalman, L., Bick, D. P., Silva, C. D., Dimmock, D. P., Funke, B. H., Gowrisankar, S., Hegde, M. R., Kulkarni, S., Mason, C. E., Nagarajan, R., Voelkerding, K. V., Worthey, E. A., Aziz, N., Barnes, J., Bennett, S. F., Bisht, H., Church, D. M., Dimitrova, Z., … Lubin, I. M. (2015). Good laboratory practice for clinical next-generation sequencing informatics pipelines. *Nature Biotechnology, 33*(7), 689–693. http://www.nature.com/nbt/index.html. doi:10.1038/nbt.3237.

Gibbs, S. N., Peneva, D., Cuyun Carter, G., Palomares, M. R, Thakkar, S., Hall, D. W., Dalglish, H., Campos, C., & Yermilov, I. (2023). Comprehensive review on the clinical impact of next-generation sequencing tests for the management of advanced cancer. *JCO Precision Oncology, 7*, e2200715. doi:10.1200/PO.22.00715.

Gibbs, S. N., Peneva, D., Cuyun Carter, G., Palomares, M. R, Thakkar, S., Hall, D. W., Dalglish, H., Campos, C., & Yermilov, I. (2025). Comprehensive review on the clinical impact of next-generation sequencing tests for the management of advanced cancer. *Oncologist, 30*(1), oyae357. doi:10.1093/oncolo/oyae357.

Grigoriadou, G. I., Esagian, S. M., Ryu, H. S., & Nikas, I. P. (2020). Molecular profiling of malignant pleural effusions with next generation sequencing (NGS): Evidence that supports its role in cancer management. *Journal of Personalized Medicine, 10*(4), 1–18. https://www.mdpi.com/2075-4426/10/4/206/pdf. doi:10.3390/jpm10040206.

Guo, Z., Xie, Z., Shi, H., Du, W., Peng, L., Han, W., Duan, F., Zhang, X., Chen, M., Duan, J., Lin, J., Chen, X., Ang Lizaso, A., Han-Zhang, H., He, J., & Yin, W. (2019). Malignant pleural effusion supernatant is an alternative liquid biopsy specimen for comprehensive mutational profiling. *Thoracic Cancer, 10*(4), 823–831. doi:10.1111/1759-7714.13006.

Gupta, A., Singh, K., Clark, J., Nuthakki, V., Aburajab, M., Aboujaoude, S., Yousef, M., Alakharas, M., Kharbanda, A., Govert, K., Flanders, V., Lefranc, M., Dowell, J., Jhajria, S., Patel, N., Kukreja, S., Leagre, C., Smith, A., Bryant, T., … Gupta, N. K. (2025). Next generation sequencing (NGS) of malignant effusions and comparison to blood liquid biopsy (ctDNA): A community experience. *Journal of Clinical Oncology, 43*(16_suppl). doi:10.1200/jco.2025.43.16_suppl.e20661.

He, W., He, X., Qiao, Z., & Wang, Y. (2026). Emerging immunotherapeutic strategies for malignant serous effusions: Current evidence and future directions. *Frontiers in Oncology, 16*. http://www.frontiersin.org/Oncology/about. doi:10.3389/fonc.2026.1757825.

Hong, M., Tao, S., Zhang, L., Diao, L. T., Huang, X., Huang, S., Xie, S. J., Xiao, Z. D., & Zhang, H. (2020). RNA sequencing: New technologies and applications in cancer research. *Journal of Hematology and Oncology, 13*(1). http://www.jhoonline.org/. doi:10.1186/s13045-020-01005-x.

Horak, P., Griffith, M., Danos, A. M., Pitel, B. A., Madhavan, S., Liu, X., Chow, C., Williams, H., Carmody, L., Barrow-Laing, L., Rieke, D., Kreutzfeldt, S., Stenzinger, A., Tamborero, D., Benary, M., Rajagopal, P. S., Ida, C. M., Lesmana, H., Satgunaseelan, L., … Sonkin, D. (2022). Standards for the classification of pathogenicity of somatic variants in cancer (oncogenicity): Joint recommendations of clinical genome resource (ClinGen), cancer genomics consortium (CGC), and variant interpretation for cancer consortium (VICC). *Genetics in Medicine, 24*(5), 986–998. https://www.journals.elsevier.com/genetics-in-medicine. doi:10.1016/j.gim.2022.01.001.

Jennings, L. J., Arcila, M. E., Corless, C., Kamel-Reid, S., Lubin, I. M., Pfeifer, J., Temple-Smolkin, R. L., Voelkerding, K. V., & Nikiforova, M. N. (2017). Guidelines for validation of next-generation sequencing–based oncology panels: A joint consensus recommendation of the association for molecular pathology and college of american pathologists. *Journal of Molecular Diagnostics, 19*(3), 341–365. http://www.sciencedirect.com/science/journal/15251578. doi:10.1016/j.jmoldx.2017.01.011.

Kamps, R., Brandão, R. D., van den Bosch, B. J., Paulussen, A. D. C., Xanthoulea, S., Blok, M. J., & Romano, A. (2017). Next-generation sequencing in oncology: Genetic diagnosis, risk prediction and cancer classification. *International Journal of Molecular Sciences, 18*(2). http://www.mdpi.com/1422-0067/18/2/308/pdf. doi:10.3390/ijms18020308.

Khranovska, N., Gorbach, O., Skachkova, O., & Klimnyuk, G. (2024). Application of next-generation sequencing to realize principles of precision therapy in management of cancer patients. *Experimental Oncology, 46*(4), 295–304. https://exp-oncology.com.ua/index.php/Exp/article/view/426/381. doi:10.15407/exp-oncology.2024.04.295.

Li, J., Nazeer Batcha, A. M., Grüning, B., & Mansmann, U. R. (2016). An NGS workflow blueprint for DNA sequencing data and its application in individualized molecular oncology. *Cancer Informatics, 15*, 87–107. http://www.la-press.com/redirect_file.php?fileId=7376&filename=5527-CIN-An-NGS-Workflow-Blueprint-for-DNA-Sequencing-Data-and-Its-Application-.pdf&fileType=pdf. doi:10.4137/CIN.S30793.

Liu, X., Xu, X., & Wang, Q. (2026). Exosomes in early lung cancer diagnostics: The current state of progress made and prospects. *Frontiers in Cell and Developmental Biology, 13*. doi:10.3389/fcell.2025.1739242.

McPhail, S., Swann, R., Johnson, S. A., Barclay, M. E., Abd Elkader, H., Alvi, R., Barisic, A., Bucher, O., Clark, G. R. C., Creighton, N., Danckert, B., Denny, C. A., Donnelly, D. W., Dowden, J. J., Finn, N., Fox, C. R., Fung, S., Gavin, A. T., Gomez Navas, E., … You, H. (2022). Risk factors and prognostic implications of diagnosis of cancer within 30 days after an emergency hospital admission (emergency presentation): An International Cancer Benchmarking Partnership (ICBP) population-based study. *The Lancet Oncology, 23*(5), 587–600. http://www.journals.elsevier.com/the-lancet-oncology/. doi:10.1016/S1470-2045(22)00127-9.

Noda, Y., Yamaka, R., Atsumi, N., Higasa, K., & Tsuta, K. (2023). Areas of crush nuclear streaming should be included as tumor content in the era of molecular diagnostics. *Cancers, 15*(6), 1910. doi:10.3390/cancers15061910.

Pérez-Barrios, C., Sánchez-Herrero, E., Garcia-Simón, N., Barquín, M., Blanco Clemente, M., Provencio, M., & Romero, A. (2021). ctDNA from body fluids is an adequate source for EGFR biomarker testing in advanced lung adenocarcinoma. *Clinical Chemistry and Laboratory Medicine (CCLM), 59*(7), 1221–1229. doi:10.1515/cclm-2020-1465.

Qi, C., Li, Y., Zeng, H., Wei, Q. i, Tan, S., Zhang, Y., Li, W., & Tian, P. (2024). Current status and progress of PD-L1 detection: Guiding immunotherapy for non-small cell lung cancer. *Clinical and Experimental Medicine, 24*(1). doi:10.1007/s10238-024-01404-1.

Schilter, K. F., Dubay, S., Stachowiak, M., Kaplan, A., Stauffenger, A., & Reddi, H. V. (2023). Implementation of a customized tertiary analysis platform for the reporting of somatic variants. *Biomarker Insights, 18*. https://journals.sagepub.com/home/BMI. doi:10.1177/11772719231178618.

Sharma, S., Arora, R. D, & Boster, J. (2025). Malignant pleural effusion. *StatPearls [Internet]*. StatPearls Publishing https://www.ncbi.nlm.nih.gov/books/NBK574541/.

Shin, S., Woo, H. I., Kim, J. W., Kim, Y., & Lee, K. A. (2021). Clinical practice guidelines for pre-analytical procedures of plasma epidermal growth factor receptor variant testing. *Annals of Laboratory Medicine, 42*(2), 141–149. http://annlabmed.org/journal/view.html?doi=10.3343/alm.2022.42.2.141. doi:10.3343/alm.2022.42.2.141.

Stenzinger, A., Vogel, A., Lehmann, U., Lamarca, A., Hofman, P., Terracciano, L., & Normanno, N. (2024). Molecular profiling in cholangiocarcinoma: A practical guide to next-generation sequencing. *Cancer Treatment Reviews, 122*, 102649. doi:10.1016/j.ctrv.2023.102649.

Suver Stević, M., Holik, H., Periša, V., Marczi, S. aš, Kolobarić, N., & Samardžija, M. (2025). From sample to sequencing: The importance of pre-analytical sample treatment in NGS analysis of patients with chronic lymphocytic leukemia. *Cancers, 17*(22), 3668. doi:10.3390/cancers17223668.

Wang, H.-Y., & Shih, J.-Y. (2025). Lung cancer driver mutation testing using pleural effusion [Internet]. pleural diseases - anatomy, physiology, and pathophysiology. *IntechOpen*. https://doi.org/10.5772/intechopen.1012553.

Whitfield, H. J., Berthelet, J., Mangiola, S., Bell, C., Anderson, R. L., Pal, B., Yeo, B., Papenfuss, A. T., Merino, D., & Davis, M. J. (2023). Single-cell RNA sequencing captures patient-level heterogeneity and associated molecular phenotypes in breast cancer pleural effusions. *Clinical and Translational Medicine, 13*(9). doi:10.1002/ctm2.1356.

Wu, H., Ji, H., Yang, W., Zhang, M., Guo, Y., Li, B., Wang, J., Chen, R., Chen, Y., & Wang, X. (2022). Liquid biopsy using ascitic fluid and pleural effusion supernatants for genomic profiling in gastrointestinal and lung cancers. *BMC Cancer, 22*(1), 1020. doi:10.1186/s12885-022-09922-5.

Yan, Z., Sun, C., Tang, W., Cao, W., Lv, J., Liang, Z., Wei, S., Zhong, W., Zhao, Z., Zhao, Z., & Li, Y. (2025). Application of the metagenomic next-generation sequencing technology to identify the causes of pleural effusion. *Frontiers in Medicine, 12.* journal.frontiersin.org/journal/medicine. doi:10.3389/fmed.2025.1525100.

Yu, Q., Jiang, M., & Wu, L. (2022). Spatial transcriptomics technology in cancer research. *Frontiers in Oncology, 12.* http://www.frontiersin.org/Oncology/about. doi:10.3389/fonc.2022.1019111.

13

Flow cytometry for body fluids and cerebrospinal fluid: Oncologist's perspective

Saksham Singh[a] and Aditya Jandial[b]

[a]DEPARTMENT OF MEDICAL ONCOLOGY, DR. RAM MANOHAR LOHIA INSTITUTE OF MEDICAL SCIENCES (DR. RMLIMS), LUCKNOW, UTTAR PRADESH, INDIA [b]DEPARTMENT OF CLINICAL HEMATOLOGY AND MEDICAL ONCOLOGY, POST GRADUATE INSTITUTE OF MEDICAL EDUCATION AND RESEARCH (PGIMER), CHANDIGARH, INDIA

13.1 Introduction

Effusion fluids, such as pleural, peritoneal, and pericardial, are commonly encountered in cancer care. Accurately identifying malignant epithelial cells in these fluids is essential for confirming metastatic disease, directing targeted therapy, and determining prognosis. While cytology remains the first-line diagnostic modality, its limited sensitivity, especially in low-cellularity specimens or morphologically equivocal cases, necessitates the use of adjunct methods. Flow cytometry (FCM) is now emerging as a promising complementary technique for detecting malignant cells in effusion fluids. It requires cells to be in suspension; effusion fluids present a biologically suitable medium for advancing its applications in the management of solid tumors. Leveraging effusion fluid analysis in FCM may enhance tumor characterization, biomarker identification, and treatment monitoring.

The central nervous system (CNS) is a distinct anatomical space that may serve as a sanctuary site for malignant cells in cancer patients. Early diagnosis of CNS involvement by malignancy is essential for the timely initiation of CNS-directed therapy and for improving overall disease control (Thastrup et al., 2022). Conventionally, CSF cytology and neuroimaging are essential diagnostic modalities used in the evaluation of cancer patients with suspected CNS involvement. For diagnosing early CNS disease, both modalities have major limitations. In this regard, CSF FCM has emerged as a valuable tool with substantial potential to improve the diagnostic yield of CSF analysis in cancer patients (Fig. 13.1). It has major implications for risk stratification, treatment, and follow-up strategies (Del Principe et al., 2020; Subirá et al., 2012). It aligns with the concept of "precision oncology" and may impact treatment decisions in contemporary oncology practice. It is also particularly relevant in the current paradigm of risk-adapted therapy and Measurable residual disease (MRD).

Advancing Body Fluid Diagnostics with Flow Cytometry: A Comprehensive Guide to Effusions and CSF.
DOI: https://doi.org/10.1016/B978-0-443-27426-8.00009-0

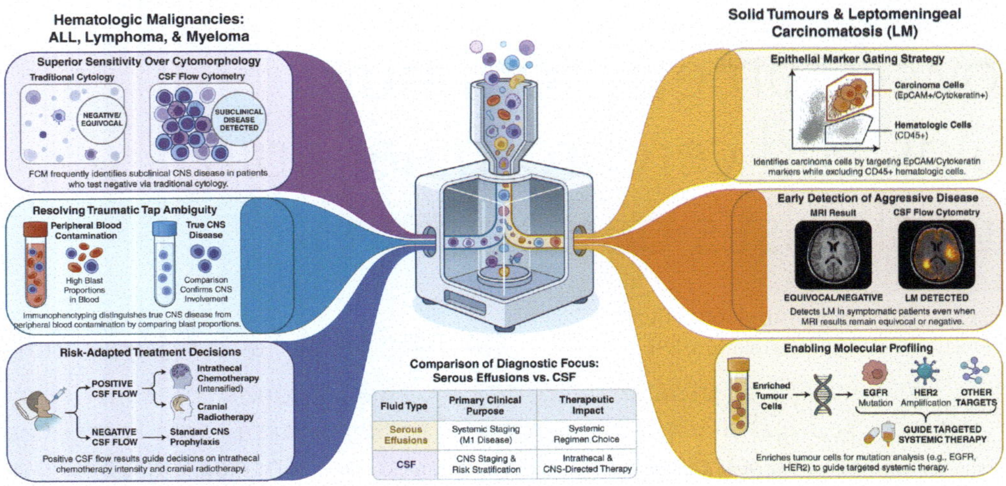

FIGURE 13.1 Role of CSF flow cytometry in hematolymphoid and solid cancers. *CSF*, Cerebrospinal fluid.

From an oncologist's perspective, any new diagnostic technique should have the following attributes:

- Potential to change disease stage or risk stratification
- Ability to alter the choice, intensity, route, and duration of treatment
- Being minimally invasive
- Rapid identification of treatment failure/relapse

These attributes should serve as a benchmark against which the performance of FCM in body fluids and CSF should be determined.

13.2 Clinical value of flow cytometry in solid and hematolymphoid malignancies: An oncologist's view

As oncologists, our core concerns when assessing a patient with malignant effusion include:

- **Confirming malignancy**: Does the effusion contain tumor cells?
- **Determining molecular targets**: Can this fluid be used for Epidermal Growth Factor Receptor (EGFR), Analplastic Lymphoma Kinase (ALK), Kirsten Rat Sarcoma viral oncogene homologue (KRAS), or other targeted testing?
- **Making timely decisions**: Can we get a result fast enough to start therapy?
- **Intra-operative setting**: Whether faster results can be achieved for intra-operative peritoneal lavage as part of staging.

Likewise, the following questions are relevant in the context of CNS involvement in cancer patients:

- Is CNS disease present at initial diagnosis?
- If present, how does the presence of CNS disease impact the choice of treatment and its duration?
- Does CNS disease persist despite therapy on follow-up?
- Is CNS disease present at the time of relapse?

FCM is a valuable tool for answering these questions when used in conjunction with cytology and imaging.

13.3 Flow cytometry in body fluids

13.3.1 Diagnostic utility and improved sensitivity in body fluids

- Cytology alone has a sensitivity of around 70%–75% for detecting malignancy in effusions (Wong-Arteta et al., 2020). In studies involving cytologically "suspicious" or "atypical" effusions, FCM has demonstrated high concordance with biopsy-proven malignancy (Davidson et al., 2002). A study by Gaur et al. (2023) evaluating FCM's diagnostic accuracy in 96 serous fluid samples (86 ascitic and 10 pleural) found that using epithelial markers EpCAM and MUC-1, FCM achieved a diagnostic accuracy of 93.75%. Combining FCM with cytomorphology increased this to 96.8%, highlighting FCM's role in cases with atypia or negative cytology (Gaur et al., 2023).
- FCM can confirm malignancy in effusions where cytology is nondiagnostic, especially in low-cellularity fluids, reactive mesothelial proliferation mimicking malignancy, and poorly differentiated carcinomas.
- FCM increases detection sensitivity by enabling rapid analysis of thousands of cells, differentiation from benign mesothelial and inflammatory cells using exclusion markers like CD45, and identification of tumor-specific surface and intracellular markers, such as EpCAM (CD326), cytokeratins, and CEA. In a cohort of 570 pleural effusion samples, high-sensitivity FCM using a 3-color assay detected malignant cells in 77.5% of cases, including 55.8% of new cancer diagnoses. FCM identified a significantly lower percentage of EpCAM+ cells in cytopathology-negative cases, demonstrating its utility in early cancer detection (Subirá et al., 2023).

13.3.2 Tumor staging and prognosis

- Malignant cells in effusion fluids may upstage the cancer as well as have prognostic implications. High malignant cell burden in effusion has been associated with shorter survival and more aggressive disease biology (Sung et al., 2021).
- In lung cancer, the presence of malignant pleural effusion confirms stage IV (M1a) disease under the AJCC TNM classification, ruling out curative surgery and redirecting the patient to systemic therapy.
- In ovarian cancer, malignant ascites often signifies peritoneal carcinomatosis, upstaging the disease to FIGO stage III or IV (Sung et al., 2021).

- In gastric cancer, positive cases on peritoneal washing cytology are classified as distant metastasis. The presence of free carcinoma cells (FCC) in the peritoneal lavage fluid is an independent adverse prognostic factor in gastric carcinoma patients (Cieśla et al., 2021).

Intra-operative FCM improves the detection of FCC in peritoneal lavage fluid compared to cytology (Ramalingam et al., 2025).

- Immunophenotypic profiles (e.g., high EpCAM or CEA expression) may correlate with poor prognosis in certain cancers. In a systematic review and meta-analysis, EpCAM overexpression was associated with clinicopathological significance and poor prognosis in patients with hepatocellular carcinoma. EpCAM was assessed by Immunohistochemistry (IHC), real-time polymerase chain reaction (PCR), ELISA, or microarray (Zhou & Zhu, 2018).

Studies have demonstrated that higher EpCAM levels predicted poor survival outcomes in ovarian, colon, squamous cell carcinoma of the head and neck region, as well as breast cancers (Munz et al., 2009; Osta et al., 2004; Pauli et al., 2003; Tas et al., 2014).

There is a strong likelihood that FCM detection of epithelial antigens in effusion fluids may be employed for staging and prognostication of solid tumors in the near future.

13.3.3 Monitoring treatment response

Effusion cytology can be used serially to monitor therapy response. FCM enhances this application by providing quantitative data on tumor cell population size over time. A study published in *Cytometry Part B: Clinical Cytometry* demonstrated that FCM could accurately quantify tumor cells in effusion specimens. The research highlighted the utility of FCM in detecting malignant cells, even in samples with low-cellularity, thereby providing quantitative data that can be tracked over multiple time points to assess treatment efficacy (Wong-Arteta et al., 2020). A systematic review in *Cancer Treatment Reviews* examined circulating blood biomarkers for MRD in Hepatocellular Carcinoma (Galli et al., 2025). The review found that higher levels of EpCAM+/CK+ circulating tumor cells (CTCs) were associated with worse recurrence-free survival, underscoring the role of these markers in detecting MRD and predicting relapse. A comprehensive liquid biopsy analysis in *Breast Cancer Research* demonstrated that the presence of EpCAM+/CK+ CTCs was associated with disease progression and shorter survival. The study highlighted the prognostic value of these markers in monitoring therapy response and detecting early relapse (Nicolò et al., 2024).

In a study focused on the diagnostic role of CTCs in gallbladder cancer, authors observed that CTC levels correlated significantly with clinico-pathological parameters of cases in terms of known prognostic indicators. A significant difference in CTC levels between T stage (T1+T2 vs. T3+T4), systemic metastasis (M0 vs. M1), tumor stage (I+II vs. III+IV), and nodal metastasis (N0 vs. N1+N2) was reported (Awasthi et al., 2017).

These studies provide a foundation for future research to underscore the significance of detecting EpCAM+/CK+ cell populations in effusion fluids as markers of measurable residual disease (MRD) and early relapse, ultimately facilitating timely therapeutic interventions.

13.3.4 Utility in targeted mutation testing and targeted therapies

Effusion fluids often represent a minimally invasive, high-yield source of tumor cells—particularly valuable when tissue biopsy is risky or inaccessible. FCM can isolate tumor cell-enriched populations using epithelial markers (EpCAM+ and CK+) while gating out leukocytes (CD45+) and mesothelial cells. A technique called cell sorting, often referred to as fluorescence-activated cell sorting (FACS) (Telford, 2023), is an extension of FCM that not only analyses but also physically separates cells of interest. After labeling cells with fluorescent markers, the flow cytometer directs individual cells into droplets. Each droplet is charged based on the cell's characteristics, and an electric field deflects these charged droplets into separate collection tubes. This process allows for the isolation of specific cell populations with high purity and also facilitates molecular profiling, including EGFR/ALK/ROS1 mutations in NSCLC, Her2 (ErbB2) in breast and gastric cancers, BRCA1/2 mutations in ovarian cancer, PD-L1 expression, MSI, and TMB analysis. Tumor heterogeneity and therapy-induced selection can result in resistant populations (e.g., **EGFR T790M, HER2 loss, and KRAS mutations**). FCM coupled with molecular assays like PCR or NGS on sorted cells, can identify these clones in real time from effusion samples. In practice, studies show equivalent mutation detection rates using effusion-derived tumor cells compared to tissue biopsies. For Instance, a study involving 192 patients with advanced nonsmall cell lung cancer (NSCLC) compared EGFR mutation detection in primary tumor tissue and pleural effusion samples. The EGFR mutation rate was 61.98% in tumor tissue and 58.85% in pleural effusion samples, with an overall concordance rate of 86.98% between the two sources (Liu et al., 2018). FCM-enrichment before molecular testing improves yield and minimizes contamination by nontumor cells.

By enabling detailed **immunophenotyping** and **cell sorting**, FCM plays several critical roles in the era of targeted and personalized cancer therapy. FCM can directly detect overexpression of key therapeutic targets on tumor cells such as **EGFR** in lung adenocarcinoma, **HER2 (ErbB2)** in breast and gastric cancers, and **PD-L1** in NSCLC and other malignancies. This facilitates prompt initiation of targeted therapies even when tissue biopsies are not available or are inadequate (Table 13.1). Recent research on intra-operative FCM describes a rapid, sensitive and objective technique which can significantly improve the detection of free cancer cells in peritoneal washings with concurrent evaluation of PD-L1 expression for immunotherapy in gastric cancer patients (Ramalingam et al., 2025).

13.3.5 Real-world case scenarios for the role of flow cytometry in diagnostic utility and targeted therapies

Case 1: Atypical cells in pleural effusion—diagnostic challenge

Background: A 65-year-old smoker presents with dyspnea. Chest X-ray shows pleural effusion. Cytology is reported as "suspicious for malignancy." Cell block immunocytochemistry (CB-ICC) awaited. FCM reveals a subpopulation of cells negative for CD45 and CD200 and expressing EpCAM.

Usefulness: Given a positive FCM result alongside a suspicious cytological diagnosis, a malignant effusion can be inferred with a high degree of confidence. FCM aids in prompt oncologic intervention in such scenarios.

Table 13.1 Summary of targeted therapy markers via effusion-based testing.

Marker	Cancer type	FCM role	Targeted therapy
PD-L1	NSCLC	Direct detection on EpCAM+ cells	Pembrolizumab and nivolumab
HER 2	Breast cancer	Direct detection on EpCAM+ cells	Trastuzumab
BRAF V600E	Melanoma and CRC	Sort cells → PCR/NGS	Dabrafenib + Trametinib
KRAS	CRC, NSCLC, and Pancreatic	Sort → PCR/NGS	Sotorasib (KRAS G12C) and avoid EGFR mAbs
ALK	NSCLC	Sort → FISH/RT-PCR	Crizotinib, alectinib, and lorlatinib
EGFR (exon 19 deletion)	NSCLC	Sort → PCR/NGS	Osimertinib
BRCA	Breast cancer	Sort → PCR/NGS	PARP inhibitors

CRC, colorectal cancer; *NGS*, next generation sequencing; *NSCLC*, nonsmall cell lung cancer; *PARP*, poly(ADP-ribose) polymerase; *PCR*, polymerase chain reaction.

Case 2: Peritoneal effusion in ovarian cancer—monitoring and staging

Background: A 54-year-old female with a history of high-grade serous carcinoma of the ovary, posttreatment. Peritoneal fluid collected during staging surgery. FCM revealed a small population of EpCAM+, MUC 1+ cells. A separate population of CD45 + leucocytes and histocytes, and CD45- cells (likely mesothelial cells) were identified.

Usefulness: FCM detects measurable residual disease not seen on cytology and helps guide adjuvant treatment decisions

Case 3: Peritoneal effusion in a known case of lymphoma to rule out a second malignancy or a reactive cause

Background: A patient of diffuse large B-cell lymphoma (DLBCL) on follow-up presents with ascites. Cytology shows mixed inflammatory cells admixed with large lymphoid cells. FCM reveals 28% cells that are CD45+, CD19+, CD20+, and kappa-restricted. No evidence of epithelial cells (negative for EpCAM).

Usefulness: Diagnosis of lymphomatous infiltration suggests a relapse, supporting prompt oncologic intervention.

Case 4: Mesothelioma versus adenocarcinoma in pleural fluid

Challenge: Reactive mesothelial cells vs. metastatic adenocarcinoma

FCM panel reveals cells expressing CD200 and CD71 (suggesting proliferating mesothelial cells), which were negative for EpCAM and MUC1.

Interpretation: Pleural effusion rich in mesothelial cells with no evidence of epithelial cells, favoring mesothelioma. Clinical and radiological correlation is advised.

Case 5: Case of HER2+ breast cancer

- **HER2 overexpression** on **EpCAM+ cells** in pleural or peritoneal effusion is identified using FCM with HER2-specific antibodies.
- Confirms metastatic disease and reinforces use of **HER2-targeted therapies** (e.g., **trastuzumab and T-DXd).**

Table 13.2 Comparison of conventional cytology, tissue biopsy and flow cytometry for tumor detection and molecular profiling.

Method	Turnaround time	Cost range	Accessibility	Invasive	Tumor enrichment for molecular testing	Diagnostic sensitivity
Cytology alone	2–3 days	Low	High	Minimally	Yes	~60%–70%
Tissue biopsy (CT-guided)	5–7 days	High	High	Moderate	Yes	~90%
Flow cytometry (FCM)	Same-day to 24 h	Moderate	Moderate	Minimally	Yes (EpCAM+ selection possible by FACS)	~80%–90%

- May also help in assessing **HER2 status conversion** posttreatment, guiding changes in therapy.

Case 6: Metastatic colorectal cancer (mCRC) presents with ascites

- EpCAM+/Ber-EP4+ tumor cells are **isolated from the effusion fluid** via FCM or FACS.
- These sorted cells are then subjected to **molecular testing** (e.g., NGS or real-time PCR) to detect **KRAS mutations** (commonly G12D and G12V).
- In **colorectal cancer,** detection of **KRAS mutations** rules out eligibility for **EGFR-targeted monoclonal antibodies** (e.g., cetuximab and panitumumab).

13.3.6 Cost-effectiveness: Flow cytometry versus cytology and biopsy

While FCM requires investment in equipment and trained personnel, it may reduce overall diagnostic costs with a definite advantage of shorter turnaround time, in selected patients (Table 13.2).

13.3.7 Clinical implications

FCM may offer a faster and safer route to diagnosis and molecular profiling in patients with the following attributes:

- Unable to undergo invasive procedures
- Unfit for biopsy due to poor performance status
- With inaccessible lesions
- Awaiting rapid confirmation for treatment decisions
- Cytology-negative or atypical but radiologically suspicious effusions

FCM can identify malignancy in cytology-negative effusions, avoiding repeat invasive procedures.

Table 13.3 Advantages and constraints of key diagnostic techniques in tumor evaluation.

Modality	Strength	Limitation
Cytology	Morphology-based, fast, and inexpensive	Subjective; lower sensitivity
Immunocytochemistry (ICC)	Phenotypic confirmation	Time-consuming; semiquantitative
FCM	Rapid, qualitative, quantitative, and sensitive	Requires technical setup; limited morphologic data
Biopsy/histology	Gold standard for diagnosis	Invasive; not always feasible

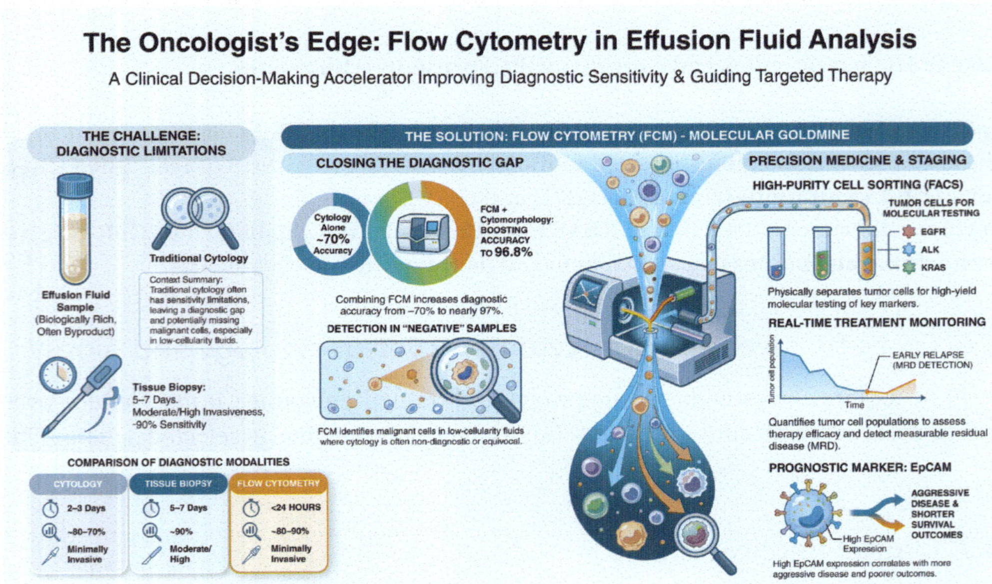

FIGURE 13.2 Role of flow cytometry in enhancing diagnostic sensitivity and clinical utility in effusion fluid analysis.

13.3.8 Complementary role with other diagnostic modalities

FCM does not replace cytology or histopathology but functions best when integrated into a multimodal diagnostic workflow (Table 13.3).

Combined use (e.g., **cytology, FCM, and ICC**) increases diagnostic confidence, especially in **low-cellularity** or **indeterminate** cases (Fig. 13.2).

13.3.9 Limitations

- **False negatives** in tumors lacking epithelial marker expression (e.g., sarcomatoid tumors and mesothelioma).
- May miss **poorly differentiated tumors** with low EpCAM expression.
- Requires **logistical support**, including trained cytometrists and marker panels tailored to cancer type.
- Requires fresh samples (viable cells).
- Loss of architectural context compared to cytology/histology.

However, these are offset by FCM's rapidity and utility in **molecular oncology workflows**.

13.4 Conclusion

From a medical oncologist's perspective, FCM **is not just a lab tool—it is a clinical decision-making accelerator**. It helps identify malignancy and has the potential to stratify patients for targeted therapy. It often avoids the need for invasive procedures in select patients. In the age of precision medicine, FCM can transform effusion fluid from a diagnostic byproduct into a **real-time molecular diagnostic goldmine**, especially when cytology is equivocal or biopsy is not feasible. It is time we consider **routine incorporation of** FCM into the diagnostic algorithm for effusion analysis in oncology—selectively, strategically, and collaboratively.

13.5 Flow cytometry for cerebrospinal fluid

13.5.1 Central nervous system dissemination of cancer—The sanctuary effect

The CNS represents an extraordinary anatomical sanctuary site for malignant cells, mainly due to the protective effect of the blood-brain and blood-CSF barriers, which are crucial for maintaining neural homeostasis. These barriers prevent the CNS penetration of systemic anticancer drugs, thereby creating a sanctuary site where tumor cells may persist and achieve a disproportionate survival advantage despite clearance of cancer from the rest of the body. Once the tumor cells gain access to the subarachnoid space, they can circulate freely and seed distant sites within the neuraxis (Weston et al., 2011)

13.5.2 Cerebrospinal fluid flow cytometry for hematolymphoid malignancies

CSF behaves as a distinct biological territory compared to other body compartments. Whereas evaluation of the pleural, pericardial, and peritoneal compartments informs about malignant dissemination and disease burden, CSF evaluation also informs about disease biology, prognosis, and potential treatment decisions, which may differ from those of the systemic disease (Del Principe et al., 2020). Application of FCM to CSF has potential implications for the choice and intensity of systemic and CNS-directed treatment.

13.5.3 Common clinical scenarios and unmet needs in hematolymphoid malignancies

- **Acute lymphoblastic leukemia (ALL)**: CSF examination is routinely performed at diagnosis and on follow-up (during induction, consolidation, re-induction, and maintenance phases of treatment) to establish CNS involvement in pediatric and adult ALL patients as part of standard treatment protocols. Baseline CNS involvement puts these patients at risk of CNS as well as systemic relapse. Application of FCM to CSF has the potential to detect "occult" leukemic involvement that is not detected by conventional CSF cytology or neuroimaging (Thastrup et al., 2025).
- **Aggressive nonHodgkin lymphomas and CNS lymphoma**: CSF FCM is usually performed in patients with overt neurological signs and symptoms, high CNS-international prognostic index (CNS-IPI), and/or in the presence of risk factors well-known to be associated with CNS involvement (testicular, renal, adrenal, and breast involvement) (Gini et al., 2021). These include DLBCL, Burkitt lymphoma, high-grade B-cell lymphoma, primary CNS lymphoma (PCNSL), and some T-cell lymphomas (Wilson et al., 2014). Here, early diagnosis of CNS involvement not only alters the curative trajectory of the underlying disease but also supports a more aggressive CNS-directed prophylaxis and treatment (Singh et al., 2018). Bing-Neel syndrome (BNS) is characterized by direct CNS involvement by clonal lymphoplasmacytic cells in the setting of Waldenstrom macroglobulinemia. CSF FCM can confirm the diagnosis of BNS by identifying clonal B-lymphocytes, especially in low-level leptomeningeal disease and in cases with equivocal neuroimaging findings (Kotsos et al., 2022). It also helps in monitoring the response to CNS-directed treatment, as clearance of CSF clone often precedes radiological resolution.
- **Acute myeloid leukemia (AML):** Overall, CNS involvement is uncommon in AML. Hence, CSF FCM is more selective in this setting. AML patients with hyperleukocytosis at presentation, monocytic or myelomonocytic variants, and overt neurological clinical features at diagnosis are at high risk for CNS involvement. Moreover, new-onset neurological signs and symptoms in an AML patient during treatment may arise due to metabolic derangements, opportunistic infections, CNS disease, and treatment-related complications. CSF FCM can guide the management in such scenarios by rapidly identifying CNS involvement.
- **Plasma cell dyscrasias:** CNS involvement in plasma cell disorders like multiple myeloma (MM) is rarely seen (<1% of cases) (Egan et al., 2020). Conventional CSF cytology and neuroimaging may yield equivocal findings. In suspected cases, CSF FCM can confirm the presence of clonal plasma cells in CSF and justify treatment escalation.

13.5.4 Preanalytical considerations

Despite its sensitivity, certain fundamental preanalytical variables affect the diagnostic accuracy of CSF FCM (Table 13.4). All the steps before actual analysis of the CSF sample in the FCM laboratory are crucial—how the sample is collected, how it is stored, how it is transported to the laboratory, and whether there was a traumatic lumbar puncture. A clear understanding of these variables is crucial for oncologists to accurately interpret test results.

Table 13.4 Preanalytical challenges and ESCCA/ISCCA recommendations for multiparameter flow cytometry in CSF (Del Principe et al., 2021).

Preanalytical challenge	Recommendation
Paucicellularity	Acquire a minimum of 1000 events
Diagnostic threshold	Sample should be labeled positive if at least 10 phenotypically abnormal events are noted
Cellular fragility	Process the CSF sample within 1 h of lumbar puncture
Traumatic lumbar puncture	Evaluate simultaneous peripheral blood sample to rule out contamination. Even if contamination is suspected, presence of blasts in CSF should be reported
Flow panel	Eight-or more-color panels should be preferred

CSF, cerebrospinal fluid.

- **Paucicellularity**: Unlike other body fluids, where abundant cells are available for analysis, normal CSF has very few leucocytes per microlitre. Even in pathological samples, there may be only a few hundred cells, or even fewer. Hence, each sample is precious, and errors in sample collection and handling can significantly compromise the accuracy of FCM.
- **Cellular fragility**: The cells in CSF samples are vulnerable to degradation by apoptosis as soon as they are removed from their physiologic milieu. Hence, the sample should reach the laboratory as soon as possible—ideally within an hour (Del Principe et al., 2021; Kraan et al., 2008). Delay in sample transportation and processing can significantly decrease the sensitivity of FCM. The coordination between oncology and laboratory teams is essential for optimal results. The addition of stabilizing media may temporarily reduce the cellular loss for up to a few hours (Thastrup et al., 2022; Nguyen-Them et al., 2023).
- **Sample volume**: A common avoidable cause of false-negative test results is inadequate sample volume. Rare malignant cells may be missed in a small-volume sample. Higher CSF sample volumes improve the sensitivity, especially in the presence of low-level disease. There is no universally accepted definition of the ideal CSF sample volume for FCM. At least a 3–5 mL sample should be collected, and wherever clinically feasible, a 5–10 mL sample should be collected (Kim et al., 2021).
- **Traumatic lumbar puncture**: CSF contamination by peripheral blood during lumbar puncture ($\geq$ 10 erythrocytes/μL) may lead to false isolation of circulating blasts or lymphomatous cells in CSF and an overestimation of CNS disease. This is particularly bothersome in acute leukemias with high circulating blast counts (Gold et al., 2013). In such cases, comparing peripheral blood and CSF immunophenotypes and assessing the relative proportion of malignant cells in CSF may help differentiate contamination from true CNS involvement.

Takeaway for oncologists: If the CSF FCM report truly drives major treatment decisions (protocol intensity, transplant eligibility, etc.), prioritize proper collection of an adequate CSF volume. Expecting a high-sensitivity CSF FCM report is unrealistic when a small CSF sample is sent to the flow laboratory late in the evening, when immediate processing is not possible. Exercise caution while reducing the treatment intensity based on a "negative" CSF flow report derived from a delayed and inadequate sample.

13.6 Approach to immunophenotyping in hematolymphoid malignancies

- **ALL:** The immunophenotype of diagnostic bone marrow [leukemia-associated immunophenotype (LAIP)] serves as the benchmark for the interpretation of CSF samples (Del Principe et al., 2021). In B-lineage ALL, CD19 expression is typically seen on blasts, accompanied by variable expression of CD10, CD34, and dim CD45 (abnormal cells may be missed if the CD45 gate is too restrictive). There may be aberrant expression of myeloid markers (CD13 and CD33), and some markers may show an unusual intensity pattern (too bright or too dim). In T-lineage ALL, leukemic blasts typically express CD3, CD5, CD7, CD2, and CD34 on the cell surface or in the cytoplasm (Del Principe et al., 2021).
- **Aggressive nonHodgkin lymphomas and CNS lymphomas:** The primary goal is to confirm the clonality of B cells. Hence, the typical flow panels focus on CD19, CD20, or CD22; light chain restriction is the definitive marker. The expression of related markers (CD10, BCL6, and MUM1) may have prognostic relevance but is not truly essential for establishing clonality. Immunophenotype is usually similar between PCNSL and systemic lymphoma with secondary CNS involvement. Diagnosis of T-cell lymphoma in CSF may be challenging. Abnormal clonal T-cells may show loss of pan-T-cell markers (CD3, CD5, CD2, and CD7), abnormal CD4/CD8 ratio, or unusual expression patterns. In some situations, reactive T-lymphocyte populations in CSF (inflammatory conditions, infections, etc.) may mimic a similar immunophenotype. For accurate diagnosis, the full clinical context should be considered. In BNS, clonal B-lymphocytes express CD19, CD20, surface Immunoglobulin M (IgM), and light chain restriction.
- **AML:** Like ALL, myeloid leukemic blasts in CSF recapitulate the immunophenotype of the bone marrow myeloid clone. They usually express CD13, CD33, CD117, CD34, HLA-DR and lineage-specific markers (CD14, CD64, CD11c, and CD36).
- **Plasma cell dyscrasias:** Clonal plasma cells express bright CD38 and CD138, with light chain restriction.

13.6.1 Diagnostic performance in hematolymphoid malignancies—Cerebrospinal fluid conventional cytology versus multiparametric flow cytometry

Conventional cytological (CC) examination of CSF is considered the "gold standard" for detecting CNS infiltration by hematologic neoplasms. However, despite the high specificity of CC, its sensitivity is low in low-cellularity CSF samples (<5 cells/μL), resulting in a high false-negative rate (20%–60%) (Bento et al., 2018; Del Principe et al., 2021). Hence, relying solely on conventional cytology can underestimate the CNS involvement, potentially delaying the diagnosis and appropriate treatment of such patients (Table 13.5). Rather than merely being an optional adjunct, multiparametric flow cytometry (MFC) plays a complementary role to CC in diagnosing CNS disease. It can detect abnormal clonal cells as low as 0.1% in CSF samples, particularly those considered negative by CC (Nguyen-Them et al., 2023; Šúri & Mociková, 2025). In the context

Table 13.5 Comparison between conventional cytology and multiparameter flow cytometry in CSF.

Variable	Conventional cytology (CC)	Multiparameter flow cytometry (MFC)
Specificity	High ("gold standard" for diagnosis of CNS disease)	High (relies on multiantigen gating and clonality)
Sensitivity	Limited	High (can detect scarce tumor cells in paucicellular samples)
False-negative rate	20% to 60%	Markedly low (due to objective immunophenotypic identification)
Interpretation challenges	Morphological similarities between tumor cells and benign reactive cells	Bypasses morphology; utilizes objective fluorescence intensity

CSF, cerebrospinal fluid.

of ALL, multiple studies have demonstrated the superiority of MFC over CC in diagnosing CNS disease. Positive yield of MFC at diagnosis (16.3% to 40.5%) has been roughly 2 to 3 times higher than CC (2.8% to 18%) (Gabelli et al., 2019; Thastrup et al., 2020).

In a systematic review of 27 studies directly comparing MFC with CC in CSF samples from patients with lymphoid neoplasms (ALL/lymphoma, DLBCL, and other lymphomas), 24 out of 27 studies (89%) reported cases with positive MFC but negative CC (MFC+/CC-). There was a wide variation in the proportion of such cases (MFC+/CC-), ranging from 3.5% to 61.5% of all samples. Conversely, positive CC but negative MFC (MFC-/CC+) was reported in only 13 out of 27 studies (48%). And, when this happened (MFC-/CC+), it affected a relatively small proportion of cases (0.5% to 10% of total samples) (Canovi & Campioli, 2016). The evidence clearly shows that relying only on CC leads to significant underestimation of CNS disease in hematolymphoid malignancies (especially ALL). Therefore, MFC should be performed in conjunction with CC to improve the detection of CNS disease.

Takeaway for oncologists: The evidence clearly shows that CNS involvement may be underestimated by CSF conventional cytology alone. Conversely, CSF FCM has much lower false-negative rates. This supports the routine execution of FCM in combination with conventional cytology on CSF samples. A positive FCM result should be acknowledged even if cytology is negative. In cases with positive cytology but negative flow, repeat testing should be considered.

13.6.2 Prognostic significance of cerebrospinal fluid flow cytometry in hematolymphoid malignancies

Evolving evidence supports the prognostic role of MFC in hematologic malignancies, especially ALL. Across multiple studies, positive MFC at diagnosis has been reported as an independent predictor of relapse and inferior survival in ALL (Cancela et al., 2017; Popov et al., 2019; Thastrup et al., 2020; Thastrup et al., 2022). The level of CSF blasts (20 blasts/μL) estimated by MFC at diagnosis has also been shown to have an independent prognostic role in predicting ALL relapse in some studies (Popov et al., 2019; Thastrup et al., 2020). Underestimation of CNS disease by relying solely on CC may put patients at risk of receiving suboptimal treatment, which may ultimately lead to inferior survival.

13.6.3 Real-world case scenarios for the role of flow cytometry in cerebrospinal fluid

Case 1: A 14-year-old boy presents with anemia and thrombocytopenia. He is diagnosed with B-lineage acute lymphoblastic leukemia (B-ALL) on bone marrow examination and FCM. He denies any neurologic symptoms, and the nervous system examination is unremarkable. CSF examination shows no cells, with normal protein and sugar. Conventional CSF cytology at diagnosis is reported as negative for malignant cells. CSF FCM with a panel reflecting the bone marrow immunophenotype reveals a small abnormal population of B-lymphoblasts, representing 0.8% of all CSF events. Based solely on cytology, this case would have been labeled CNS1. However, because of positive FCM, the treating team decides to escalate the CNS risk in this case, and the patient is treated with intensified CNS-directed therapy [including intensified (more frequent) intrathecal injections and inclusion of systemic CNS-directed blocks].

Usefulness: This case demonstrates the utility of CSF FCM in identifying the "occult" CNS disease in an otherwise asymptomatic B-ALL patient, thereby revising CNS risk stratification and CNS-directed therapy. A positive CSF flow should move a patient from "standard CNS risk" to "high CNS risk".

Case 2: A 45-year-old gentleman presents with generalized lymphadenopathy and weight loss. Lymph node histopathology reveals DLBCL. Imaging shows extranodal involvement in the form of renal and adrenal deposits. He is labeled as DLBCL, stage IV with B symptoms (Stage IVB) with high CNS-IPI (5/6). He denies any neurologic symptoms. MRI brain is reported as normal. Baseline CSF cytology is reported as normal. However, CSF FCM reveals a small abnormal population of clonal B cells. Instead of R-CHOP alone, given the positive CSF flow, he is treated with R-CHOP plus intrathecal methotrexate in each cycle, followed by two cycles of high-dose methotrexate after completion of six cycles of R-CHOP. Subsequent CSF FCM reports are reported as negative, and the end-of-treatment Positron Emission Tomography (PET) scan and MRI are normal.

Usefulness: In this case, a positive CSF flow study guided the intensification of CNS-directed therapy (intrathecal and systemic) in a patient who would have been labeled "CNS-negative" based on cytology alone.

Case 3: A 35-year-old lady with intermediate risk AML (FLT3-ITD positive and NPM1 positive) achieves morphological remission with bone marrow MRD-negativity after intensive chemotherapy. She undergoes an allogeneic hematopoietic stem cell transplant (HSCT). Day 90 assessment shows bone marrow in remission with flow MRD negative and 100% donor chimerism. On day 140 posttransplant, she develops intermittent headache and gait abnormalities. The MRI brain is reported as grossly normal. CSF cytology shows no malignant cells. CSF flow identifies a small population of abnormal myeloblasts that matches the immunophenotype of baseline marrow. A repeat bone marrow (BM) examination reveals marrow in remission with flow MRD negative. She was diagnosed with AML with isolated CNS relapse and started on azacytidine with gilteritinib (second generation FLT3 inhibitor) and intrathecal therapy (with cytarabine and methotrexate), and subsequently underwent a second allogeneic HSCT.

Usefulness: A positive CSF FCM guides the prompt initiation of CNS-directed therapy before the onset of overt neurological deficits, thereby preserving the fitness for a second transplant.

Case 4: A 49-year-old gentleman with relapsed DLBCL is treated with CD19-targeted CAR T-cell therapy and achieves a complete response on PET scan (Deauville 1) at 3 months. At 4 months, he presents with cognitive impairment. MRI brain shows equivocal leptomeningeal enhancement. CSF cytology is nondiagnostic. CSF FCM reveals clonal CD20-positive, CD19-negative B cells (indicating immune escape). PET scan suggests isolated CNS relapse after CD19-directed CAR T-cell therapy. He is started on CNS-directed therapy (high-dose methotrexate, rituximab, and lenalidomide) and planned for CD20-directed bispecific antibody (Glofitamab).

Usefulness: CSF flow cytometry identifies post-CART immune escape and guides initiation of salvage treatment with CNS-penetrant drugs.

13.6.4 Cerebrospinal fluid flow cytometry for solid tumors

Leptomeningeal metastases (LM) are the third most common catastrophic complication affecting the CNS in patients with solid tumors, following brain metastases and epidural metastases. It is seen in approximately 10% of patients with solid tumors, the most common being lung cancer NSCLC, breast cancer, and melanoma (Boire et al., 2019). Adenocarcinomas are the most common solid tumors that metastasize to the leptomeninges (Nayar et al., 2017). Occasionally, sarcomas can metastasize to the leptomeninges. As with hematologic neoplasms, neuroimaging may be inconclusive in early disease, and CC may yield false-negative results, especially in low-cellularity CSF samples. Early identification of leptomeningeal disease via MFC provides an opportunity to initiate appropriate CNS-focused systemic therapy or radiotherapy promptly. With the availability of effective novel anticancer therapies, cancer patients are expected to have longer survival due to the effective control of systemic disease. However, the incidence of LM is expected to increase due to the limited CNS penetration of novel drugs (Nayar et al., 2017; Pellerino et al., 2021).

CSF FCM shall be particularly useful for the following patients with solid tumors:

- There is high clinical suspicion of LM, but imaging is inconclusive
- Discordance between imaging findings and conventional cytology findings
- Scenarios where confirmation of LM will alter the intent of treatment (from curative to palliative intent), and/or support the use of CNS-penetrant targeted drugs

13.6.5 Approach to immunophenotyping in solid tumors

The principles of CSF FCM in solid tumors are derived from those of effusion FCM. However, technical challenges are compounded by the paucicellularity of CSF samples. The core methodology is to exclude CD45-positive hematopoietic cells from the analysis and focus on CD45-negative cells, which are positive for epithelial markers such as EpCAM (epithelial cell adhesion molecule, CD326) for epithelial tumors (absent on normal leukocytes) and tumor-specific markers such as cytokeratin for carcinoma (Milojkovic Kerklaan et al., 2016; van Bussel et al., 2018). Human high molecular weight-melanoma-associated-antigen (HMW-MAA) or melanoma chondroitin sulfate proteoglycan (MSCP) and CD146 are the markers used for melanoma (Campoli et al., 2004). The Cell Search is a highly sensitive and specific platform that utilizes an EpCAM antibody-conjugated ferroparticle to quantify tumor cells in CSF samples (Pellerino et al., 2021).

The flow panel is guided by clinical suspicion (i.e., the type of systemic malignancy), and the gating strategy is like that used in effusion FCM.

- Breast cancer: Epithelial markers (EpCAM and cytokeratin) with/without HER2
- Lung cancer: Epithelial markers (EpCAM and cytokeratin)
- Melanoma: Melanocytic markers (HMW-MAA/MSCP or CD146)

13.6.6 Diagnostic performance in solid tumors—Cerebrospinal fluid conventional cytology versus multiparameter flow cytometry

CSF conventional cytology is the "gold standard" for diagnosing LM in solid tumors (Le Rhun et al., 2023). Its sensitivity for LM at the first lumbar puncture is 50%–70%. Sensitivity improves by 2% to 5% with each additional lumbar puncture (Nayar et al., 2017).

Quantification of CTCs in CSF by EpCAM-based FCM has prognostic value in predicting survival in patients with LM (van Kessel et al., 2025). A recent meta-analysis comprising seven studies on CTCs in CSF reported pooled sensitivities and specificities of 87% and 94%, respectively (Nakasu et al., 2023).

13.6.7 Role in molecular profiling and targeted therapy

CNS progression may reflect clonal evolution of the systemic cancer. A discordance may be observed between the primary site and LM in terms of response to targeted anticancer drugs. In such situations, in addition to diagnosis, CSF FCM provides an opportunity for mutational analysis of enriched tumor cells (Congur et al., 2023). Using a strategy similar to the effusion analyses provides valuable information on resistance mutations and guides the choice of further targeted therapy.

- In breast cancer with LM, identification of strong HER2 expression on EpCAM-positive tumor cells may guide intensification of CNS-directed and HER2-directed treatment [intrathecal or systemic or combined treatment with Trastuzumab, or Tucatinib (a tyrosine-kinase inhibitor that specifically targets HER2)] (Lin et al., 2020).
- In an EGFR-mutant lung cancer patient on targeted therapy, identification of an acquired resistance mutation may inform treatment modification. For example, a patient with an EGFR T790M mutation who was initially treated with a first-or second generation TKI (such as Gefitinib) may benefit from Osimertinib-based treatment (Yun et al., 2025)

- Integration with advanced modalities like CSF cell-free DNA analyses fosters a "liquid CNS biopsy" paradigm, especially in cases where tissue sampling is not feasible.

13.6.8 Significance of ploidy status on cerebrospinal fluid flow cytometry

It serves as an additional objective marker that confirms the truly malignant nature of a cell population in CSF. In epithelial cancers with LM, malignant cells identified on EpCAM-based CSF flow assays have hyperdiploid or aneuploid DNA content. This feature differentiates them from normal CSF leucocytes, which form diploid DNA peaks. Integration of ploidy with EpCAM and cytokeratin has been shown to improve the diagnostic accuracy, particularly in paucicellular samples (Milojkovic Kerklaan et al., 2016; Subirá et al., 2012).

Apart from diagnosis, abnormalities in ploidy and DNA content correlate with the burden of disease in the CNS. A higher proportion of hyperdiploid cells in CSF has been shown to be associated with biologically aggressive disease and rapid neurological deterioration, ultimately leading to inferior overall survival (Subirá et al., 2015).

Although immunophenotyping and matching the clone to blood or bone marrow have a central role in hematolymphoid malignancies, the ploidy information may be useful in challenging cases. From an oncologist's viewpoint, adding ploidy information to CSF FCM may provide semiquantitative information to an otherwise qualitative "positive/negative" interpretation, with implications on neurologic trajectory.

13.6.9 Conclusion

In hematolymphoid malignancies and solid tumors, CSF FCM has emerged as an indispensable tool bridging the gap between conventional cytology and advanced molecular diagnostics. It allows early and accurate diagnosis of scarce tumor cells in a technically challenging fluid compartment. From an oncologist's viewpoint, this translates into better risk stratification of patients and prompt initiation of appropriate CNS-focused therapy, ultimately improving overall cancer control. In fact, detection of CNS involvement via MFC in hematologic and solid neoplasms represents a proactive, biology-driven approach rather than a mere reactive approach in a patient with established neurological deficits. Forging ahead, standardization of flow panels across centers and integration of CSF FCM with advanced molecular diagnostic modalities will further boost its clinical utilization.

References

Awasthi, N. P., Kumari, S., Neyaz, A., Gupta, S., Agarwal, A., Singhal, A., & Husain, N. (2017). EpCAM-based flow cytometric detection of circulating tumor cells in Gallbladder Carcinoma cases. *Asian Pacific Journal of Cancer Prevention, 18*(12), 3429–3437. http://journal.waocp.org/article_54072_714de445c775c73e68ec566250a0515a.pdf. doi:10.22034/APJCP.2017.18.12.3429.

Bento, L. C., Correia, R. P., Alexandre, A. M., Nosawa, S. T., Pedro, E. d. C., Vaz, A. d. C., Schimidell, D., Fernandes, G. B. P., Duarte, C. A. S., Barroso, R. d. S., & Bacal, N. S. (2018). Detection of central nervous system infiltration by myeloid and lymphoid hematologic neoplasms using flow cytometry analysis:

Diagnostic accuracy study. *Frontiers in Medicine, 5*(MAY). https://www.frontiersin.org/articles/10.3389/fmed.2018.00070/full. doi:10.3389/fmed.2018.00070.

Boire, A., Brandsma, D., Brastianos, P. K., Rhun, E. L., Ahluwalia, M., Junck, L., Glantz, M., Groves, M. D, Lee, E. Q., Lin, N., Raizer, J., Rudà, R., Weller, M., Bent, M. JV., Vogelbaum, M. A, Chang, S., Wen, P. Y, & Soffietti, R. (2019). Liquid biopsy in central nervous system metastases: A RANO review and proposals for clinical applications. *Neuro-Oncology, 21*(5), 571–584. doi:10.1093/neuonc/noz012.

Campoli, M. R., Chang, C. C., Kageshita, T., Wang, X., McCarthy, J. B., & Ferrone, S. (2004). Human high molecular weight-melanoma-associated antigen (HMW-MAA): A melanoma cell surface chondroitin sulfate proteoglycan (MSCP) with biological and clinical significance. *Critical Reviews in Immunology, 24*(4), 267–296. doi:10.1615/CritRevImmunol.v24.i4.40.

Cancela, C. S. P., Murao, M., Assumpção, J. G., Souza, M. E. d. L., de Macedo, A. V., Viana, M. B., & De Oliveira, B. M. (2017). Immunophenotyping of the cerebrospinal fluid as a prognostic factor at diagnosis of acute lymphoblastic leukemia in children and adolescents. *Pediatric Hematology and Oncology, 34*(2), 53–65. doi:10.1080/08880018.2017.1313920.

Canovi, S., & Campioli, D. (2016). Accuracy of flow cytometry and cytomorphology for the diagnosis of meningeal involvement in lymphoid neoplasms: A systematic review. *Diagnostic Cytopathology, 44*(10), 841–856. doi:10.1002/dc.23539.

Cieśla, S., Lisiecki, R., Ławnicka, A., Kudliński, B., Ostrowska, P., Davì, A., Veroux, M., & Murawa, D. (2021). Clinical significance of peritoneal fluid examination for free cancer cells in patients qualified for surgery for gastric cancer. *Frontiers in Surgery, 8*. journal.frontiersin.org/journal/surgery. doi:10.3389/fsurg.2021.685868.

Congur, I., Koni, E., Onat, O. E, & Keskin, Z. T (2023). Meta-analysis of commonly mutated genes in leptomeningeal carcinomatosis. *PeerJ, 11*, e15250. doi:10.7717/peerj.15250.

Davidson, B., Dong, H. P., Berner, A., Christensen, J., Nielsen, S., Johansen, P., Bryne, M., Asschenfeldt, P., & Risberg, B. (2002). Detection of malignant epithelial cells in effusions using flow cytometric immunophenotyping: An analysis of 92 cases. *American Journal of Clinical Pathology, 118*(1), 85–92. doi:10.1309/M877-QABM-D9GB-FJAX.

Del Principe, M. I., Buzzatti, E., Piciocchi, A., Forghieri, F., Bonifacio, M., Lessi, F., Imbergamo, S., Orciuolo, E., Rossi, G., Fracchiolla, N., Trappolini, S., Neri, B., Sarlo, C., Zappasodi, P., Dargenio, M., Cefalo, M., Irno-Consalvo, M. A, Conti, C., Paterno, G., … Guarini, A. R (2020). Clinical significance of occult central nervous system disease in adult acute lymphoblastic leukemia. A multicenter report from the Campus All Network. *Haematologica, 106*(1), 39–45. doi:10.3324/haematol.2019.231704.

Del Principe, M. I, Gatti, A., Johansson, U., Buccisano, F., & Brando, B. (2021). ESCCA /ISCCA protocol for the analysis of cerebrospinal fluid by multiparametric flow-cytometry in hematological malignancies. *Cytometry Part B: Clinical Cytometry, 100*(3), 269–281. doi:10.1002/cyto.b.21981.

Egan, P. A., Elder, P. T., Deighan, W. I., O'Connor, S. J. M., & Alexander, H. D. (2020). Multiple myeloma with central nervous system relapse. *Haematologica, 105*(7), 1780–1790. http://www.haematologica.org/content/105/7/1780. doi:10.3324/haematol.2020.248518.

Gabelli, M., Disarò, S., Scarparo, P., Francescato, S., Zangrando, A., Valsecchi, M. G., Putti, M. C., Basso, G., & Buldini, B. (2019). Cerebrospinal fluid analysis by 8-color flow cytometry in children with acute lymphoblastic leukemia. *Leukemia and Lymphoma, 60*(11), 2825–2828. https://www.tandfonline.com/loi/ilal20. doi:10.1080/10428194.2019.1602269.

Galli, E., Patelli, G., Villa, F., Gri, N., Mazzarelli, C., Mangoni, I., Sgrazzutti, C., Ghezzi, S., Sartore-Bianchi, A., Saverio Belli, L., De Carlis, L., Vanzulli, A., Siena, S., & Bencardino, K. (2025). Circulating blood biomarkers for measurable residual disease in hepatocellular carcinoma: A systematic review. *Cancer Treatment Reviews, 135*, 102908. doi:10.1016/j.ctrv.2025.102908.

Gaur, G., Awasthi, N. P., Gupta, A., Agarwal, A., Sachan, R., Malhotra, K. P., Shukla, S., Singh, A. K., Singh, P., & Husain, N. (2023). Diagnostic accuracy of flow cytometry in detecting malignant epithelial cells in serous effusions. *Journal of the American Society of Cytopathology, 12*(6), 423–435. https://www.sciencedirect.com/science/journal/22132945. doi:10.1016/j.jasc.2023.09.003.

Gini, G., Di Rocco, A., Nassi, L., Arcari, A , Tisi, M. C., Loseto, G., Olivieri, A., Gentile, M., Annibali, O., Cabras, M. G., Chiappella, A., Rusconi, C., Ferreri, A. J. M., & Balzarotti, M. (2021). CNS prophylaxis: How far is routine practice from the guidelines? focus on a nationwide survey by the Fondazione Italiana Linfomi (FIL). *Frontiers in Oncology, 11*. http://www.frontiersin.org/Oncology/about. doi:10.3389/fonc.2021.730194.

Gold, D. R., Nadel, R. E., Vangelakos, C. G., Davis, M. J., Livingston, M. Y., Heath, J. E., Reich, S. G , Gojo, I., Morales, R. E., & Weiner, W. J. (2013). Pearls and oy-sters: The utility of cytology and flow cytometry in the diagnosis of leptomeningeal leukemia. *Neurology, 80*(14), e156–e159. doi:10.1212/WNL.0b013e31828ab295.

Kim, D., Alperstein, S. A., & Siddiqui, M. T. (2021). Standardizing a volume benchmark for cerebrospinal fluids for optimal diagnostic accuracy. *Diagnostic Cytopathology, 49*(2), 258–266. http://onlinelibrary.wiley.com/journal/10.1002/(ISSN)1097-0339. doi:10.1002/dc.24635.

Kotsos, D., Chatzileontiadou, S., Apsemidou, A., Xanthopoulou, A., Rapi, A., Frouzaki, C., & Hatjiharissi, E. (2022). Bing-neel syndrome: Real-life experience in personalized diagnostic approach and treatment. *Frontiers in Oncology, 12*. doi:10.3389/fonc.2022.891052.

Kraan, J., Gratama, J. W., Haioun, C., Orfao, A., Plonquet, A., Porwit, A., Quijano, S., Stetler-Stevenson, M., Subira, D., & Wilson, W. (2008). Flow cytometric immunophenotyping of cerebrospinal fluid. *Current Protocols in Cytometry*, (45). http://onlinelibrary.wiley.com/book/10.1002/0471142956. doi:10.1002/0471142956.cy0625s45.

Le Rhun, E., Weller, M., van den Bent, M., Brandsma, D., Furtner, J., Rudà, R., Schadendorf, D., Seoane, J., Tonn, J.-C., Wesseling, P., Wick, W., Minniti, G., Peters, S., Curigliano, G., & Preusser, M. (2023). Leptomeningeal metastasis from solid tumours: EANO–ESMO clinical practice guideline for diagnosis, treatment and follow-up. *ESMO Open, 8*(5), 101624. doi:10.1016/j.esmoop.2023.101624.

Lin, N. U., Borges, V., Anders, C., Murthy, R. K., Paplomata, E., Hamilton, E., Hurvitz, S., Loi, S., Okines, A., Abramson, V., Bedard, P. L., Oliveira, M., Mueller, V., Zelnak, A., DiGiovanna, M. P., Bachelot, T., Jo Chien, A., O'Regan, R., Wardley, A., … Winer, E. P. (2020). Intracranial efficacy and survival with tucatinib plus trastuzumab and capecitabine for previously treated HER2-positive breast cancer with brain metastases in the HER2CLIMB trial. *Journal of Clinical Oncology, 38*(23), 2610–2619. https://ascopubs.org/doi/10.1200/JCO.20.00775. doi:10.1200/JCO.20.00775.

Liu, N., Sun, R. Z., Du, J., Dong, Q. Z., Fan, C. F., Li, Q. C., Wang, E. H., & Liu, Y. (2018). Comparison of epidermal growth factor receptor gene mutations identified using pleural effusion and primary tumor tissue samples in non-small cell lung cancer. *Applied Immunohistochemistry and Molecular Morphology, 26*(4), e44–e51. http://www.appliedimmunohist.com. doi:10.1097/PAI.0000000000000543.

Milojkovic Kerklaan, B., Pluim, D., Bol, M., Hofland, I., Westerga, J., Tinteren, H., Beijnen, J. H., Boogerd, W., Schellens, J. H. M., & Brandsma, D. (2016). EpCAM-based flow cytometry in cerebrospinal fluid greatly improves diagnostic accuracy of leptomeningeal metastases from epithelial tumors. *Neuro-Oncology, 18*(6), 855–862. doi:10.1093/neuonc/nov273.

Munz, M., Baeuerle, P. A., & Gires, O. (2009). The emerging role of EpCAM in cancer and stem cell signaling. *Cancer Research, 69*(14), 5627–5629. http://cancerres.aacrjournals.org/cgi/reprint/69/14/5627. doi:10.1158/0008-5472.CAN-09-0654Germany.

Nakasu, Y., Deguchi, S., Nakasu, S., Yamazaki, M., Notsu, A., Mitsuya, K., & Hayashi, N. (2023). Diagnostic accuracy of cerebrospinal fluid liquid biopsy and MRI for leptomeningeal metastases in solid cancers: A systematic review and meta-analysis. *Neuro-Oncology Advances, 5*(1). doi:10.1093/noajnl/vdad002.

Nayar, G., Ejikeme, T., Chongsathidkiet, P., Elsamadicy, A. A., Blackwell, K. L., Clarke, J. M., Lad, S. P., & Fecci, P. E. (2017). Leptomeningeal disease: Current diagnostic and therapeutic strategies. *Oncotarget, 8*(42), 73312–73328. doi:10.18632/oncotarget.20272.

Nguyen-Them, L., Alentorn, A., Ahle, G., Soussain, C., Mathon, B., Garff Tavernier, M. L. e, Houillier, C., & Hoang-Xuan, K. (2023). CSF biomarkers in primary CNS lymphoma. *Revue Neurologique, 179*(3), 141–149. doi:10.1016/j.neurol.2022.06.014.

Nicolò, E., Gianni, C., Pontolillo, L., Serafini, M. S., Munoz-Arcos, L. S., Andreopoulou, E., Curigliano, G., Reduzzi, C., & Cristofanilli, M. (2024). Circulating tumor cells et al.: Towards a comprehensive liquid biopsy approach in breast cancer. *Translational Breast Cancer Research, 5*. https://tbcr.amegroups.org/article/download/85637/62651. doi:10.21037/tbcr-23-55.

Osta, W. A., Chen, Y., Mikhitarian, K., Mitas, M., Salem, M., Hannun, Y. A., Cole, D. J., & Gillanders, W. E. (2004). EpCAM is overexpressed in breast cancer and is a potential target for breast cancer gene therapy. *Cancer Research, 64*(16), 5818–5824. doi:10.1158/0008-5472.CAN-04-0754.

Pauli, C., Münz, M., Kieu, C., Mack, B., Breinl, P., Wollenberg, B., Lang, S., Zeidler, R., & Gires, O. (2003). Tumor-specific glycosylation of the carcinoma-associated epithelial cell adhesion molecule EpCAM in head and neck carcinomas. *Cancer Letters, 193*(1), 25–32. doi:10.1016/S0304-3835(03)00003-X.

Pellerino, A., Brastianos, P. K., Rudà, R., & Soffietti, R. (2021). Leptomeningeal metastases from solid tumors: Recent advances in diagnosis and molecular approaches. *Cancers, 13*(12). https://www.mdpi.com/2072-6694/13/12/2888/pdf. doi:10.3390/cancers13122888.

Popov, A., Henze, G., Verzhbitskaya, T., Roumiantseva, J., Lagoyko, S., Khlebnikova, O., Streneva, O., Bidanov, O., Tsaur, G., Inaba, H., Karachunskiy, A., & Fechina, L. (2019). Absolute count of leukemic blasts in cerebrospinal fluid as detected by flow cytometry is a relevant prognostic factor in children with acute lymphoblastic leukemia. *Journal of Cancer Research and Clinical Oncology, 145*(5), 1331–1339. doi:10.1007/s00432-019-02886-3.

Ramalingam, T. R., Marimuthu, B., Rasheed, H. N., Lakshmanan, A., Narla, S. L., Vaidhyanathan, L., & Pai, A. (2025). Intraoperative flow cytometry in detecting free carcinoma cells in peritoneal lavage fluid of gastric carcinoma cases. *Cytometry Part B - Clinical Cytometry*. http://onlinelibrary.wiley.com/journal/10.1002/(ISSN)1552-4957. doi:10.1002/cyto.b.22238.

Singh, C., Lakshman, A., Jandial, A., Sharma, S., Nampoothiri, R., Prakash, G., & Malhotra, P. (2018). Pachymeningeal Involvement with Blindness as the Presenting Manifestation of Non-Hodgkin Lymphoma. *Turkish Journal of Hematology, 35*(1), 73–74. doi:10.4274/tjh.2016.0404.

Subirá, D., Barriopedro, F., Fernández, J., Martínez, R., Chara, L., Castelao, J., & García, E. (2023). High sensitivity flow cytometry immunophenotyping increases the diagnostic yield of malignant pleural effusions. *Clinical and Experimental Metastasis, 40*(6), 505–515. https://www.springer.com/journal/10585. doi:10.1007/s10585-023-10236-4.

Subirá, D., Serrano, C., Castanón, S., Gonzalo, R., Illán, J., Pardo, J., Martínez-Garciá, M., Millastre, E., Aparisi, F., Navarro, M., Dómine, M., Gil-Bazo, I., Segura, P. P., Gil, M., & Bruna, J. (2012). Role of flow cytometry immunophenotyping in the diagnosis of leptomeningeal carcinomatosis. *Neuro-Oncology, 14*(1), 43–52. doi:10.1093/neuonc/nor172.

Subirá, D., Simó, M., Illán, J., Serrano, C., Castañón, S., Gonzalo, R., Granizo, J. J., Martínez-García, M., Navarro, M., Pardo, J., & Bruna, J. (2015). Diagnostic and prognostic significance of flow cytometry immunophenotyping in patients with leptomeningeal carcinomatosis. *Clinical & Experimental Metastasis, 32*(4), 383–391. doi:10.1007/s10585-015-9716-3.

Sung, H., Ferlay, J., Siegel, R. L., Laversanne, M., Soerjomataram, I., Jemal, A., & Bray, F. (2021). Global cancer statistics 2020: GLOBOCAN estimates of incidence and mortality worldwide for 36 cancers in 185 countries. *CA Cancer Journal for Clinicians, 71*(3), 209–249. http://onlinelibrary.wiley.com/journal/10.3322/(ISSN)1542-4863. doi:10.3322/caac.21660.

Šúri, A., & Mociková, H. (2025). Diagnostic approach to leptomeningeal involvement in diffuse large B-cell lymphoma. *Expert Review of Molecular Diagnostics, 25*(6), 209–215. doi:10.1080/14737159.2025.2485211.

Tas, F., Karabulut, S., Serilmez, M., Ciftci, R., & Duranyildiz, D. (2014). Clinical significance of serum epithelial cell adhesion molecule (EPCAM) and vascular cell adhesion molecule-1 (VCAM-1) levels in patients with epithelial ovarian cancer. *Tumor Biology, 35*(4), 3095–3102. doi:10.1007/s13277-013-1401-z.

Telford, W. G. (2023). Flow cytometry and cell sorting. *Front Med (Lausanne), 10*, 1287884.

Thastrup, M., Marquart, H. V., Levinsen, M., Grell, K., Abrahamsson, J., Albertsen, B. K., Frandsen, T. L., Harila-Saari, A., Lähteenmäki, P. M., Niinimäki, R., Pronk, C. J., Ulvmoen, A., Vaitkevičienė, G., Taskinen, M., Schmiegelow, K., Wehner, P., Frost, B. M., Norén-Nyström, U., Behrendtz, M., … Wojcik, D. M. (2020). Flow cytometric detection of leukemic blasts in cerebrospinal fluid predicts risk of relapse in childhood acute lymphoblastic leukemia: A Nordic society of pediatric hematology and oncology study. *Leukemia, 34*(2), 336–346. http://www.nature.com/leu/index.html. doi:10.1038/s41375-019-0570-1.

Thastrup, M., Marquart, H. V., Levinsen, M., Ulvmoen, A., Albertsen, B. K., Pronk, C. J., Olsen, M., Wehner, P., Taskinen, M., Als-Nielsen, B., & Schmiegelow, K. (2025). Flow cytometric analysis of cerebrospinal fluid identifies patients with occult central nervous system involvement in relapsed pediatric acute lymphoblastic leukemia. *Haematologica, 111*(1), 355–360. www.haematologica.org. doi:10.3324/haematol.2025.287392.

Thastrup, M., Marquart, H. V., & Schmiegelow, K. (2022). Flow cytometric detection of malignant blasts in cerebrospinal fluid: A biomarker of central nervous system involvement in childhood acute lymphoblastic leukemia. *Biomolecules, 12*(6). https://www.mdpi.com/2218-273X/12/6/813/pdf?version=1654786927. doi:10.3390/biom12060813.

van Bussel, M. T. J., Pluim, D., Bol, M., Beijnen, J. H., Schellens, J. H. M., & Brandsma, D. (2018). EpCAM-based assays for epithelial tumor cell detection in cerebrospinal fluid. *Journal of Neuro-Oncology, 137*(1). www.wkap.nl/journalhome.htm/0167-594X. doi:10.1007/s11060-017-2691-6.

van Kessel, E., Woerdman, N., Pluim, D., Kerklaan, B. M, Broek, D., Huitema, A. D. R., & Brandsma, D. (2025). Added prognostic value of circulating tumor cell numbers in CSF of patients with leptomeningeal metastasis from epithelial tumors. *European Journal of Cancer, 220*, 115377. doi:10.1016/j.ejca.2025.115377.

Weston, C. L., Glantz, M. J., & Connor, J. R. (2011). Detection of cancer cells in the cerebrospinal fluid: Current methods and future directions. *Fluids and Barriers of the CNS, 8*(1). http://www.fluidsbarrierscns.com/content/8/1/14. doi:10.1186/2045-8118-8-14United States.

Wilson, W. H., Bromberg, J. E. C., Stetler-Stevenson, M., Steinberg, S. M., Martin-Martin, L., Muñiz, C., Sancho, J. M., Caballero, M. D., Davidis, M. A., Brooimans, R. A., Sanchez-Gonzalez, B., Salar, A., González-Barca, E., Ribera, J. M., Shovlin, M., Filie, A., Dunleavy, K., Mehrling, T., Spina, M., … Orfao, A. (2014). Detection and outcome of occult leptomeningeal disease in diffuse large B-cell lymphoma and Burkitt lymphoma. *Haematologica, 99*(7), 1228–1235. http://www.haematologica.org/content/99/7/1228.full.pdf. doi:10.3324/haematol.2013.101741.

Wong-Arteta, J., Rey, M., Aragón, L., Gil-Rodríguez, E., & Bujanda, L. (2020). The utility of flow cytometry in the diagnostic work up of malignant effusions due to nonhematopoietic neoplasms. *Cytometry Part B: Clinical Cytometry, 98*(6), 504–515. doi:10.1002/cyto.b.21886.

Yun, W., Liu, A., Wen, J., Zhao, K., & Meng, X. (2025). Treatment options for EGFR mutation-positive NSCLC patients with leptomeningeal metastases after third-generation TKI resistance: A retrospective real-world study. *Therapeutic Advances in Medical Oncology, 17*. doi:10.1177/17588359251403396.

Zhou, L., & Zhu, Y. (2018). The EpCAM overexpression is associated with clinicopathological significance and prognosis in hepatocellular carcinoma patients: A systematic review and meta-analysis. *International Journal of Surgery, 56*, 274–280. doi:10.1016/j.ijsu.2018.06.025.

Index

Page numbers followed by "*f*" and "*t*" indicate, figures and tables respectively.

Printed and bound by CPI Group (UK) Ltd, Croydon, CR0 4YY
17/09/2026
02218533-0011